This volume is dedicated to the memory of our dear echinologist friend, Lucienne Fenaux. (Left to right: John Lawrence, Lucienne Fenaux, Michael Jangoux. 4éme Seminaire International sur les Echinodermes. Saint-Aubin sur Mer, France, 1985).

ECHINODERM STUDIES

Edited by
MICHEL JANGOUX
Université Libre de Bruxelles, Belgium

JOHN M. LAWRENCE
University of South Florida, Tampa, FL, USA

VOLUME 6

A.A. BALKEMA / LISSE / ABINGDON / EXTON (PA) / TOKYO

Library of Congress Cataloging-in-Publication Data

Library of Congress no. 84643702.

Typesetting: Grafische Vormgeving Kanters, Sliedrecht, The Netherlands
Printed by: Krips, Meppel, The Netherlands

Published by
A.A. Balkema Publishers, a member of Swets & Zeitlinger Publishers
www.balkema.nl and www.szp.swets.nl

ISBN HB 90 5809 301 8

Contents

Larval supply, settlement and recruitment in echinoderms

TOBY BALCH*, ROBERT E. SCHEIBLING
Department of Biology, Dalhousie University, Halifax, Nova Scotia, Canada, B3H 4J1

CONTENTS

*E-mail: tbalch@is2.dal.ca

1

ABSTRACT

Over the past 2 decades, numerous studies have addressed settlement and recruitment patterns of echinoderms, particularly species (mainly echinoids and asteroids) of ecological or economic importance. However, the processes which regulate the transitions from early life stages through to adulthood, and thereby determine the demography, distribution and abundance of echinoderm populations, remain poorly understood. The supply of planktonic larvae of echinoderms to benthic populations is regulated by a complex interaction of biotic and abiotic factors such as hydrodynamics, sea temperature, predation and starvation. Echinoderm larvae have been induced to settle on various substrata in laboratory studies and specific chemical inducers have been isolated in some cases. However the importance of settlement induction and substratum preferences in determining settlement patterns in natural habitats is not clear. A salient feature of most echinoderm populations is a high degree of spatial and temporal variability in settlement and recruitment. Settlement variation exists at spatial scales ranging from metres between adjacent habitats to hundreds of kilometres between regions. Seasonal settlement is common among echinoderm species but interannual variation in settlement and/or recruitment is often high, and several years may elapse between successful recruitment events. Some of this variability is likely attributable to measurement inaccuracies associated with sampling individuals which are small, cryptic and transient, and procedural inconsistencies among different studies. A variety of post-settlement processes including predation, migration, disease and starvation may alter observed patterns of settlement and play an important role in regulating recruitment rates and patterns. This review attempts to consolidate our understanding of the importance of early life history events in determining population structure and dynamics of echinoderms in order to identify deficiencies and suggest fruitful avenues for future research.

KEYWORDS: Settlement, Recruitment, Echinodermata, Echinoidea, Asteroidea, Ophiuroidea, Holothuroidea, Crinoidea.

1 INTRODUCTION

The role of settlement and recruitment in determining population and community structure, or what has been coined "supply-side ecology" (Lewin 1986), has been the focus of intensive study for over a decade (reviewed by Underwood & Fairweather 1989, Ólafsson et al. 1994, Booth & Brosnan 1995, Caley et al. 1996, Hunt & Scheibling 1997). Settlement of benthic marine invertebrates is generally defined as the attachment of larvae to the substratum and attendant metamorphosis into the juvenile form (reviewed by Pawlik 1992). Recruitment is usually defined operationally as occurring some time after settlement when individuals can be reliably counted and some post-settlement mortality or migration may have occurred (*sensu* Keough & Downes 1982). Consequently, methodological differences often complicate comparisons of recruitment within and among taxa and habitats. Most studies on recruitment have involved sessile species, such as barnacles and mussels, in rocky intertidal habitats (e.g., Connell 1985, Gaines & Roughgarden 1985, Sutherland 1987, 1990, Raimondi 1990, Menge 1991, Minchinton & Scheibling 1991). Fewer studies have involved mobile species, such as echinoderms, for which patterns of distribution and abundance are influenced both by post-settlement mortality and migration (Hunt & Scheibling 1997).

Echinoderms are ubiquitous in all marine benthic habitats from the intertidal zone to the deep-sea. They are significant components of the trophic structure in many communities and serve important ecological roles as both predators and prey. In some cases, their impact as predators or grazers can be catastrophic. Population outbreaks of the asteroid *Acanthaster planci*, for example, have devastated coral reefs in the south Pacific (Moran 1986, Johnson 1992a) and intensive grazing by strongylocentrotid echinoids has destroyed kelp forests throughout the north Atlantic and Pacific (Harrold & Pearse 1987, Vadas & Elner 1992). Consequently, these species have been viewed as pests to be controlled or eradicated. In contrast, some holothuroids and echinoids, are valued food resources and form the basis of major fisheries around the world (Sloan 1984). Yet others, such as ophiuroids and crinoids, are suspension feeders comprising a large proportion of the macrofaunal biomass of sedimentary environments, particularly in the deep-sea (Haedrich et al. 1980).

In view of their considerable ecological and economic importance, the population ecology of many echinoderms, particularly asteroids and echinoids, has been extensively studied. However, the importance of early life-history events in determining population structure and dynamics remains poorly understood for most species. The larval phase is a critical component of the life-history of echinoderms and other marine invertebrates with meroplanktonic larvae. Factors determining larval supply to benthic habitats include abiotic (e.g. currents, temperature, salinity) and biotic factors (e.g. larval behaviour, food availability, predation) which regulate larval production, development and survival (reviewed by Young & Chia 1987, Rumrill 1990). Settlement can be induced by a variety of biological, physical and chemical factors (reviewed by Rodriguez et al. 1993) but their importance relative to pre- and post-settlement processes is not clear for echinoderms (reviewed by Chia et al. 1984, Pearce 1997). Ebert (1983) reviewed studies of recruitment in echinoderms and found that recruitment was spatially and temporally variable both within and among species. He concluded that many factors contribute to recruitment variability in ways that remain poorly understood. In many studies, researchers have made inferences about patterns of settlement and recruitment based on population size structure or distributional patterns of older individuals (Ebert 1983, Chia et al. 1984), although sampling constraints and the small scale of such observations have limited the strength or generality of the conclusions.

In this review, we summarise recent progress (much of it over the past 14 years since Ebert's 1983 review) in the search for links between larval supply, settlement, and recruitment in echinoderms, which ultimately determine the distribution and abundance of populations. The text is divided into four major sections based on different stages of the early life-history of echinoderms and the processes that influence them: 1) Factors affecting larval supply, 2) Induction of settlement, 3) Spatial and temporal patterns of settlement and recruitment, and 4) Post-settlement processes. The section on patterns of settlement and recruitment is central to the review and studies on this topic from the four most studied echinoderm classes are summarised in tabular form to facilitate comparisons within and between species, classes and geographic regions. Our intent is to provide a comprehensive synthesis and critical evaluation of the literature on patterns and processes of larval supply, settlement and recruitment in echinoderms, and to identify gaps in our knowledge which may guide future research.

2 FACTORS AFFECTING LARVAL SUPPLY

2.1 *Hydrodynamics*

While in the water column, invertebrate larvae generally have little control over horizontal movement, although they may actively migrate vertically (reviewed by Young & Chia 1987). Echinoderm larvae usually are found near the sea surface (Rumrill 1988a, Pedrotti & Fenaux 1992) where ocean currents and wind may enhance larval dispersal (reviewed by Ebert 1983, Harrold & Pearse 1987, Pearse & Cameron 1991). Larval dispersal is also common in deep-sea echinoderms (Eckelbarger 1994, Pearse 1994, Tyler et al. 1994, Young 1994a, b) where larvae can be retained locally or advected away from spawning areas by currents (Mullineaux 1994). To fully understand the effect of hydrodynamics on larval supply to the benthos, various spatial scales must be considered. Because the arrival of planktonic larvae at suitable settlement sites is primarily dependant on advective transport, both local hydrodynamics and large-scale oceanographic features are important determinants of recruitment success (reviewed by Shanks 1995). Ebert and coworkers (Ebert 1983, Ebert et al. 1994) documented the existence of a latitudinal cline in settlement for *Strongylocentrotus purpuratus* and *S. franciscanus* and correlated settlement with general oceanographic processes along the California coast. They found settlement was more variable along the northern coast, which is subjected to greater offshore advection rates, than in the southern California Bight, which is thought to have a longer residence period and therefore retain larvae. Ebert and Russel (1988), however, did not observe this latitudinal cline in settlement of *S. purpuratus* in intertidal populations from central California to Oregon. They proposed that local topographic features such as capes and headlands reduced recruitment due to upwelling and cold water plumes that advect larvae away from the coast. At sites between headlands with no predictable upwelling, there was substantial annual recruitment. Further north in Washington, Paine (1986) found that recruitment of *S. purpuratus* occurred only four times in 22 years suggesting sporadic supply of larvae to the area. Although Paine (1986) found no consistent correlation with El Niño events at this latitude, recruitment was associated with above average sea temperatures which he assumes is an indication of transport of larvae in northward flowing currents.

Most echinoderm larvae have relatively long planktonic periods and are capable of delaying metamorphosis in the absence of suitable substrata (Strathmann 1978a, b, Bosch et al. 1989). Echinoid, asteroid and ophiuroid larvae have been found in offshore plankton tows in the central Pacific where currents may enable long distance dispersal between distant islands (Scheltema 1986). Once larvae are advected far offshore, however, the likelihood of settling in a suitable habitat is greatly diminished unless currents deliver them to coastal areas when they are competent to settle (Jackson & Strathmann 1981). In the absence of a settlement substrate, ophiuroids may metamorphose in the plankton or settle indiscriminately resulting in major losses (Hendler 1991). For example, Mileikovsky (1968) found post-larval ophiuroids in plankton tows down to 4000 m in the Oyashio Current (northwest Pacific) and attributed a decrease in numbers with depth to increasing mortality. Local hydrodynamic processes may act to retain larvae within the coastal region. Pedrotti and Fenaux (1992) found that ophiuroid and echinoid larvae remained in the surface layer in the Bay of Villefranche, Mediterranean and rarely occurred beyond a divergence zone 30 km offshore. At a smaller spatial scale, Sewell and Watson (1993) found high densities of asteroid larvae in plankton tows in an enclosed bay in Nootka Sound, British Columbia, where there was substantial recruitment of *Pisaster ochraceus*. They proposed that the larvae are spawned and retained within the bay where they settle and then disperse over time. Similarly, Lubchenco-Menge and Menge (1974) found atypically high densities of recruits of *P. ochraceus* at one site on San Juan Island, Washington, which they considered a nursery area.

By far the largest body of research on asteroid early life-history is based on one species, the crown-of-thorns starfish *Acanthaster planci*. This research has been motivated by population outbreaks of *A. planci* which have had catastrophic effects on coral reefs throughout the western tropical Pacific (reviewed by Moran 1986, Birkeland & Lucas 1990). There is general recognition that larval supply, settlement and recruitment are critical aspects of this problem, although the relative importance of each of these processes in initiating population outbreaks remains speculative (Johnson 1992b). The genetic relatedness of populations of *A. planci* throughout the Pacific suggests that there is widespread dispersal of larvae and considerable gene flow within reef systems connected by ocean currents (Benzie 1992). Populations on the Great Barrier Reef are genetically homogeneous, suggesting that outbreak populations arise from a single source (Nash et al. 1988, Benzie 1992). Although the primary source of larvae is not known, the two series of outbreaks of *A. planci* recorded on the Great Barrier Reef during the past four decades have progressed from north to south (Johnson 1992b, Moran et al. 1992) presumably by advection of larvae (Nash et al. 1988). Hydrodynamic models of larval dispersal, correlated with data on reported outbreaks, showed that the severity of the outbreak decreases, and the proximity to the mainland increases, as larvae disperse southward (James & Scandol 1992, Scandol & James 1992). Black and Moran (1991) developed a numerical model of current patterns and larval supply to six reefs in the central Great Barrier Reef. They found a clear correspondence between observed and predicted distributions of populations of *A. planci* in all six simulations and concluded that hydrodynamics and larval supply are largely responsible for recruitment patterns. Recruitment of *A. planci* probably occurs in deeper water on the reef slope; subsequent migration up the reef slope results in outbreaks (Johnson et al. 1991).

2.2 *Temperature and salinity*

Thorson (1950) proposed that increased temperature may enhance settlement by accelerating larval development and reducing the period that larvae are exposed to planktonic predators. Evidence for a correlation between temperature and recruitment of echinoids was reviewed by Ebert (1983) who found a positive relationship for echinoids (species not stated) in Japan, an inverse relationship for *Strongylocentrotus purpuratus* in southern California, and no obvious relationship for *S. purpuratus* in Oregon. In a laboratory study, Hart & Scheibling (1988) showed that temperature has a strong positive influence on larval development of *S. droebachiensis*. By comparing sea temperature patterns off Nova Scotia in the early 1980s to recruitment events during the same period, they found that recruitment tended to occur in years of relatively warm spring sea temperature and not in colder years. Long term records for the same area show abnormally warm spring sea temperatures in 1960 and 1983, which preceded echinoid population outbreaks in the late 1960s and the early 1990s respectively (Hart & Scheibling 1988, Scheibling 1996). In accordance with Thorson's hypothesis, Hart & Scheibling (1988) proposed that increased larval survival during warm years results in recruitment pulses which lead to population outbreaks several years later. Other studies of this species, however, have suggested an inverse relationship between recruitment and sea temperature during larval development. Foreman (1977) attributed heavy recruitment of *S. droebachiensis* in British Columbia in 1969 to record low spring temperatures and Himmelman (1986) related weak recruitment at sheltered sites in Newfoundland to relatively high water temperatures. These equivocal findings suggest that temperature alone does not reliably explain recruitment patterns. Changes in sea temperature may simply reflect shifts in local hydrodynamics or other environmental factors, such as salinity, food availability or predator abundance, which may either enhance or limit larval survival.

The temperature dependence of developmental rate also has been proposed as a mechanism influencing settlement rates in asteroids. Laboratory experiments with *Acanthaster planci* have shown the length of the larval period ranges from 9 to 28 days and decreases with increasing temperature between 25°C and 32°C (reviewed by Moran 1986, Johnson 1992b). Larvae of *A. planci* are also found to respond to differences in salinity. Wide salinity ranges (21 to 33‰) can be tolerated but survival is greatest at 30‰ (reviewed by Moran 1986, Brodie 1992). However, Brodie (1992) concluded that fluctuations in temperature and salinity are relatively minor on the Great Barrier Reef and are probably not responsible for outbreaks of *A. planci* in that region.

2.3 *Predation*

Although predation in the plankton has long been recognised as a major component of larval mortality (Thorson 1950), few studies have examined predation of echinoderm larvae (reviewed by Ebert 1983, Harrold & Pearse 1987, Young & Chia 1987, Rumrill 1990, Scheibling 1996). Laboratory experiments on predation of embryos and larvae of *Dendraster excentricus*, *Strongylocentrotus franciscanus* and *S. purpuratus* showed that planktonic invertebrate predators, such as crustaceans, chaetognaths, medusae and ctenophores, selectively feed on embryos and early larval stages, whereas small planktivorous fish choose the larger plutei (Rumrill & Chia 1985, Pennington et al. 1986). Possible mechanisms which have been proposed to explain stage-specific predation include selection of prey size by predators, escape swimming behaviour of larvae, and larval structural defences (Rumrill & Chia 1985, Pennington et al. 1986).

As larvae approach the bottom they become exposed to a suite of suspension-feeding benthic invertebrates including mussels, ophiuroids, ascidians, anemones, and tunicates (Hooper 1980, Cowden et al. 1984). In the laboratory, Tegner and Dayton (1981) offered larvae of *Strongylocentrotus purpuratus* to three species of bryozoans and a serpulid polychaete which are common suspension feeders inhabiting kelp fronds. They found that only the polychaete could consume young plutei and that it had difficulty consuming larger, mature plutei. They concluded that it is unlikely that these filter feeders reduce larval abundance within a kelp forest, and suggested that planktivorous fish are more likely responsible (see also Gaines & Roughgarden 1987).

The coral *Pocillopora damicornis* and several species of fish have been observed to feed on eggs and/or larvae of *Acanthaster planci* (Yamaguchi 1973, and reviewed by Yamaguchi 1975, Moran 1986). However, laboratory studies have shown that some species of fish reject *A. planci* eggs and larvae, suggesting a chemical defence (reviewed by Yamaguchi 1975, Moran 1986). Lucas et al. (1979) found that pomacentrid fish rejected gelatin food particles with extracts of saponins from larval and adult *A. planci* and concluded that the concentrations of saponins in eggs and larvae are sufficient to limit predation. A chemical defence is also proposed for larvae of *Pisaster ochraceus* and for three species of echinoids which had higher survival rates than polychaete, gastropod or cirriped larvae when offered to mussels or ascidians in the laboratory (Cowden et al. 1984, but see also Young & Chia 1987).

2.4 *Starvation*

Starvation is a possible cause of larval mortality, although in reviewing the literature on nutrition of larval echinoids, Pearse and Cameron (1991) concluded that starvation is probably less important than predation in limiting larval survival. Strathmann (1996) reviewed several studies on feeding rates and natural densities of echinoderm larvae and concluded that death through starvation is unlikely under most conditions, although food limitation may occur in nutrient poor waters. Laboratory studies on larvae of the asteroid *Asterina miniata* showed that food limitation slows development, extends the time to metamorphosis, and reduces survival (Allison 1994, Basch & Pearse 1996). Basch and Pearse (1996) found that larvae of *A. miniata* reared in field enclosures developed faster than those in the laboratory where different hydrodynamic or light regimes may have reduced food capture or quality. In laboratory feeding experiments with larvae of *Acanthaster planci*, Lucas (1982) found that food limitation prevented development to the late brachiolaria stage. He suggested that starvation and/or increased predation associated with a longer larval period may reduce survival of larvae of

A. planci on the Great Barrier Reef under food-limited conditions. However, Olson (1985) pointed out that food used in laboratory studies differs in composition from natural food resources and proposed that a mixed diet may not be limiting, even at the same particle concentrations. He used *in situ* culture chambers to examine larval survival of *A. planci* under natural food levels on the Great Barrier Reef. Surviving larvae did not appear to be food limited, although survivorship was relatively low (40 to 58%). Olson (1985) attributed this to handling error but it is clear from discrepancies between laboratory and field studies that more experimentation is needed to resolve the issue of food limitation.

Birkeland (1982) correlated outbreaks of *Acanthaster planci* with heavy rainfall events and proposed that terrestrial runoff, which results in phytoplankton blooms, enhances larval survival, assuming the larvae normally are food limited. However, Moran (1986) pointed out some of the weaknesses with this correlational hypothesis and concluded that it is not well supported for some regions, such as the Great Barrier Reef (but see also Shanks 1995). Brodie (1992) reviewed several hypotheses relating enhanced larval survival of *A. planci* to various effects of runoff, including increased nutrient supply, increased mortality of predators, a reduction in salinity to an optimal level and increased temperature. Although he found the existing evidence for each of these hypotheses to be inconclusive, he suggested that biotic and abiotic effects of riverine input should not be discounted, and that further investigation may elucidate their roles in causing primary outbreaks.

2.5 *Summary and conclusions*

Although much research has been done on larval behaviour in relation to the dynamics of dispersal, most studies on larval echinoderms have been conducted in the laboratory. Field studies have tended to focus on distribution in the water column and have not followed larvae through to settlement. Despite major advances during the past decade in methods of marking and tracking larvae (reviewed by Levin 1990), there remain substantial deficiencies in our present understanding of larval dispersal and dynamics. New methods such as the use of genetic markers for larval tracking may enable researchers to determine the discreteness of populations and identify potential sources of larvae (Palumbi 1995, Medeiros-Bergen et al. 1995). Hydrodynamic models of dispersal, which have been proven effective in predicting recruitment of *Acanthaster planci* (Black & Moran 1991), could be applied to other species to obtain a more general understanding of the relationship between physical factors and larval supply to benthic habitats. There is strong evidence for regulation of settlement by currents and hydrodynamic forcing. However, these factors can act at different spatial scales, all of which must be considered before fully understanding settlement patterns. The roles of larval predation and starvation in determining settlement and recruitment also remain unclear. Laboratory studies have identified some predators of echinoderm larvae and have shown that larvae possess chemical, structural, and behavioural defences to resist predation. Although starvation seems to be less important than predation in regulating larval abundance, this needs to be tested experimentally under realistic conditions of food availability and predator abundance, preferably in the field (e.g. Olson 1985).

3 INDUCTION OF SETTLEMENT

Numerous laboratory experiments (mostly with echinoids) have documented various cues for induction of settlement and metamorphosis of echinoderm larvae (reviewed by Strathmann 1978b, Chia et al. 1984, Chia 1989, Rodriguez et al. 1993, Pearce 1997). Larvae of regular echinoid species, including *Anthocidaris crassispina*, *Arbacia punctulata*, *Lytechinus pictus*, *Pseudocentrotus depressus*, *Strongylocentrotus droebachiensis* and *S. purpuratus*, have been found to settle in response to a variety of microbial and/or algal films that occur on natural substrata (Cameron & Hinegardner 1974, Cameron & Schroeter 1980, Rowley 1989, Pearce & Scheibling 1990a, 1991, Kitamura et al. 1993, reviewed by Morse 1992). Although specific inducers have been isolated, studies which have tested a variety of substrata suggest that larvae are responding less to a single cue than to a suite of signals which indicate the suitability of a habitat (Cameron & Schroeter 1980, Pearce & Scheibling 1991, but see also Rowley 1989). In contrast, irregular echinoids, such as *Dendraster excentricus* and *Echinarachnius parma* have been shown to settle selectively in the presence of a chemical cue associated with conspecifics, which largely restricts settlement to within adult populations (Highsmith 1982, Pearce & Scheibling 1990b).

Studies of settlement induction and substrate preferences in asteroids have involved only a few species. In Washington, Birkeland et al. (1971) found that larvae of the asteroid *Mediaster aequalis* were highly selective, settling only on the tubes of the polychaete *Phyllochaetopterus prolifica*. In New Zealand, Barker (1977) found that larvae of *Stichaster australis* settled only on the encrusting coralline alga *Mesophyllum insigne*, but that larvae of *Coscinasterias calamaria* were non-selective, provided there was a microbial film on the substrate. In a subsequent study in England, he also found that larvae of *Asterias rubens* and *Marthasterias glacialis* showed no marked substrate preference in laboratory experiments, but tended to settle on the undersides of various substrata, as was observed in the field (Barker & Nichols 1983). *Acanthaster planci* has been shown to settle on a wide

10

variety of substrata, although many studies report only qualitative results (reviewed by Moran 1986) or are compromised by flawed methodology such as inadequate controls for spontaneous settlement (reviewed by Johnson 1992b). In laboratory experiments, Johnson et al. (1991) showed that some crustose coralline algae such as *Lithothamnium pseudosorum* and/or associated bacteria are highly inductive to larvae of *A. planci*, whereas other coralline algal species are not. The authors proposed that *A. planci* on the Great Barrier Reef settle in deeper water at the base of reefs where these inductive substrata occur (see also Johnson 1992b). Alternatively, Zann et al. (1987) suggested that *A. planci* in Fiji settle on coral in shallow water. However, because of the limited number of settlers found in most studies (see part 4, hereafter) spatial patterns of settlement of *A. planci* remain unresolved.

Very little is known about possible cues to settlement in holothuroids (Smiley et al. 1991), crinoids (Holland 1991) and ophiuroids (Hendler 1991). Some species exhibit gregarious settlement, although the chemical or physical factors inducing metamorphosis and settlement have not been experimentally examined. Young & Chia (1982) showed that larvae of the holothuroid *Psolus chitonoides* settle gregariously on or near adult conspecifics in laboratory and field studies in the San Juan Islands, Washington. In contrast, Hamel & Mercier (1996) found no evidence of gregarious settlement in another holothuroid *Cucumaria frondosa* in laboratory and field studies in the St. Lawrence Estuary. Rather, the larvae tended to settle on the undersides of rocks and avoided mud and sand bottoms. Mladenov & Chia (1983) showed aggregated settlement of *Florometra serratissima* on the bottom of culture dishes and concluded that gregarious settlement may account for the adult aggregations that they observed in the field. Larval ophiuroids often appear to settle indiscriminately or metamorphose in the water column in the absence of a substrate more so than other echinoderms (reviewed by Hendler 1991).

3.1 *Summary and conclusions*

Although various cues for induction of settlement of echinoderms have been isolated, the evidence is often based on findings from laboratory studies or from anecdotal evidence from the field. Chia et al. (1984) and Pearce (1997) concluded that larval preferences for settlement substrata are generally less important than post-settlement processes, such as migration and mortality, in determining population distribution and abundance. This brings into question the relative importance of induction for most species of echinoderms and underscores the need to follow settlers in the field through to early juvenile stages to better understand the consequences of preferential settlement (e.g. Highsmith 1982, Young & Chia 1982).

4 SPATIAL AND TEMPORAL PATTERNS OF SETTLEMENT AND RECRUITMENT

4.1 *Sampling methods*

A variety of techniques have been used to record settlement and recruitment rates of echinoderms (Tables 1 to 4), but sampling accuracy remains a considerable methodological challenge. Conventional sampling methods, such as quadrat and grab sampling, are labour intensive and must be repeated frequently if recent recruits are to be enumerated. Harris et al. (1994) have used photographic sampling to measure recruitment of *Strongylocentrotus droebachiensis* in the Gulf of Maine, although the small size and cryptic nature of early juveniles limit the applicability of this approach. Typically, these methods either fail to detect or to accurately census recently settled individuals. Variation in sampling method and frequency among studies may significantly affect measures of recruitment rate, as has been shown for barnacles (Minchinton & Scheibling 1993, Miron et al. 1995), suggesting caution when interpreting data and comparing across studies using different methods.

Of the fifty-two studies of echinoids that we reviewed, ten used artificial collectors to measure settlement or recruitment rates (Table 1). Bak (1985) monitored recruitment of *Diadema antillarum* on plastic collectors (light diffuser panels) suspended vertically 20 cm above coral reefs in Curaçao. He found that submergence times of more than 2 months resulted in reduced settlement due to fouling. Harrold et al. (1991) used plastic pipes containing either light diffuser panels or coralline algae (*Calliarthron* and *Bossiella*), and suspended 1 m above the seabed, to monitor settlement of *Strongylocentrotus purpuratus* and *S. franciscanus* in a kelp forest in California. They found that settlement over 30 to 41 day sampling intervals was greater in the plastic-filled than in the coralline-filled collectors, conflicting with their laboratory observations which showed greater settlement on corallines. However, there were more juvenile crabs and polychaetes in the coralline-filled collectors, which may have preyed upon the newly settled echinoids. Keesing et al. (1993) used plastic bio-filter spheres suspended 1 m off the bottom in mesh bags to compare settlement of several species of echinoderms on the Great Barrier Reef. They found a significant correlation between settler densities on collectors and those on the natural substrata for species from all five echinoderm classes. They also compared various sorting techniques and found that recovery rates of settlers varied from 52 to 100%, and that the most time consuming method was required to collect all echinoderm settlers. Harris et al. (1994) used panels of plastic turf mounted on racks on the bottom to collect settlers of *S. droebachiensis* in the Gulf of Maine. They found much higher rates of recruitment (3 months after settlement) on the turf than on natural substrata such as bare rock and coralline algae. Ebert et al. (1994) and Schroeter et al. (1996) used scrub brushes suspended vertically 1 to 1.4 m off the bottom to monitor settlement of *Strongylocentrotus* spp. in California. Schroeter et al. (1996) concluded that patterns of recruitment on natural substrata observed during benthic surveys were similar to the patterns of settlement measured at weekly intervals in their collectors.

Many studies of ophiuroid recruitment have been done in the deep-sea where sampling is logistically difficult and confined to conventional methods (Table 3). Constraints on sampling frequency and the accuracy of site relocation limit conclusions about recruitment dynamics in such remote communities (Grassle 1994). Ophiuroids also are ubiquitous in shallow waters but there they typically are cryptic and hard to sample. In Denmark, Muus (1981) collected recent settlers (0.325 mm) of *Amphiura filiformis* in sediment samples sieved through a 0.265 mm mesh screen. She observed high rates of settlement and concluded that previous studies using larger mesh sizes did not sample settlers. Estimates of ophiuroid settlement often have involved back-calculation based on modal analysis of size distributions. These methods measure recruitment at best and the stacked age classes that usually occur seriously limit the accuracy of this technique (Gage 1985). Even methods of measuring the size of an ophiuroid have been debated and the reliability of some methods has been questioned (O'Connor & McGrath 1980, O'Connor et al. 1983, Duineveld & Van Noort 1986, Bosselmann 1989, Munday & Keegan 1992). In an effort to circumvent many of the problems with conventional sampling methods, artificial collectors have been used to measure ophiuroid settlement and have detected high rates of settlement in differing shallow water habitats (Keesing et al. 1993, Balch et al. 1999, Balch & Scheibling in press).

Methods used to detect patterns of spatial and temporal variability in recruitment of holothuroids have varied among studies (Table 4). Compared to other echinoderms, holothuroids are particularly difficult to measure accurately due to their soft and flexible body wall. Cameron & Fankboner (1989) suggested that a single measurement of length or wet weight can be unreliable. Different methods of measurement could account for some of the variability observed both within and between studies, and may influence conclusions that are drawn from them.

4.2 Echinoidea

Many studies have documented large spatial and temporal variability (over tens of metres to thousands of kilometres and over months to years) in settlement and recruitment rates of echinoids (Table 1 and reviews by Ebert 1983, Harrold & Pearse 1987, Pearse & Cameron 1991). Pearse &

Table 1. A summary of studies examining spatial and temporal patterns of settlement and/or recruitment of echinoids.

Species	Location	Habitat, Depth	Sampling Method	Size*	Sampling Period, Frequency	Patterns and Conclusions	Source
North Temperate							
Strongylocentrotus droebachiensis	Vega Island, Norway	barrens and kelp, 2.5-10 m	0.25 m² quadrats	≥ 10 mm	Jul 1990, Apr and Jul 1991	Age/size-frequency analysis showed regular annual recruitment in barren areas; kelp beds inhibit recruitment.	Leinaas & Christie 1996
Strongylocentrotus droebachiensis	Isles of Shoals, New Hampshire, USA	barrens and kelp, 0-30 m	artificial collectors (plastic turf) and photo quadrats	0.49 mm	1990, 1992 and 1993, once in Jun-Jul	Substantial annual settlement decreasing with depth (after 9 m). Recruitment greater on artificial turf than rock.	Harris et al. 1994
Strongylocentrotus droebachiensis	Gulf of Maine, USA, Bay of Fundy and Atlantic coast of Nova Scotia, Canada	barrens, kelp, silt, 10 cm-8 m off bottom at 5-30 m	artificial collectors (plastic turf and scrub brush)	0.150 mm or cheese cloth mesh size	May to Sep 1994, 2-8 week intervals	Order of magnitude differences in maximum settlement between regions. Highest settlement in the Gulf of Maine (34008 m⁻²), lowest in the Bay of Fundy (32 m⁻²), and intermediate in Nova Scotia (1066 m⁻²). Within regions, settlement differed between sites but was within the same order of magnitude. Settlement greater in the barrens than in adjacent kelp beds in the Gulf of Maine.	Balch et al. 1998
Strongylocentrotus droebachiensis	St. Margaret's Bay, Nova Scotia, Canada	habitat not stated, intertidal-20 m	0.25 m² quadrats	2 mm	Jul 1968	A size-frequency distribution suggested annual recruitment in May and Jul, following Nov and Apr spawnings.	Miller & Mann 1973

* see page 28.

14

Table 1. (Continued).

Species	Location	Habitat, Depth	Sampling Method	Size*	Sampling Period, Frequency	Patterns and Conclusions	Source
Strongylocentrotus droebachiensis	St. Margaret's Bay, Nova Scotia, Canada	barrens	0.1 m² quadrat	> 5 mm	May-Sep 1975, monthly	Evidence for high recruitment in the year after destructive grazing of kelp.	Lang & Mann 1976
Strongylocentrotus droebachiensis	Atlantic coast of Nova Scotia, Canada	barrens and kelp, 0.2 and 2.3 m off bottom at 5-10 m	artificial collectors (plastic turf) and 1.0 m² quadrat for recruits	0.150 mm and 1 mm mesh size	Jun-Nov 1992-1994, biweekly and recruits once per year 1993-1995	Settlement pulse in Jul of each year. Low settlement in 1992 and '93, high settlement in '94. Settlement greater in barrens than in kelp beds but not significant. Recruitment reflects settlement.	Balch & Scheibling in press
Strongylocentrotus droebachiensis	Avalon Peninsula, Newfoundland, Canada	barrens, 0-24 m	0.2, 0.8 or 1.0 m² quadrat (depending on density)	2-3 mm	1968-1969, once or twice in summer	Decreased recruitment with depth as reduced food and slower growth results in increased predation. Suggested increased recruitment with exposure to wave action (fewer predators) and lower temperature (increased larval survival).	Himmelman 1986
Strongylocentrotus droebachiensis	Conception Bay, Newfoundland, Canada	barrens and macro-algae, 0-18 m	diver-operated air dredge in 0.1 m² quadrats	1 mm	summer 1979-summer 1983, once or twice per year	High recruitment in 1979 and '80 and low from '81-83. Juveniles most abundant in barren zone (6-9 m) particularily on rough coralline algae and rare in the macro-algal zones (0-3 m and 12-18 m). Suggested that macro-algae inhibits recruitment but rough coralline algae in barren areas acts as a refuge from predation.	Keats et al. 1985

15

Table 1. (Continued).

Species	Location	Habitat, Depth	Sampling Method	Size*	Sampling Period, Frequency	Patterns and Conclusions	Source
Strongylocentrotus droebachiensis	St. Lawrence Estuary, Québec, Canada	rock and cobble, intertidal-18 m	diver-operated "scraping devices and an air lift aparatus" in 0.25 m² quadrats	0-1 mm	once in 1978 or 1979 at 8 sites and from 1978-1980/81 at 2 sites, every 2-8 months	Single pulse of recruitment in 1977 and then none from '78-81. High recruitment at all depths in the lower estuary, but limited to below the fresh layer (0-2 m) further upstream. Suggested that low survival of juveniles in the upper estuary related to low salinity.	Himmelman et al. 1983
Strongylocentrotus droebachiensis	Strait of Georgia, British Columbia, Canada	urchin grazed and ungrazed bedrock in estuarine flow 30 km from river, 7.5-9 m	0.25 m² airlift	2.5 mm mesh size	1972-1975, once per year	First record of an urchin outbreak in the region. Recruitment in 1973 and '69 and none in '74 or '75. Recruitment success related to plankton bloom and cold winters which enhance larval survival.	Foreman 1977
Strongylocentrotus droebachiensis, *S. purpuratus*, *S. franciscanus* (Alaska), *S. polyacanthus* (Aleutians)	Aleutian Islands and SE Alaska, USA	rock/boulder bottom with or without kelp, 6-13 m	0.25 m² quadrats	≤ 2-3 mm	1972-1990, every 1-7 years (sites sampled for 3-15 years)	Heavy recruitment in Aleutians in most of the 19 years; low and episodic recruitment in SE Alaska. Regional differences related to larval transport and supply.	Estes & Duggins 1995

16

Table 1. (Continued).

Species	Location	Habitat, Depth	Sampling Method	Size*	Sampling Period, Frequency	Patterns and Conclusions	Source
Strongylocentrotus purpuratus	Tatoosh Island area, Washington, USA	habitat and depth not stated	method not stated	1-2 mm "can be found"	1963-1985	Recruitment only observed in 1963, '69, '82 and '83. No consistent association with El Niño events affecting these waters but some association with warmer water temperatures suggested northward movement of water enhances recruitment in this area.	Paine 1986
Strongylocentrotus purpuratus	Sunset Bay, Oregon, USA	habitat and depth not stated	method not stated	2.5-5 mm	1964-1978, once per year	Good recruitment in 1963. Little or no recruitment during remainder of study.	Ebert 1983
Strongylocentrotus franciscanus, *S. purpuratus*	southern Oregon, USA	habitat not stated, 1 m off bottom and 4 m below surface at 18 and 22 m	plankton tows; artificial collectors (Astroturf)	0.202 mm mesh size (tow); 0.3 mm mesh size (collector)	Jan-Aug 1994 and Mar-Aug 1995, ~ every 2 days- 4 weeks	Larvae more abundant and occur earlier in the year in 1995 than '94. Variable settlement between sites, species and years occurring from Mar-Aug with peak in mid-June of each year and greater in 1995 than '94. More settlement in bottom collectors. Aging indicated settlement events occurred over several days within a sampling period during relaxation of upwelling.	Miller & Emlet 1997

17

Table 1. (Continued).

Species	Location	Habitat, Depth	Sampling Method	Size*	Sampling Period, Frequency	Patterns and Conclusions	Source
Strongylocentrotus franciscanus / S. purpuratus (not distinguished)	Bodega Head, California, USA	granitic rock and attached algae, 5 and 11 m	1 m² quadrats	5-20 mm (< 5 mm "not accurately sampled")	Oct 1988-Oct 1992, ~ quarterly	Juveniles more abundant, smaller and more cryptic at 5 m than 11 m. Greater proportion of adult S. franciscanus provided shelter under spine canopies at 5 m (28 %) than 11 m (2%). Suggested recruitment is greater in shallow water and in association with adults.	Rogers-Bennett et al. 1995
Strongylocentrotus franciscanus, S. purpuratus	northern California, USA	habitat not stated (1992); "rocky bottom" (1993), 0.5-1 m off bottom at 5-20 m (1992), 10-12 m (1993)	artificial collectors (scrub brush)	0.06 mm mesh size	Apr-Sep 1992 and 1993, weekly	No significant vertical (5-20 m) or horizontal (10-100's m) variation in settlement between collectors within a site. Low settlement of both species in 1992; mainly during an unusual upwelling relaxation event in Jul. Minimal settlement at 3 sites in 1993 but high at a 4th site where settlement of S. purpuratus during a relaxation event in Apr.	Wing et al. 1995a, b
Strongylocentrotus purpuratus	central California to central Oregon, USA	tide-pools	"thorough search"	0-2.5 mm	1985-1986, once in spring	Recruitment driven by physical events: low recruitment at upwelling sites (capes and headlands); high annual recruitment at sites between. No trend with latitude.	Ebert & Russell 1988

Table 1. (Continued).

Species	Location	Habitat, Depth	Sampling Method	Size*	Sampling Period, Frequency	Patterns and Conclusions	Source
Strongylocentrotus franciscanus, S. purpuratus	northern and southern California, USA	habitat and depth not stated but variable, 1-1.4 m off bottom	artificial collectors (scrub brush)	0.436 mm mesh size	1990-1993, weekly with some gaps	Seasonal settlement (Feb-Jul) was higher and more regular in the south. Settlement correlated with general oceanographic processes.	Ebert et al. 1994
Strongylocentrotus purpuratus / S. franciscanus (not distinguished)	Pacific Grove, California, USA	kelp forest, 1 m off bottom at 10 m	artificial collectors (plastic light diffuser and coralline algae)	0.25 mm mesh size	Jul 1988-Jul 1989, ~ monthly	Two settlement peaks (Dec-Feb and Apr-Jul). Greater settlement in plastic collectors, probably due to fewer predators.	Harrold et al. 1991
Strongylocentrotus purpuratus, S. franciscanus	Pacific Grove, California, USA	kelp forest, 10 m	10 m² circular plots	> 10 mm (5-10 mm was variable)	Oct 1972-Aug 1981, every 2-12 months with a gap in 1973/74	*S. franciscanus* densities low throughout. Recruitment pulse of *S. purpuratus* in 1975-76. A massive reduction of urchins in 1976 due to predation or disease.	Pearse & Hines 1987
Strongylocentrotus purpuratus, S. franciscanus	Carmel Bay, central California, USA	kelp forest and deforested rock reef, 16-29 m	1 m² quadrats	2-3 mm "some <1 cm overlooked"	1986-1989, ~ twice per year	Evidence of recruitment in kelp in 1987-88. Suggested that heavy recruitment in 1984 was responsible for population increase and deforestation at both sites (but only one site recovered).	Watanabe & Harrold 1991

Table 1. (Continued).

Species	Location	Habitat, Depth	Sampling Method	Size*	Sampling Period, Frequency	Patterns and Conclusions	Source
Strongylocentrotus purpuratus, S. franciscanus	Santa Barbara, California, USA	"patchy" kelp forest and coral-line barren, 8-12 m	collected shale with resident organisms	0.243 mm mesh size	1984-1986, mostly during summer	Few settlers in Apr 1984/85 and many in May '86; similar patterns for both species but lower densities of S. franciscanus. No significant difference between habitats; concluded that post-settlement mortality is greater in the kelp.	Rowley 1989, 1990
Strongylocentrotus purpuratus, S. franciscanus	Point Loma, California, USA	inner, middle and outer kelp forest, 12, 15 and 18 m	1 m² quadrat "haphazardly placed over rock piles where urchins were abundant"	5-7.5 mm	1974-1977, ~ monthly	Substantial annual recruitment at all sites. S. franciscanus recruitment lower at inner (12 m) site and much higher outside the kelp canopy. Concluded that kelp and/or resident predators reduce recruitment.	Tegner & Dayton 1981
Strongylocentrotus purpuratus, S. franciscanus	Point Loma, California, USA	outer, middle and inner kelp forest, 18, 15 and 12 m	1 m² quadrats "haphazardly placed over aggregations of urchins in boulder piles"	"< 10 mm not quantitative-ly sampled"	1983-1987, ~ twice per year	Recruitment on outer edge of kelp forest was low from 1983-85 and increased in '86 and '87 to previous levels (1970s, see above); recruitment remained low within forest. Low recruitment related to El Niño conditions leading to reduced production/survival of larvae and altered currents.	Tegner & Dayton 1991

Table 1. (Continued).

Species	Location	Habitat, Depth	Sampling Method	Size*	Sampling Period, Frequency	Patterns and Conclusions	Source
Strongylocentrotus franciscanus, S. purpuratus	San Diego, California, USA	inshore, offshore and within 3 kelp forests, 1 m off bottom at 12-15 m	artificial collectors (scrub brush)	0.436 mm mesh size	Mar-May 1991 and Jan, Apr-Jul 1992, ~ weekly (range = 5-20 d)	Low settlement (0-10 per collector) in both years. Kelp forest had no effect on settlement of S. purpuratus. Variable settlement of S. franciscanus with some evidence for higher settlement offshore. Concluded that kelp forests do not affect larval supply or settlement.	Schroeter et al. 1996
Strongylocentrotus purpuratus	La Jolla, California, USA	habitat not stated	method not stated	0-2.5 mm	1970-1978, 6 months-3 years apart	Good recruitment in 1969, '71, '74 and substantial recruitment in '72 and '76.	Ebert 1983
Strongylocentrotus purpuratus	Papalote Bay, Baja California, Mexico	habitat not stated	method not stated	0-2 mm	1962-1969, once or twice per year	Substantial annual recruitment with more than one pulse in some years and variability in timing between years.	Pearse 1970 in Ebert 1983
Dendraster excentricus	Monterey Bay, California, USA	inshore sand dollar bed to offshore, 8-30 m	plankton tows; grab and core samples	0.202 mm mesh size (tow); 0.3 mm mesh size (grab)	1978-1981, monthly with gaps	Abundance of competent larvae peaked in summer. Variable interannual settlement within and outside adult sand dollar bed.	Cameron & Rumrill 1982
Echinarachnius parma	Middle Atlantic Bight and Georges Bank, northwest Atlantic	habitat not stated, 20-70 m	0.1 m² Smith-McIntyre grab	1 or 0.5 mm mesh size	Jul 1977-Jul 1985, every 2-13 months	Regular annual recruitment at 3 of 4 sites from Dec-Apr in Middle Atlantic Bight, from Nov-Jul on Georges Bank. No recruitment at the fourth site after 1978 attributed to anoxic conditions.	Steimle 1990

21

Table 1. (Continued).

Species	Location	Habitat, Depth	Sampling Method	Size*	Sampling Period, Frequency	Patterns and Conclusions	Source
Echinus affinis	Rockall Trough, north-east Atlantic	habitat not stated, 1632-2300 m	3 m wide Agassiz trawl, epibenthic sledge, single-warp trawl or semi-balloon otter trawl	20, 16, 12, 10, 1 or 0.5 mm mesh size	Jun 1973-Apr 1985, 1 day-37 months apart	Mixed unimodal and bimodal size-frequency distributions, but variation in sampling methods and possible bias may have obscured patterns. Samples using small mesh sizes suggested recruitment is rare.	Gage & Tyler 1985
Echinocardium cordatum	German Bight, North Sea	fine sand, 25 and 35 m	vertical plankton tows with Nansen net; 10 cm² subsample from 0.017m² Reineck box core	0.15 mm mesh size (tow); 0.1 mm mesh size (core)	Apr 1985-Dec 1986, monthly	Larvae and settlers present from Jun-Aug in both years. Maximum settlement (26000 m⁻²) in Aug; heavy post-settlement mortality eliminated recruits by Sep of both years. Suggested dense settlement enables population recovery in years following heavy adult mortality.	Bosselmann 1989
Echinocardium cordatum	Terschelling and Texel Islands, North Sea, Netherlands	sand, 7-18 m	0.2 m² Van Veen grab sampler	1 mm mesh size	1972-1982, ~ once per year from Apr-Jun	High recruitment (individuals < 1.5 cm) in 1973, 1977 and 1980; low or no recruitment in other years. Recruitment greatest at depths of 8-12 m and low above 8 m and below 12 m. Suggested that recruitment occurs infrequently but simultaneously over large areas (hundreds of km), and may be related to colder water temperatures.	Beukema 1985

Table 1. (Continued).

Species	Location	Habitat, Depth	Sampling Method	Size*	Sampling Period, Frequency	Patterns and Conclusions	Source
Echinocardium cordatum	Bay of Seine, Normandy, France	sand, 0–25 m	0.1 m² grab samples	3–6 mm	1986–1987, once per year	Recruitment increased with depth; no recruits in littoral zone. Suggested that 2–4 year olds migrate to the littoral zone.	DeRidder et al. 1991
Sphaerechinus granularis	Penfret Island, Brittany, France	algae-encrusted shell debris, mud, Zostera marina and rock, < 5 m	"1 m wide dredge" or diver collected	20 mm mesh size or not stated	Dec 1988–Dec 1991, monthly with an 8 month gap in 1989/90	Consistent recruitment in Aug/Sep each year but of variable magnitude among years. Recruits found on algae whereas adults occur only on rocks. Suggested post-settlement migration from juvenile to adult habitats occurs both passively (with currents) and actively (with change in diet).	Guillou & Michel 1993a, b, see also Glémarec & Guillou 1996
Paracentrotus lividus, Arbacia lixula	Bay of Villefranche, Provence, France	6–52 km offshore, 10 m and 200–0 m	plankton tows	0.2 mm mesh size	1984–1988, every 2 weeks–3 months	Synchronous spawning twice per year (spring and fall). Larvae restricted to the surface layer, decreased with distance from the coast and rarely occurred beyond a divergence zone 30 km offshore. Timing of recruitment of P. lividus related to larval supply.	Pedrotti & Fenaux 1992, Pedrotti 1993
Clypeaster ravenelii	northern Gulf of Mexico	sand, 110 m	otter trawl	10 mm mesh size	Dec 1988–Jan 1991, bimonthly with gaps	Low numbers of juveniles (1–4 cm) noticed in Sept 1989, and May and Aug '90 but little evidence of substantial recruitment. Suggested that if recruitment occurs, it is episodic.	McClintock et al. 1994

23

Table 1. (Continued).

Species	Location	Habitat, Depth	Sampling Method	Size*	Sampling Period, Frequency	Patterns and Conclusions	Source
South Temperate							
Evechinus chloroticus	Goat Island Marine Reserve, New Zealand	barrens, 5-8 m	0.25 m² quadrats "carefully searched"	> 5 mm "reliably found"	1975-1977, yearly; 1979-1981, monthly with gaps	Juveniles found at all sampling dates. Smaller proportions of juveniles in 1975-77 than in '79-81 suggested variable recruitment among years. Concluded that recruitment maintains populations in barrens.	Andrew & Choat 1982
Centrostephanus rodgersii	central coast New South Wales, Australia	barrens	5 m² transects	> 3 mm	Jan 1985-Jan 1988, 1-3 times per year	Recruitment varied between 4 sites spanning 300 km from no recruitment over 4 years to high annual recruitment. Suggested differences in availability of spatial refugia (rock types) may result in variable rates of juvenile survival.	Andrew & Underwood 1989
Centrostephanus rodgersii	Botany Bay, New South Wales, Australia	barrens and foliose algae, < 6 m	"haphazard swim"	5-10 mm	Oct 1986-Jun 1988, ~ bimonthly	Recruitment observed at all 7 sites from Jan-Apr in both years. Suggested that growth of foliose algae (at 5 sites following mass mortalities of C. rodgersii) did not affect recruitment.	Andrew 1991

Table 1. (Continued).

Species	Location	Habitat, Depth	Sampling Method	Size*	Sampling Period, Frequency	Patterns and Conclusions	Source
Tropical							
Araeosoma fenestratum, Archaeopneustes hystrix, Aspidodiadema jacobyi Brissopsis sp., Cidaris blakei, Conolampas sigsbei, Linopneustes longispinus, Lytechinus euerces, Paleobrissus hilgardi, Paleopneustes cristatus, Paleopneustes tholoformis, Phormosoma placenta, Salenia goesiana, Stylocidaris lineata	northern Bahamas	habitat not stated, 100-930 m	individually collected using suction device or scoop on submersible	≥ 5 mm	Oct 1985-Feb 1990, ~ twice per year (9 cruises)	Only Phormosoma placenta showed direct evidence of recruitment; small indivduals of other species not detected. Suggested recruitment of those species is sporadic and patchy, based on adult size distributions.	Young 1992
Diadema antillarum	Curaçao, Caribbean	coral reef terrace and slope, 3-30 m	transect survey	size not stated	May-Oct 1983, twice	After mass mortality due to disease, ≤ 1 mm recruits settled throughout the year. Larvae may have originated from an island 52 km up current not affected by disease.	Bak et al. 1984

25

Table 1. (Continued).

Species	Location	Habitat, Depth	Sampling Method	Size*	Sampling Period, Frequency	Patterns and Conclusions	Source
Diadema antillarum	Curaçao, Caribbean	2 coral reefs 24 km apart, 20 cm off bottom at 8 m	artificial collectors (plastic light diffuser panels)	≤ 3 mm	1982-1984, biweekly with gaps	Continuous settlement with spring and fall peaks. Settlement similar between years but differed between sites.	Bak 1985
Diadema antillarum	Barbados, Caribbean	fringing coral reef	10-250 m² search area	< 10 mm = settlers; 10-15 mm = recruits	Oct 1984-Dec 1985, monthly	Settlers found in Jun in cryptic habitats and aggregated with adults on offshore region of reefs. Populations recovered to 57% of pre-mortality levels within 2 years after disease. Suggested recruitment is greater on reefs with higher adult densities.	Hunte & Younglao 1988
Diadema antillarum	St. Croix, US Virgin Islands, Caribbean	coral reef, 2-10 m	1m² quadrats	0-10 mm	Dec 1983-Mar 1986, 2-15 months apart	Low recruitment at 2 of 4 sites in Feb, Apr 1984 only. Paucity of larvae/settlers after mass mortalies from 1983-85 suggested population recovery is recruitment-limited.	Carpenter 1990
Diadema antillarum	St. John, US Virgin Islands, Caribbean	habitat previously stated	method previously stated	recruit = < 50 mm	1984-1988, ~ yearly	Annual recruitment rates were low (0.017-0.534 m⁻²yr⁻¹) after mass mortalities from 1983-84. Significant interannual variation but no significant variation between sites in the same bay. Highest recruitment in 1985, lowest in 1988. Concluded that recruitment rate was density-independent but too low to enable populations to recover to pre-disturbance levels.	Karlson & Levitan 1990

26

Table 1. (Continued).

Species	Location	Habitat, Depth	Sampling Method	Size*	Sampling Period, Frequency	Patterns and Conclusions	Source
Diadema antillarum	Molasses Reef, Florida Keys reef tract, USA	offshore reef, 1-2 m	140 m^2 quadrat	3-4 mm	Jul 1991-Aug 1992, every 1-4 months	Mass mortality in Jan/Feb 1991 reduced densities by 97%. Low recruitment from Jul-Oct 1991, none in 1992. Suggested populations will not recover because mortality rate exceeds recruitment.	Forcucci 1994
Diadema antillarum	Panama, Caribbean	sand, coral and seagrass between 2 reefs, 0.5-2.5 m	3-2500 m^2 permanent quadrats	< 15 mm	1982-1987, biweekly-bimonthly	Recruitment immediately after mass mortality in 1983 but none during remainder of study. Experiments showed that predation, conspecific protection, settlement cues, and competition by other echinoids were not responsible for recruitment failure. Low fertilisation success and poor larval supply were likely the cause.	Lessios 1988
Diadema antillarum	Panama, Caribbean	coral reef	method not stated	size not stated	1983-1993, "continuously"	Since mass mortality in 1983, population density has been < 10% of pre-mortality level. Recruitment was very low over study period.	Lessios 1994
Anthocidaris crassispina	Hong Kong	habitat and depth not stated	diver-collected	5-10 mm	Jan 1983-Dec 1984, monthly	Recruitment from Feb-Jun in both years at 3 sites; no recruitment at fourth site probably due to adverse environmental conditions (low salinity).	Chiu 1986

27

Table 1. (Continued).

Species	Location	Habitat, Depth	Sampling Method	Size*	Sampling Period, Frequency	Patterns and Conclusions	Source
several species including *Echinometra mathaei, Mespilia globulus* (could not be distinguished)	Davies Reef, Great Barrier Reef, Australia	windward and leeward coral reef, 1 m off bottom at ~ 15 m	artificial collectors (100 plastic bio-filter spheres in a net bag), boxes of rubble	0.5 mm mesh size	Nov 1991-Jan 1992, rubble collected in Jan	Settlement was significantly greater on the windward edge than on the leeward edge of the reef which corresponds with observed recruitment to rubble. Within a given reef habitat there was no significant difference in settlement rate between collectors placed 1-100's of meters apart.	Keesing et al. 1993
Echinometra mathaei	Rottnest Island and other sites along Western Australia's coast	habitat and depth not stated	used 0.25 m² quadrats to collect size frequency data and genetic traits to distinguish groups	5-10 mm	1985 (Rottnest Island) and 1987 (all other sites)	Populations within 4 km of each other had as much genetic variance as populations sampled over 1300 km of coast. Suggested that, over time, variable recruitment supplies different sites with larvae from separate populations.	Watts et al. 1990
Polar							
Sterechinus neumayeri	McMurdo Sound, Antarctica	mud, sand and gravel, 6-18 m	diver-collected	30-31 mm	1989, 1990 and 1992, Dec of each year	Size-frequency analysis and ageing failed to detect recruitment between 1985 and 1992. Suggested that recruitment in McMurdo Sound is rare and controlled by physical factors (current and ice activity) that vary interannually.	Brey et al. 1995

* Measurements are test diameters unless specified. Size is either the smallest individual recorded or deemed detectable, the smallest size-frequency range containing individuals, and/or the mesh size used to filter or collect samples.

Table 2. A summary of studies examining spatial and temporal patterns of settlement and/or recruitment of asteroids.

Species	Location	Habitat, Depth	Sampling Method	Size*	Sampling Period, Frequency	Patterns and Conclusions	Source
North Temperate							
Asterias forbesi	Long Island Sound, Connecticut, USA	habitat not stated, 0-33 m	collectors (100 clean oyster shells in wire mesh bags)	size not stated	1937-1961, twice per week	Settlement increased with depth to 10 m and then decreased to 33 m. Timing and magnitude of settlement varied between sites and years, occurring between Jun and Sep and ranging in duration from 1-91 days (mean = 52 days). Settlement intensity varied from a single peak in 1 week to constant settlement throughout the period. Annual settlement increased from NE to SW and varied by 4 orders of magnitude (0-1700 collector⁻¹) with no pattern of good and bad years. Settlement not correlated with adult density.	Loosanoff 1964
Asterias vulgaris and/or A. forbesi	Atlantic coast of Nova Scotia, Canada	barrens and kelp, 0.2 and 2.3 m off bottom at 5-10 m	artificial collectors (plastic turf) and 1.0 m² quadrat for recruits	0.150 mm and 1 mm mesh size	Jun-Nov 1992-1994, biweekly and recruits once per year 1993-1995	Settlement pulse in Aug/Sep of each year. Highest settlement in 1993 and 1994, intermediate settlement in 1992. Settlement greater in kelp beds than in barrens. Recruitment reflects settlement.	Balch & Scheibling in press
Asterias rubens (= A. vulgaris)	Torbay, southwest England	rocky intertidal shore	"carefully searched"	2-5 mm	1980-1981, monthly	Recruits first appeared in Jul 1980 and Sep 1981 beneath boulders and in crevices. No recruits found in adjacent areas with similar topography.	Barker & Nichols 1983, Nichols & Barker 1984

29

Table 2. (Continued).

Species	Location	Habitat, Depth	Sampling Method	Size*	Sampling Period, Frequency	Patterns and Conclusions	Source
Asterias vulgaris, Leptasterias polaris, Crossaster papposus	Mingan Islands, Gulf of St. Lawrence, Canada	bedrock, boulders, cobble, sand/mud, intertidal-20 m	2 m² quadrat	≥ 5 mm	Aug-Sep 1984	A. vulgaris juveniles most abundant at 4-7 m and found only in boulders or cobble. L. polaris juveniles most abundant at 0-1 m and found only in boulders, cobble or bedrock. Small C. papposus juveniles found only on sediment bottoms > 11 m.	Himmelman & Dutil 1991
Pisaster ochraceus, Pycnopodia helianthoides, Dermasterias imbricata	Nootka Sound, British Columbia, Canada	bare and algae-covered rock, 1-9 m; 1-2 m (tows)	100 cm² quadrat for recruits; plankton tows	0-5 mm "ray length" of recruits; 0.125 mm mesh size (tows)	1987-1991, recruitment data once per year; plankton tows in May and Sep 1989	Recruitment varied between species, sites and years. Recruitment of Pisaster ochraceus was high in all years but mortality of recruits after 1 year was > 97%. Pycnopodia helianthoides recruited in 1987-89 and 1991. D. imbricata recruited in low densities in 1988. Highest densities of bipinnaria larvae found in plankton tows in a semi-enclosed bay. Proposed this area acts as a source of larvae.	Sewell & Watson 1993
Pisaster ochraceus	San Juan Island, Washington, USA	rocky intertidal	0.1-1 m² quadrats	1-10 mm	May 1969, Aug 1970 and Apr 1971	Annual recruitment at Point Caution over all 3 years. This site proposed as a nursery area because Gull Reef and other sites in region showed few recruits.	Lubchenco-Menge & Menge 1974

Table 2. (Continued).

Species	Location	Habitat, Depth	Sampling Method	Size*	Sampling Period, Frequency	Patterns and Conclusions	Source
Pisaster ochraceus	Pacific north-west	habitat not stated (includes intertidal)	method not stated	size not stated	1984-1987	Size-structure varied markedly over 20 sites. Predictable and intense recruitment at sheltered sites. Density of recruits declined rapidly and survivors were restricted to crevices.	Rumrill 1988b (abstract)
Asterina miniata	Barkley Sound, British Columbia, Canada	mud and cobble, 4-6 m	0.25 m² quadrats	2-4 mm "ray length"	Mar 1985-Aug 1987, bimonthly-yearly	Recruitment at 1 of 2 sites in 1983 and 1984; none at either site from 1985 to 1987. Evidence that migration and juvenile mortality are not as important as pre- and early post-settlement events in determining recruitment.	Rumrill 1989
Mediaster aequalis, Luidia foliolata, Crossaster papposus, Henricia leviuscula, Solaster stimpsoni, S. dawsoni, Pteraster tesselatus	San Juan Islands, Washington, USA	sand and tube worm (Phyllo-chaetopterus prolifica) beds, 20 m	diver-collected	> 2 mm	"fall and winter, 1968-1969", 4 dives	Juvenile asteroids of all species were commonly found on poly-chaete tubes and rare elsewhere. Suggested tube worm beds are a nursery ground for juveniles, which subsequently migrate to sandy areas to feed on larger prey.	Birkeland et al. 1971
unidentified species	northern California, USA	habitat not stated, 0.5-1 m off bottom at 5-20 m	artificial collectors (scrub brush)	0.06 mm mesh size	Apr-Sep 1992, weekly	Low levels of settlement occurred throughout most of the sampling period at both sites, with a peak in settlement in Jul during an unusual upwelling relaxation event.	Wing et al. 1995a

31

Table 2. (Continued).

Species	Location	Habitat, Depth	Sampling Method	Size*	Sampling Period, Frequency	Patterns and Conclusions	Source
Patiria miniata	Southern California Bight, USA	rock reefs and boulders	method not stated	5-10 mm	period not stated	Juveniles found under boulders and adults on reef surface. Suggested juveniles are either excluded by adults or subject to predation when exposed.	Day & Osman 1981
Astropecten bispinosus, *A. aranciacus*	Sardinia, Italy	sand bottom (5-10 m) and seagrass (*Zostera marina*) beds (10-20 m)	25 m² grid	10-15 mm radius	Jul/Aug 1980	Recruitment never observed on sandy bottom where adults of both species coexist; suggested migration of juvenile *A. bispinosus* from seagrass beds compensates for loss via predation by *A. aranciacus*.	Jost & Rein 1985
Anseropoda placenta	Bay of Brest, Brittany, France	"muddy sand, sediments with shells and sandy gravel", 17-35 m	"0.8 m wide 'Charcot' dredge"	20 mm mesh size	Dec 1983-Nov 1986, monthly with a 6 month gap in 1984	Annual recruitment from Aug-May. Higher recruitment at deeper site (where shells provide a refuge); possible migration to shallower sites (where adults are more abundant).	Guillou & Diop 1988, see also Glémarec & Guillou 1996
South Temperate							
Stichaster australis, Coscinasterias calamaria	Maori Bay, and other sites around North Island, New Zealand	rock reefs and boulders, intertidal-5 m	"searched exhaustively on many occasions"	≥ 7 mm	1974-1976, "on many occasions"	*S. australis* settlers found only on encrusting coralline alga (*Mesophyllum insigne*) at all sites. *C. calamaria* settlers not found in Maori Bay (where juveniles and adults existed) but occurred on various algae at other sites.	Barker 1977

Table 2. (Continued).

Species	Location	Habitat, Depth	Sampling Method	Size*	Sampling Period, Frequency	Patterns and Conclusions	Source
Stichaster australis	Maori Bay, New Zealand	a large (2 m2) intertidal boulder	measured each juvenile found on the boulder	2-3 mm	1974-1979, monthly	Recruitment density varied by an order of magnitude between years. Timing of recruitment also varied between years.	Barker 1979
Tropical							
Oreaster reticulatus	Grenadines and St. Croix, Caribbean	seagrass beds and sand bottoms, 1-13 m	25-100 m² quadrats along a transect	> 20 mm radius	1974-1977, ~ every 1-4 months	Juveniles mainly observed adjacent to or within dense seagrass beds. Suggested that they settle in seagrass beds and migrate to adult populations in open sand areas once they reach a size refuge from predation.	Scheibling 1980a, b
Acanthaster planci	Iriomote-jima, Ryukyu Islands, Japan	fringing coral reef (flat and slope, floor of groove or base of coral), depth not stated (imply 5.5-18.5 m)	0.25 m² quadrats (20 or 10 cm layer of substrate collected by hand)	0.35 or 2 mm mesh size	Mar-Nov 1985, frequency and allocation of 191 samples not stated	Found 13 recruits in Sep and Nov 1985 (9 in 1984). 12 of 13 found in 5.5-7 m and one at 18.5 m depth. 11 of 13 found on the reef floor and the other 2 on the reef slope.	Yokochi & Ogura 1987
Acanthaster planci	Suva Reef, Fiji Islands	coral reef crest, intertidal-? (imply 2 m)	1 m² quadrats	10-11 mm	1979-1987, monthly-yearly	Low recruitment for most of the 9 years except 1984 when massive recruitment (3 order of magnitude increase) over most of the reef flat resulted in a population outbreak and subsequent migration down the reef slope.	Zann et al. 1987

33

Table 2. (Continued).

Species	Location	Habitat, Depth	Sampling Method	Size*	Sampling Period, Frequency	Patterns and Conclusions	Source
Acanthaster planci	Suva, Nukubuco and other reefs, Fiji Islands	coral reef and rubble, intertidal-0.5 m	1 m wide belt transects, rubble searches, random 0.25 m² quadrats, spot dives and reef users' reports	10-30 mm	1979-1989, yearly	Massive recruitment in 1977 and 1984, large recruitment in 1987 and little or no recruitment in other years, over thousands of hectares. Intense recruitment over only a few hectares in 1982 and 1983. No correlation between recruitment and terrestrial runoff associated with increased rainfall.	Zann et al. 1990
Acanthaster planci	Green Island, Great Barrier Reef, Australia	live coral and coralline algal-covered rubble, 2-12 m	rubble searches	≥ 2 mm	1986-1990, once or twice per year	Only 2 recruits (2 and 4 cm) found in 4 years, suggesting low or very patchy recruitment.	Fisk 1992, see also Fisk et al. 1988
Acanthaster planci	16 reefs in central Great Barrier Reef, Australia	coral reef base, crest and flat; 15 m below, 2-5 m below and on top of reef respectively	destructive sampling of 10 m² belt transects	> 10 mm	Jul-Nov 1986 and Nov 1987, monthly	Greater recruitment in the lower zones of the reef. Reduced recruitment with increasing distance downstream of the primary outbreak population. Recruitment in 1985 was an order of magnitude greater than in 1986 and 1987. Suggested that a single year of successful recruitment could result in outbreaks.	Doherty & Davidson 1988

Table 2. (Continued).

Species	Location	Habitat, Depth	Size*	Sampling Method	Sampling Period, Frequency	Patterns and Conclusions	Source
Acanthaster planci, *Choriaster granulatus*, *Culcita novaeguineae*	Davies Reef, Great Barrier Reef, Australia	windward and leeward coral reef, 1 m off bottom at ~15 m	0.5 mm mesh size	artificial collectors (100 plastic bio-filter spheres in a net bag), boxes of rubble	Nov 1991-Jan 1992, rubble collected in Jan	Minimal settlement of asteroids (n = 11) on collectors with corresponding low recruitment to rubble. 7 of the 11 settlers, including all 3 *A. planci* found on the windward reef slope. Suggested collectors could be a useful tool in predicting outbreaks of *A. planci* 3 years in advance.	Keesing et al. 1993
Acanthaster planci	Lord Howe Island, Australia	coral reef and unstated habitat, 1-40 m	> 15 cm	500-5000 m^2 "searched systematically"	1987 and 1989, once each year	Suggested that range of observed sizes (15-52 cm) indicates annual recruitment since 1985.	DeVantier & Deacon 1990
Polar							
Odontaster validus	McMurdo Sound, Antarctica	volcanic sediments, 10, 20 and 30 m	0-1 g wet weight	diver-collected	Oct 1984-Aug 1985, every 2-7 months	Size-frequency distributions showed no obvious recruitment over the year. Suggested that recruitment is low and temporally stable.	McClintock et al. 1988

* Measurements are diameters unless specified. Size is either the smallest individual recorded or detectable, the smallest size class containing individuals and/or the mesh size used to filter or collect samples.

35

Table 3. A summary of studies examining spatial and temporal patterns of settlement and/or recruitment of ophiuroids.

Species	Location	Habitat, Depth	Sampling Method	Size*	Sampling Period, Frequency	Patterns and Conclusions	Source
North Temperate							
Acrocnida brachiata	Douarnenez Bay, Brittany, France	sand, intertidal and 20 m	0.25 m² core (intertidal), 0.062 m² suction sample (subtidal)	0.16 mm mesh size (intertidal), 1 or 0.2 mm mesh size (subtidal)	Mar 1984–Jul 1986, ~monthly with some gaps	Annual recruitment at both sites in June with high mortality in the subsequent 2 months. Higher recruitment in the subtidal zone is attributed to less harsh hydrodynamic conditions.	Bourgoin et al. 1990, see also Glémarec & Guillou 1996
Amphiura filiformis, Ophiothrix fragilis, Ophiopluteus bimaculatus, O. compressus ophioplutei	Bay of Ville-franche, Pro-vence, France	6-52 km offshore, 10 m and 200-0 m	zooplankton tows	0.2 mm mesh size	1984-1988, every 2 weeks-3 months	Larvae restricted to the surface layer, decreased with distance from the coast and rarely occurred beyond a divergence zone 30 km offshore. A mixture of larval stages at various times of the year suggested variable recruitment over most of the year.	Pedrotti & Fenaux 1992, Pedrotti 1993
Amphiura filiformis	Concarneau Bay, Brittany, France	muddy sand, 17 and 28 m	0.1 m² Smith McIntyre grab sampler	1 mm mesh size	May 1972–Oct 1973 and Jun 1977-Feb 1979, every 3-7 months	Recruitment in spring and fall each year. Suggested high post-spawning mortality of adults facilitates recruitment, particularly under unstable evironmental conditions.	Bourgoin & Guillou 1988, see also Glé-marec & Guillou 1996

36

Table 3. (Continued).

Species	Location	Habitat, Depth	Sampling Method	Size*	Sampling Period, Frequency	Patterns and Conclusions	Source
Amphiura filiformis	North Sea, off the Dutch coast	fine sand with silt, 30 m	0.06 m² Reineck box corer	1.0 and 0.2 mm nested sieves (measured disk size: "distance between alternating radial shields")	Sept 1982-May 1984, ~ quarterly	Stable size structure with large adult mode and seasonal recruitment from Jul-Sep each year. Density of recruits varied annually (mean: 3000-15000 m⁻²) but high mortality in their first year resulted in a low but steady rate of renewal of adult population.	Duineveld & Van Noort 1986
Amphiura filiformis	German Bight and the central North Sea	muddy fine sand, 38 and 54 m	0.1 m² van Veen grab; 0.017m² Reineck box sampler	0.5 mm mesh size; 0.125 mm mesh size (0.2-0.3 mm)	1983, 1984 and 1986-1988, every Mar/Apr (and Jun, Jul/Aug and Nov in 1983 and 1987); Mar and Dec 1987 and Apr 1988	Size-frequency analysis suggested annual recruitment with settlement beginning Jul/Aug. Data from several studies indicated a SW to NE progression in timing of settlement in the North Sea.	Künitzer 1989

Table 3. (Continued).

Species	Location	Habitat, Depth	Sampling Method	Size*	Sampling Period, Frequency	Patterns and Conclusions	Source
Amphiura filiformis	Galway Bay, west coast of Ireland	silty sand, 20 m	0.12 m^2 van Veen grab or suction sampler	0.5 or 1 mm mesh size	Oct 1974-Sept 1976 and Nov 1978- Apr 1982, ~ every 1-7 months	High densities found at a permanent site (mean = 290-2226 m^{-2}) and throughout the bay, but only ~ 5 % were juveniles (< 4 mm). Very low and patchy recruitment observed over the 8 year period. Settlers were not sampled due to large sieve size, but peak settlement assumed to occur in the fall (Sep-Nov).	O'Connor et al. 1983, see also O'Connor & McGrath 1980
Amphiura filiformis, A. chiajei	the Øresund, off Denmark	muddy sand, 27 m	0.02 m^2 "mouse-trap" sampler	0.265 mm mesh size (settler = 0.325 mm)	Oct 1963-Oct 1965, biweekly	79% of A. filiformis sampled were recruits (0.3-0.6 mm). Peak settlement (6-7000 m^{-2}) occurred annually during a 2-6 week interval in Sep-Nov. Few recruits survived their first year. Few juveniles (0.7-4 mm) were present and adult cohorts overlapped with relatively stable densities (mean = 575 m^{-2}). A. chiajei were present in very low numbers and settlement started in Nov (~ 3 mo after A. filiformis).	Muus 1981

38

Table 3. (Continued).

Species	Location	Habitat, Depth	Sampling Method	Size*	Sampling Period, Frequency	Patterns and Conclusions	Source
Amphiura chiajei	Killary Harbour, west coast of Ireland	soft mud, 14 m	0.02 m² "mousetrap" sampler and 0.12 m² van Veen grab	0.5 or 1.4 mm mesh size (measured oral width and disk diameter)	Nov 1985-Oct 1988, every 1-2 months with a 12 month gap in 1987/88	Limited and variable recruitment occurred over the sampling period. Suggested that low survival of recruits is due to competition with a dense adult population (~ 700 m⁻²). Adult mortality during severe winters may allow for occasional heavy recruitment.	Munday & Keegan 1992
Amphipholis squamata	Firth of Forth, Scotland	"boulder-strewn rocky shore", low water	"200 ml of muddy gravel" from under boulders	0.5-1.0 mm	Dec 1975-Nov 1976, monthly	Juveniles recruited to the study population from Jun-Sep but declined in numbers over winter and spring.	Jones & Smaldon, 1989
Amphipholis squamata	Firth of Forth, Scotland; South Devon, England	"loose muddy gravel between boulders", low water; tidepools, high water	200 ml of gravel; 150-200 ml of fringing turf	0-0.99 mm	Dec 1975-Nov 1976, monthly; Jul 1986-Jun 1987, monthly	Recruitment occurred from May-Aug in Scotland, but occurred in a pulse (in Aug) in England. Differences in recruitment patterns attributed to differences in adult survival: adults die in fall in England whereas they continue to brood through a second winter in Scotland.	Emson et al. 1989

39

Table 3. (Continued).

Species	Location	Habitat, Depth	Sampling Method	Size*	Sampling Period, Frequency	Patterns and Conclusions	Source
Ophiomusium lymani	various sites in northeast and northwest Atlantic and northeast Pacific	habitat not stated, 1100-2300 m	epibenthic sled, Agassiz trawl, Megatrawl, Otter trawl, young fish trawl, Blake trawl	0-1 mm (various mesh sizes used)	1910-1981, variable	Re-analysed size distributions from various studies. Presence of several juvenile modes in all regions suggested seasonal and annual recruitment in spring in the Atlantic and late summer in the NE Pacific. Juvenile modes dominated most of the populations but survivorship low; variable presence of adult modes attributed to different rates of mortality with the NE Atlantic region showing lowest adult mortality.	Gage 1982, see also Gage & Tyler 1982a, Gage et al. 1980, Tyler 1988
Ophiomusium lymani	Rockall Trough, northeast Atlantic	silt, clay, and sand mixture, 2200 m	Agassiz trawl, epibenthic sled	0-2 mm (10 mm or 0.5 mm mesh size and/or "sorted on deck")	Apr 1978-Sep 1980, 1 day-7 months apart	Polymodal size distributions suggested annual recruitment with a settlement pulse in May (due to large mesh size, settlers were only effectively measured on 2 trawls in 1980). This conflicts with the observed lack of seasonal reproduction but may be explained by regulation of larval survival by seasonal inputs of detritus from surface waters. Concluded that high mortality of juveniles < 4 mm results in low rate of increase of adult population.	Gage & Tyler 1982a, b, see also Gage et al. 1980, Tyler 1988

Table 3. (Continued).

Species	Location	Habitat, Depth	Sampling Method	Size*	Sampling Period, Frequency	Patterns and Conclusions	Source
Ophiocten gracilis	Rockall Trough and Outer Hebrides continental slope, north-east Atlantic	silt, clay, and sand mixture, 2200, 2900 and 600-1200 m	epibenthic sled or 0.25 m² box-core	0.75-1.0 mm (0.5 mm and unstated mesh size)	May 1975-Sep 1980, ~ every 2-7 months	Heavy settlement in early summer (May/Jun) at two sites in the Rockall Trough (2200 and 2900 m) where adults absent. Mortality at that time was high and no settlers survived the winter. Found a population on the Hebridean continental slope (600-1200 m) with a polymodal size distribution of juveniles suggesting annual and seasonal recruitment.	Gage & Tyler 1981a, see also Gage et al. 1980, Gage & Tyler 1981b, 1982b, c, Tyler & Gage 1980, Tyler 1988
Ophiocten gracilis, *Ophiura ljungmani*	Rockall Trough and surrounding slopes, north-east Atlantic	silt, clay, and sand mixture, 704-2900 m	epibenthic sled and box core	0.25-0.5 mm (0.5 mm mesh size)	Jun 1973-Sep 1980, 1-43 months apart	Suggested uniform annual settlement over the observed depth range for both species. Adult breeding populations of *Ophiocten gracilis* (range = 704-2900 m) concentrated at ~ 1000 m with a high proportion of juveniles at all depths but decreased survivorship with depth. Adult *Ophiura ljungmani* (range = 1632-2900 m) concentrated at ~ 2900 m and juveniles dominated the shallower populations with no clear trend in survivorship. Concluded that both species showed considerable variability with depth and suggested that depth is not a direct controlling factor.	Gage and Tyler 1982c, see also Gage & Tyler 1981a, 1981b, Gage et al. 1980

41

Table 3. (Continued).

Species	Location	Habitat, Depth	Sampling Method	Size*	Sampling Period, Frequency	Patterns and Conclusions	Source
Ophiura ljungmani	Rockall Trough, north-east Atlantic	silt, clay, and sand mixture, 2200 and 2900 m	epibenthic sled	0.25-0.5 mm (0.5 mm mesh size)	Nov 1975-Sep 1980, ~ 3 times per year	Time series of size distributions showed that high densities of recruits dominated the population. Suggested settlement occurs annually in summer but the timing and magnitude varies from year to year.	Gage & Tyler 1981b, see also Gage et al. 1980, Gage & Tyler 1982c, b, Tyler & Gage 1980, Tyler 1988
Ophiura robusta and/or O. sarsi, Ophiopholis aculeata	Atlantic coast of Nova Scotia, Canada	barrens and kelp, 0.2 and 2.3 m off bottom at 5-10 m	artificial collectors (plastic turf) and 1.0 m² quadrat for recruits	0.150 mm and 1 mm mesh size	Jun-Nov 1992-1994, biweekly and recruits once per year 1993-1995	Settlement pulse of both species in late Jul-early Aug of each year. Highest settlement in 1992 for *Ophiura* and in 1993 for *Ophiopholis aculeata* where settlement occurred over a 3-day period and was associated with minor oceanographic and meteorologic fluctuations. Settlement of both species greater in barrens than in kelp beds. Recruitment same between habitats.	Balch et al. 1999, Balch & Scheibling in press
Ophiura sarsi	Gulf of Maine, USA	poorly sorted silt-clay, 148-156 m	Blake trawl	1 mm mesh size	Jul 1985-Aug 1986, every 1-4 months	Followed multiple cohorts over time. Small individuals (<3 mm) in all sample dates with high numbers of recent recruits (< 1 mm) in Jan. Suggested continuous recruitment with an annual peak around Jan.	Packer et al. 1994

42

Table 3. (Continued).

Species	Location	Habitat, Depth	Sampling Method	Size*	Sampling Period, Frequency	Patterns and Conclusions	Source
Ophiura sarsi	northeast Japan	mud, 200-600 m	trawl or dredge	1 or 5 mm mesh size	May 1987-Mar 1989, 2-7 times per site	Settlement not detected due to large mesh size. Size distribution showed strong recruitment at shallow depths and poor recruitment at deeper sites. Multiple modes at 250 m site suggested annual recruitment with a peak in Mar-May.	Fujita & Ohta 1990
Ophioplocus esmarki, Ophionereis annulata	La Jolla, California, USA	habitat and depth not stated	method not stated	2-3 mm	1971-1974, 1-6 months apart	Found few settlers of either species but conceded that they may have been missed. Size distributions showed that *Ophioplocus esmarki* (brooder) has higher or more frequent recruitment due to its positive skew than *Ophionereis annulata* (spawner) with its more normal distribution. Concluded that both species are at their geographic extreme and suggested they might show different recruitment patterns elsewhere.	Muscat 1975 in Ebert 1983
Hemipholis elongata, Microphiopholis atra	northern Gulf of Mexico	sand, silt and clay, 13 m	0.09 m² box-corer	0.5 mm mesh size	Sep 1981	Predominance of small individuals (< 1 mm) suggested heavy recruitment in both species. Based on a single sample of size distribution and no growth data, concluded that periods of recruitment overlap.	Turner & Miller 1988

43

Table 3. (Continued).

Species	Location	Habitat, Depth	Sampling Method	Size*	Sampling Period, Frequency	Patterns and Conclusions	Source
Hemipholis elongata, Microphiopholis atra	Mississippi Sound, Gulf of Mexico	habitat and depth not stated	0.4 m² Shipek grab sampler	0.5 mm mesh size (measured oral diameter)	Apr 1986-Aug 1987, ~ biweekly	Recruitment of both species occurred year round with peak levels in Sep/Oct 1986 and Jul/Aug 1987 for *M. atra*, and May-Sep 1986 and Jun-Aug '87 for *H. elongata*. Only 5 and 3% of recruits of *M. atra* and *H. elongata* respectively survived the first year. Concluded that interannual variability in recruitment has little effect on density and persistence of populations.	Valentine 1991
Asteroporpa annulata	northern Gulf of Mexico	"substrate covered with rhodolith structures", 90 m	10 m semi-balloon trawl	15 mm mesh size	Oct 1988-Apr 1991, 19 times	Of 177 individuals sampled, only 4 were juveniles (3-4 mm) all of which were attached to adult aboral discs. Suggested gradual supply of juveniles that either actively or passively recruit to adult conspecifics.	McClintock et al. 1993
Tropical							
Ophiocomella ophiactoides	Discovery Bay, Jamaica	coralline algae on rock, < 2 m	hand collected samples of *Amphiroa* spp. (where *O. ophiactoides* resides)	1-1.5 mm	Jun-Aug and Dec 1981, Jul 1982, monthly	Only 6 juveniles found suggesting recruitment via larvae is rare. Concluded that continuous reproduction by fission and a high survival rate maintains large populations.	Mladenov et al. 1983

Table 3. (Continued).

Species	Location	Habitat, Depth	Sampling Method	Size*	Sampling Period, Frequency	Patterns and Conclusions	Source
Ophiactis savignyi	Wanlitung, southern Taiwan	tidepools, 0.3-0.5 m deep	500-1000 g samples of the sponge *Haliclona* sp. (where *O. savignyi* resides)	0-0.9 mm (across widest part of disk)	Feb 1991-Jan 1992, monthly	Low numbers of recruits found from May-Dec with a pulse in Jun. Regeneration by fission occurred year round with a peak in Jul.	Chao & Tsai 1995
unidentified species	Davies Reef, Great Barrier Reef, Australia	windward and leeward coral reef, 1 m off bottom at ~ 15 m	artificial collectors (100 plastic bio-filter spheres in net bag), boxes of rubble	0.5 mm mesh size	Nov 1991-Jan 1992, rubble collected in Jan	Settlement did not differ significantly between habitats which corresponds with observed recruitment to rubble. Within a given reef habitat there were near-significant differences in settlement between collectors hundreds of meters apart but no differences between those 1-2 m apart. Settlement increased from north to south between sites on the leeward reef.	Keesing et al. 1993

* Measurements are disk diameters (radial shield to opposite disk edge) unless specified. Size is either the smallest individual recorded or detectable, the smallest size-frequency range containing individuals, and/or the mesh size used to filter or collect samples.

Table 4. A summary of studies examining spatial and temporal patterns of settlement and/or recruitment of holothuroids.

Species	Location	Habitat, Depth	Sampling Method	Size*	Sampling Period, Frequency	Patterns and Conclusions	Source
North Temperate							
Cucumaria frondosa	western Gulf of Maine, USA	mussels, coralline algae and kelp holdfasts, 7.5-12.5 m	"haphazard samples of each substrate"	0.5-1.0 mm	Jun/Jul 1993, Oct 1994 and Apr-Jul 1995	Recruitment from May-Jul with peak in mid-Jun. Recruitment highest in mussel beds. Recruitment on individual mussels, and intermediate in kelp holdfasts and coralline algae.	Medeiros-Bergen & Miles 1997, see also Medeiros-Bergen et al. 1995
Cucumaria frondosa	St. Lawrence Estuary, Canada	bedrock, boulders and gravel, 0-60 m	200 m2 "transect parallel to the coast"	0-15 mm, (measured "length")	spring 1992-winter 1993, "once at the beginning of each season"	Spawned mid-June 1992 and 1993 and settled 3 weeks later. Settlers concentrated in shallow water (0-20 m). Concluded low annual recruitment and migration to deeper water at sexual maturity (~ 3 years).	Hamel & Mercier 1996
Cucumaria pseudocurata	Shell Beach, northern California, USA	rocky intertidal, + 0.3 m	"samples collected" for reproductive cycle; 16 cm² quadrats for size distributions	0-1 mg (dried weight)	Aug 1970-Jan 1972, ~ monthly for reproductive cycle; Feb and Mar, 1971 for size distributions	Seasonal reproductive cycle with spawning in Jan 1971 and 1972. Recruits first observed in Feb 1971. All mature females in the population were found brooding, suggesting annual recruitment.	Rutherford 1973

46

Table 4. (Continued).

Species	Location	Habitat, Depth	Sampling Method	Size*	Sampling Period, Frequency	Patterns and Conclusions	Source
Leptosynapta clarki	Bamfield, British Columbia, Canada	mid-intertidal mudflat	785.4 mm^2 per-spex core	0.25 mm mesh size (length measured after relaxation in MgCl2)	May 1990-Aug 1991, every 2 months	Recruitment pulse in Apr/May of both years. Noted "patchy distribution of juveniles".	Sewell 1994
Parastichopus californicus	Howe Sound, Clayoqout Sound, and Indian Arm, British Columbia, Canada and San Juan Islands, Washington, USA	algal mats, stipes or thalli, polychaete tubes, crevices in rock walls, 5-15 m (not stated for all habitats)	"qualitatively observed" or collected	0.004 "size index" = contracted length × width × scaling factor of 0.1	May 1979-Dec 1984, every 2-3 months (varied with site)	Extended spawning period suggested recruitment occured over several months at Canadian sites with no recruitment in the San Juan Islands. Followed growth of 2 cohorts that settled in 1981 and 1983. Suggested recruitment is common and regular in some areas and weak or non-existent in others.	Cameron & Fankboner 1989
Holothuria tubulosa	Ischia Island, Gulf of Naples, Italy	seagrass (*Posidonia oceanica*) bed, 6, 19 and 33 m	diver collected	0-20 g "total wet weight"	May 1988-Dec 1989, every 2-4 months	Combining all sample dates showed a depth gradient in size distribution with smaller individuals (but of different ages and/or reproductive states) at shallow depths (6 m). Concluded that recruitment occurs in the shallow part of the seagrass bed and that some individuals migrate down-slope.	Bulteel et al. 1992

47

Table 4. (Continued).

Species	Location	Habitat, Depth	Sampling Method	Size*	Sampling Period, Frequency	Patterns and Conclusions	Source
Tropical							
Holothuria atra, H. leucospilota, Actinopyga echinites, Opheodesoma grisea, Synapta maculata	Wanlitung and Nanwan, southern Taiwan	intertidal and subtidal "flats", intertidal-10 m	"search for small individuals"	recruits defined as either < 5 g or < 20 mm	Mar 1990-Feb 1992, monthly (intertidal); 1991, 3 times in summer (subtidal)	No recruits of *H. atra* at either site; recruits of other species were common on intertidal flats. Suggested that asexual reproduction by fission maintains populations of *H. atra*.	Chao et al. 1994

* Measurements are body length (anterior to posterior ends) unless specified. Size is either the smallest individual recorded or detectable, the smallest size-frequency range containing individuals, and/or the mesh size used to filter or collect samples.

48

Hines (1987) monitored populations of *Strongylocentrotus franciscanus* and *S. purpuratus* in central California from 1972 to 1981. Throughout this period there was only 1 significant pulse of recruitment of *S. purpuratus* (resulting in a 25-fold increase in population density) and little evidence of recruitment of *S. franciscanus*. However, echinoids < 10 mm were not accurately sampled in this study and the sampling frequency was inadequate to reliably distinguish and track cohorts over time. Estes and Duggins (1995) found episodic recruitment of *S. droebachiensis*, *S. purpuratus*, and *S. franciscanus* between 1972 and 1990 in southeast Alaska, although *S. polyacanthus* recruited heavily in each of these years in the Aleutian Islands. They suggest that large-scale oceanographic processes are responsible for differences in larval supply to each of the regions. Episodic recruitment also has been observed in *S. purpuratus* and *S. franciscanus* in California and Washington (Paine 1986, Watanabe & Harrold 1991) and *S. droebachiensis* in Nova Scotia (Scheibling 1986, Raymond & Scheibling 1987, Balch & Scheibling 2000). Rowley (1989) found recently settled *S. franciscanus* and *S. purpuratus* in echinoid barren grounds and kelp beds in southern California between April and July in each of 3 successive years (1984 to 1986), but only in May of the final year did he observe heavy settlement of both species (~1000 settlers m^{-2}). He noted a rapid reduction in settler densities within 10 days and suggested that other pulses during his 3 years of sampling may have gone undetected. Sloan et al. (1987) observed low overall recruitment and a high degree of variability among populations of *S. franciscanus* in southern British Columbia in 1984 and 1985. Although most studies of strongylocentrotids have shown recruitment to be quite patchy in space and time, there are some exceptions. For example, recruitment was temporally predictable and substantial over several years for *S. droebachiensis* in the Gulf of Maine (Harris et al. 1985, 1994, Harris & Chester 1996) and *S. purpuratus* in Baja California, Mexico (Pearse 1970 in Ebert 1983).

Watts et al. (1990) measured spatial and temporal variability in recruitment using the genetic characteristics of *Echinometra mathaei* in Western Australia. They found that populations within 4 km of each other had as much genetic variance as populations sampled over 1300 km of coast. The authors suggested that different populations of larvae are supplied to adjacent areas over different years, resulting in genetic heterogeneity. On Rottenest Island, Western Australia, Prince (1995a, b) measured spatial and temporal variability in recruitment of *Echinometra mathaei* at a variety of scales. She concluded that differences in both local hydrodynamics, such as eddy formation and wave action, as well as large-scale interannual variation in ocean currents determine patterns of recruitment.

Small-scale spatial variation in settlement and recruitment of echinoids has been related to differences in habitat (Pearse & Cameron 1991), particularly between kelp beds/forests and echinoid-dominated barren grounds (Pearse et al. 1970, Lawrence 1975). Lower rates of recruitment of *Strongylocentrotus franciscanus* and *S. purpuratus* have been recorded in kelp forests (*Macrocystis pyrifera*) than in adjacent barren areas in California (Tegner & Dayton 1981) and a similar pattern has been observed for *S. droebachiensis* in kelp beds (*Laminaria* spp.) in Nova Scotia (Scheibling 1986) and Norway (Leinaas & Christie 1996). Furthermore, Basch & Tegner (1995) found that recruitment of *Strongylocentrotus* spp. was lower within a kelp forest than at the seaward edge. A number of mechanisms have been proposed to account for these patterns. Various authors have suggested that kelp forests act as larval filters by harbouring species which consume larvae as they drift through the forest or settle on the bottom (Pearse et al. 1970, Bernstein & Jung 1979, Tegner & Dayton 1981, Dayton & Tegner 1984, Gaines & Roughgarden 1987, Harrold & Pearse 1987, Chapman & Johnson 1990). In addition, Jackson & Winant (1983) showed that kelp forests deflect currents which could act to reduce the number of incoming larvae (see also Dayton & Tegner 1984). Larval supply to a kelp forest also may be limited because larvae encountering a forest settle on the first suitable substrates along the forest edge (Bernstein & Jung 1979, Tegner & Dayton 1981, Harrold & Pearse 1987). Within the forest, post-settlement survival of recruits may be adversely affected by understory macroalgae which increase sedimentation or decrease water flow, light and microalgal cover, as has been shown for other benthic invertebrates (Eckman et al. 1989, Duggins et al. 1990).

Studies which have compared settlement rates of strongylocentrotids between kelp forests/beds and barren grounds have yielded equivocal results. Rowley (1989) reported no significant difference in the densities of newly settled *Strongylocentrotus franciscanus* and *S. purpuratus* between a kelp forest and an adjacent echinoid barren ground in southern California. However, the number of samples and size of the sampling units (200 cm^2 pieces of shale) may have been too small to provide a meaningful statistical comparison. Rowley also noted that the kelp canopy density was low during this study and might not have influenced larval supply to the kelp forest. Using artificial collectors (scrub brushes), Schroeter et al. (1996) found that kelp forests had no significant effect on settlement rates of *S. purpuratus*. However, settlement of *S. franciscanus* was low and variable (0 to 10 per collector) with some evidence of higher settlement 20 m offshore of the kelp forest. The authors concluded that kelp forests do not reduce larval supply or settlement but offer the caveat that their 2 year study period may have been too short (relative to natural cycles) to assess this. Using artificial collectors (plastic turf) in the Gulf of Maine, Harris and Chester (1996) found settlement of *S. droebachiensis* was greater within natural or artificial kelp beds than in adjacent barren grounds. In contrast, using similar collectors in Nova Scotia, we found the opposite pattern: settlement of this species was lower in kelp beds than in barrens (Balch & Scheibling in press). These regional disparities may be attributable to differences between the Gulf of Maine and Nova Scotia in size and growth form of individual kelps, or in characteristics of the kelp bed (e.g., bed area and shape, kelp density, understorey species) or surrounding environment (e.g., depth, topography, hydrodynamic conditions), which may effect larval supply. Such differences are even more pronounced between the *Laminaria* beds of the northwest Atlantic and the *Macrocystis* forests of the northeast Pacific (see Harrold & Pearse 1987 for a description of different kelp habitats).

Differences in echinoid settlement rates also have been recorded between habitats on a tropical coral reef. Keesing et al. (1993) found that settlement of echinoid larvae (several species were grouped together including *Echinometra mathaei* and *Mespilia globulus*) on an artificial substrate (plastic biofilter spheres) was significantly greater on the windward edge than on the leeward edge of Davies Reef, Great Barrier Reef. The authors attributed this to different water residence times. Within a given reef there was no significant difference in settlement rate between collectors placed tens to hundreds of meters apart.

Although most studies of echinoid settlement and recruitment have focussed on horizontal variability, few studies have documented variability over a depth gradient. Harris et al. (1994) found settlement of *Strongylocentrotus droebachiensis* on artificial turf to be greatest at 6 to 8 m and orders of magnitude lower at 20 and 30 m. Himmelman (1986) also found decreased recruitment of *S. droebachiensis* with depth in Newfoundland. He attributed this pattern to reduced food and slower growth resulting in increased predation of juveniles. In contrast, DeRidder et al. (1991) found recruitment of *Echinocardium cordatum* was greatest at 15 to 25 m, much less at 5 to 10 m, and absent in the littoral zone.

Seasonal patterns in settlement have been documented for a few species of echinoids. In central California, Harrold et al. (1991) sampled artificial collectors in a kelp forest at monthly intervals for a year and found clear settlement peaks in April and November for *Strongylocentrotus purpuratus* and *S. franciscanus*. In northern and southern California, Ebert et al. (1994) monitored the same species on artificial collectors at weekly intervals between 1990 and 1993. They found that settlement of both species was strongly seasonal occurring between late winter and early summer and that settlement rate varied between species, among sites and among years. Settlement at southern sites tended to be higher and more consistently annual than at northern sites. In the Mediterranean, Pedrotti (1993) suggested that seasonal recruitment patterns of *Paracentrotus lividus* were related to biannual spawning (in spring and fall), as indicated by the presence of larvae in plankton samples.

As with echinoids, asteroids show a high degree of spatial and temporal variability in settlement and recruitment throughout the distributional range of species (Table 2). The most extensive study of settlement for any echinoderm is that by Loosanoff (1964) of *Asterias forbesi* in Long Island Sound, Connecticut. From 1937 to 1961 he deployed oyster shell collectors twice weekly at 10 sites in 3 areas along 26 km of shore and at depths of 0 to 33 m. Loosanoff (1964) found that settlement increased with depth to 10 m and then decreased to 33 m (see also Ebert 1983). Settlement occurred between June and September but the settlement period varied between sites and years, ranging from 1 to 91 days (mean = 52 days). In years of heavy settlement, the settlement period was protracted; in years of light settlement, it occurred late in the year. Settlement intensity varied from a single peak in 1 week to relatively constant settlement over a period of 3 months. Total annual settlement varied by five orders of magnitude (0.3 to 1700 per collector) with no consistent pattern of high and low settlement years. Settlement varied between sites within areas but tended to increase from northeast to southwest. Both Loosanoff (1964) and Ebert (1983), who re-analysed Loosanoff's data, concluded that settlement was not correlated with preceding or subsequent adult density (but see also Burkenroad 1957). Ebert (1983) proposed that hydrodynamic conditions or planktonic predators regulating larval supply may be more important in determining settlement and subsequent recruitment of *A. forbesi* than settlement or post-settlement processes.

Despite massive sampling efforts, most studies of *Acanthaster planci* found few settlers and these were patchily distributed in space and time (reviewed by Moran 1986, Johnson 1992b). During extensive searches at Iriomote-jima in the Ryukyu Islands, Yokochi and Ogura (1987) found only 9 juveniles of *A. planci* in 1984 and 13 in 1985. Fisk (1992) used various methods to measure recruitment of *A. planci* on Green Island in the northern Great Barrier Reef (considered a possible source for larvae that seed secondary outbreaks on reefs to the south) and found only 2 recruits between 1986 and 1990 (Fisk et al. 1988, Fisk 1992). In the central Great Barrier Reef south of Green Island, Doherty and Davidson (1988) destructively searched for *A. planci* on 16 reefs in 1986 and 1987 and found only 4 individuals < 30 mm in diameter (all in 1986), which Johnson (1992b) considered new recruits. From analysis of size-frequency distributions, they inferred low settlement rates in 1986 and 1987 and an order of magnitude higher settlement rate in 1985. However, Ebert (1983) and Moran (1986) pointed out that size- and diet-specific variation in growth rates complicates identification of cohorts from size distributions. *A. planci* individuals grow slowly at the early juvenile stage, when they feed on algae, and then undergo a dramatic increase in growth rate when they shift to a diet of coral. This dietary shift is not necessarily age-related and size at a given age can vary considerably (Moran 1986) indicating that more direct methods of measuring settlement and recruitment rates than size-frequency analysis are needed to accurately detect patterns. Keesing et al. (1993) found 11 asteroid settlers on artificial collectors on Davies Reef, Great Barrier Reef (3 *A. planci*, 5 *Choriaster granulatus* and 3 *Culcita novaeguineae*) in 1992 and concluded that it was a poor settlement year. Despite these meagre results, they suggested that collectors could be used to monitor settlement of *A. planci* in various habitats and to predict the location of outbreaks 3 years in advance.

Juveniles of *Acanthaster planci* have been found in Fiji where Zann et al. (1987, 1990) reported heavy recruitment in 1977, 1984 and 1987 in the intertidal zone of several coral reefs based on size-frequency analysis. Recruitment occurred over thousands of hectares in most years between 1979 and 1989, although intense recruitment in 1982 and 1983 occurred over only a few hectares. Zann et al. (1987, 1990) concluded that there is a high degree of spatial and temporal variability in recruitment of *A. planci* and that outbreaks originate from episodic events.

Differences in recruitment patterns of several co-existing species of temperate asteroids suggest that factors influencing settlement and recruitment may not operate uniformly across species. Sewell and Watson (1993) found that *Pisaster ochraceus* in British Columbia settled in all 5 years studied, whereas *Pycnopodia helianthoides* settled in 4 of 5 years and *Dermasterias imbricata* settled in only 1 year. Himmelman and Dutil (1991) found that differences in distribution of recruits of 3 asteroid species were associated with different habitat types across a depth gradient in the northern Gulf of St. Lawrence. Recruits of *Leptasterias polaris* were most abundant at 0 to 1 m and found only in boulders, cobble or bedrock; those of *Asterias vulgaris* were most abundant at 4 to 7 m and found only in boulders or cobble; and those of *Crossaster papposus* were only found on sedimentary bottoms deeper than 11 m.

Asteroids display a wide variety of reproductive strategies (Chia & Walker 1991) which may influence patterns of recruitment, but little is known about recruitment of species that reproduce by means other than planktonic larvae. For example, Ebert (1983) contended that increased parental investment in the form of brooding should increase recruitment success and consequently decrease longevity. In support of this hypothesis, Menge (1975) found a broadcast spawner (*Pisaster ochraceus*) lived approximately 3 times longer than a brooding asteroid (*Leptasterias hexactis*) on San Juan Island, Washington. Menge (1975) proposed that brooding has co-evolved with small body size to ensure increased reproductive success and survival of *L. hexactis* in a competitive relationship with larger *P. ochraceus*. However, Himmelman et al. (1982) suggested that brooding is a fixed trait in the genus *Leptasterias* and doubts that it evolved from competitive interaction. Boivin et al. (1986) concluded that the large egg reservoir and long development time of *L. polaris* assure steady annual recruitment in the St. Lawrence Estuary.

There is indirect evidence for habitat selectivity in several asteroids based on the differential distribution of juveniles and adults. These differences may arise by selective settlement in a habitat different from that of adults (assuming that juveniles eventually migrate to the adult habitat) or because of between-habitat differences in post-settlement mortality. Migration to adult habitats may occur when juveniles reach a size refuge from predation or require an alternate food source. Birkeland et al. (1971) found various species of recently metamorphosed asteroids (*Mediaster aequalis, Luidia foliolata, Crossaster papposas, Henricia leviuscula, Solaster stimpsoni, S. dawsoni* and *Pteraster tesselatus*) on the tubes of the polychaete *Phyllochaetopterus prolifica* and not elsewhere. The authors suggested this habitat acts as a nursery ground for juveniles where there is an abundance of food.

Plant assemblages seem to be particularly attractive habitats for asteroid settlement. Scheibling (1980a) found juveniles of *Oreaster reticulatus* mainly within and adjacent to dense seagrass beds and suggested that settlement in seagrass beds provides refuge from predation by fish (Scheibling 1980a, b). Sewell and Watson (1993) found recruits of *Pisaster ochraceus, Pycnopodia helianthoides* and *Dermasterias imbricata* on various substrata including macroalgae such as *Laminaria saccharina* and *Sargassum muticum*. Rumrill (1988b) reported that *Pisaster ochraceus* in laboratory experiments preferentially settled on substrata collected from the *Laminaria* zone. Day and Osman (1981) found juvenile *Patiria miniata* under boulders in a California kelp forest whereas adults were on the exposed algal covered reef. They suggested that juveniles are either out-competed by adults on exposed reefs or removed by predation (see also Harrold & Pearse 1987).

4.4 *Ophiuroidea*

Unlike echinoids and asteroids, most studies of ophiuroids (including a number in the deep-sea) indicate consistent annual and seasonal patterns of settlement or recruitment (Table 3). Although the deep-sea has been considered an aseasonal environment, some species of ophiuroids (and other invertebrates) show seasonality in reproduction and/or recruitment (Schoener 1968, Tyler et al. 1982, Tyler 1988, Gage 1994). In the northeast Atlantic, Gage & Tyler (1981a, b, 1982b, c) found that *Ophiura ljungmani* and *Ophiocten gracilis* on the Hebridean continental slope (600 to 1200 m depth) and *Ophiura ljungmani* in the nearby Rockall Trough (2200 to 2900 m) reproduced seasonally, and that high densities of recruits dominated populations of *Ophiura ljungmani* in the Rockall Trough. From these studies, they inferred that settlement of both species occurs annually in summer. However, they also observed annual recruitment (indicating a settlement peak in May) of *Ophiomusium lymani* which exhibited continual gametogenesis in the Rockall Trough (Gage & Tyler 1982a, b). They suggest that this pattern of recruitment in *Ophiomusium lymani* is related to the seasonal input of detritus from surface waters which may regulate larval survival. Other studies of *Ophiomusium lymani* from sites at 1100 to 2300 m depth in the northeast and northwest Atlantic and the northeast Pacific (reviewed by Gage 1982) showed several juvenile modes in population size distributions, suggesting seasonal and annual recruitment in the spring (Atlantic) or late summer (northeast Pacific).

Several studies of *Amphiura filiformis* from coastal waters of various regions have shown different patterns of recruitment, although these differences may partly reflect different sampling methods (Table 3). Very low and patchy recruitment of *A. filiformis* was observed over an 8 year period in Galway Bay, Ireland (O'Connor & McGrath 1980, O'Connor et al. 1983). Based on the reproductive cycle, settlement was assumed to occur from September to November. The authors concluded that they had missed sampling settlers because of high post-settlement mortality in the first year after settlement, although it is possible that settlement rarely occurs in this population. Off the coast of the Netherlands, Duineveld and Van Noort (1986) observed high recruitment of *A. filiformis* from July to September in each of the 2 years studied. They concluded that high mortality in the first year after settlement limits the number of intermediate size animals, but enough survive to sustain a low rate of renewal of the adult population. In the Øresund off Denmark, Muus (1981) also found high recruitment of *A. filiformis* in both years of a 2 year study, with a peak from September to November. Here too, post-settlement mortality was high and few recruits survived their first year, resulting in a relatively stable adult population. In contrast, *A. chiajei*, which began settling 3 months later than *A. filiformis*, was found in very low numbers in the same area (Muus 1981). In the Mediterranean, Pedrotti (1993) found larvae of *A. filiformis* comprised 70% of the ophiuroids in plankton tows from November to February. The author concluded that mixed larval stages found at various times of the year were evidence for prolonged and variable recruitment.

Continual recruitment also has been reported for *Ophiura sarsi* at depths of 148 to 156 m in the Gulf of Maine (Packer et al. 1994). Small individuals were found throughout the year but the highest number of recruits occurred in January, suggesting a seasonal peak. Using artificial collectors at 6 to 10 m depth off Nova Scotia, we found that *Ophiura* sp. and *Ophiopholis aculeata* both settled in a pulse between late July and early August in 3 successive years (Balch & Scheibling 2000). However, highest settlement occurred in different years for each species suggesting that different processes control larval supply and settlement of the two species. The regional difference in settlement pattern of *Ophiura* between Nova Scotia and the Gulf of Maine may be due to differences in depth or geographic location. Alternatively, different species may have been sampled, since we could not distinguish between *Ophiura sarsi* and *O. robusta* (Balch & Scheibling 2000).

Ebert (1983) hypothesised that different reproductive strategies result in different recruitment patterns, and that brooders should have more predictable recruitment than spawners due to increased parental investment in brooding. He compared recruitment (based on size distributions) of *Ophioplocus esmarki* (a brooder) and *Ophionereis annulata* (a spawner) at False Point, California and found that the brooder showed higher or more frequent recruitment. Ebert (1983) also hypothesised that recruitment declines with depth, based on data from studies of the deep-sea ophiuroids *Ophiura ljungmani* and *Ophiomusium lymani*. Assuming a constant population where mortality equals recruitment, he estimated mortality rate and showed that mortality, and thus recruitment, decreases with depth. However, Ebert (1983) included data from Gage and Tyler (1982c) who found considerable variability in survivorship of these ophiuroids and suggested that depth is not a factor affecting their recruitment. Some support for Ebert's (1983) hypothesis comes from the observations of Fujita and Ohta (1990) that recruitment of *Ophiura sarsi* off the northeast coast of Japan was greater at 250 m than at deeper sites (350 to 550 m).

Many studies of ophiuroids are from deep water habitats with relatively uniform soft-bottoms; few studies have compared settlement or recruitment of the same species between different habitats. We found higher settlement of *Ophiura* sp. and *Ophiopholis aculeata* on artificial collectors in echinoid-dominated barren grounds than in kelp beds in Nova Scotia (Balch & Scheibling in press). However, recruitment of both species did not differ between habitats 1 year later, suggesting differential post-settlement mortality. Keesing et al. (1993) compared settlement between two habitats on Davies Reef in the Great Barrier Reef. Using an artificial substrate, they found no differences in settlement of several combined species of unidentified ophiuroids between the windward and leeward edge of the reef. This pattern corresponded to observed recruitment to coral rubble in these habitats but Keesing et al. (1993) suggested that identification of separate species might yield different patterns.

4.5 *Holothuroidea*

Despite the predominance of spawners among the holothuroids (Smiley et al. 1991), most studies of recruitment are based on fissiparous or brooding species (Table 4). Ebert (1983) reviewed several studies of fissiparous populations of *Holothuria atra* in the South Pacific. He concluded that recruitment can be either continual or seasonal and that rates of fission alone can be enough to sustain the population. Rutherford (1973) sampled populations of the brooding holothuroid *Cucumaria pseudocurata* in the intertidal zone in northern California and found that they recruit annually in February. He also found a strong negative correlation between recruit survival and adult density leading him to conclude that recruitment is density dependent and that recruits are space limited. Sewell (1994) showed that *Leptosynapta clarki*, another brooder inhabiting intertidal mudflats of Bamfield Inlet, British Columbia, recruited annually in April/May. Cameron and Fankboner (1989) concluded that recruitment of *Parastichopus californicus* (a spawner) occurs over several months in fall and winter, based on an extended spawning period in southern British Columbia and Washington. They found that recruitment over a 6 year period varied markedly among sites, with regular recruitment in some areas and weak or no recruitment in others.

Differential settlement over a depth gradient has been observed for some holothuroid species which settle in the shallow range of their habitat. In the St. Lawrence Estuary, Hamel and Mercier (1996) found settlers of *Cucumaria frondosa* concentrated in shallow water (0 to 20 m) compared to adults which were more common in deeper water (40 and 60 m). Bulteel et al. (1992) found similar results for *Holothuria tubulosa* in 3 depth zones (6, 19 and 33 m) of a seagrass bed off Ischia Island, Gulf of Naples, where small individuals dominated the shallowest zone.

4.6 *Crinoidea*

There is little information about settlement or recruitment of crinoids on natural substrata in the field. Mladenov and Chia (1983) were unable to find any settlers of *Florometra serratissima* in 2 years of study in Barkley Sound, British Columbia, suggesting that recruitment is low or sporadic. On Davies Reef (Great Barrier Reef), Keesing et al. (1993) found 15 crinoid settlers (unidentified species) on artificial collectors. 12 of the settlers were on the windward edge of the reef and only 3 on the leeward, but the sample size is too small to draw any conclusions about spatial trends in settlement.

4.7 *Summary and conclusions*

Ebert (1983) identified spatial and temporal variability in recruitment as a salient feature of most echinoderm life histories. While this review supports this contention, it also demonstrates that various scales of spatial and temporal variability must be considered in establishing patterns of settlement and recruitment. Although most echinoderm species exhibit seasonal patterns of settlement, there is large inter-annual variation. Spatial variability in settlement occurs at a variety of scales including habitat, site and region, all of which introduce variables that are difficult to isolate. Although Ebert (1983) presents a number of "clines" that may explain large-scale patterns of recruitment, these are generally based on correlative data and the causative factor(s) are not known. Large spatial and temporal variability necessitates close monitoring of populations to detect ecologically relevant patterns of settlement and recruitment, and suggests caution in interpretation and generalisation of patterns from any particular study. Periodicity may operate on spatial and temporal scales greater than those sampled and thus go undetected. Alternatively, patterns may emerge that are of little consequence to the overall recruitment to a population. For example, low levels of recruitment observed during a particular study may become irrelevant when a single large recruitment event occurs before or after the sampling period. Of 88 studies which examined temporal variation in settlement and/or recruitment of echinoderms, 57 (65%) were 3 years in duration (Fig. 1). Unfortunately, studies of this duration yield little information on long-term patterns. The length of most studies likely reflects funding periods for research grants, rather than a biologically meaningful time scale.

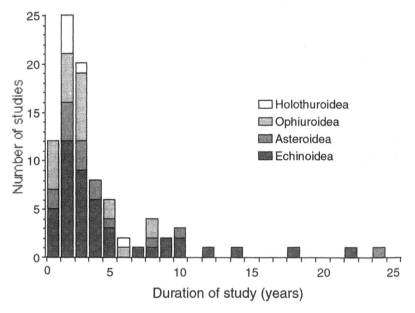

Figure 1. Frequency distribution of sampling duration for 88 studies which monitored temporal patterns of settlement and/or recruitment in four classes of echinoderms (Tables 1 to 4; studies using larval distribution or genetic analysis to predict patterns are excluded).

Traditional techniques of monitoring settlement and recruitment are time consuming, labour intensive, and often inaccurate in sampling and identifying small individuals. Identification of settlers to species level is often not done because of the lack of suitable descriptions of early post-metamorphic forms and the difficulty in discriminating taxonomic characteristics at microscopic scales. However, species identification recently has been facilitated by molecular genetic techniques (for ophiuroid species) which have broad applicability to future studies (Medeiros-Bergen et al. 1998). Of the 108 studies on settlement and/or recruitment patterns that we reviewed (Tables 1 to 4), only 57 detected individuals that might be considered recent recruits (< 1 mm for ophiuroids and < 2 mm for other classes).

Settlement and recruitment patterns often are inferred from infrequent measures of size distributions which are difficult to interpret (Botsford et al. 1994). Estimates of juvenile abundance can vary greatly between studies due to differences in sampling method, frequency and efficiency, which limit direct comparisons between studies. Of 82 studies which provide information on sampling frequency, 30 (37%) sampled at intervals < 1 month

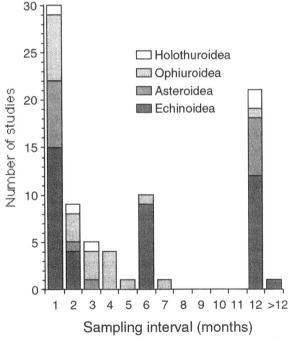

Figure 2. Frequency distribution of modal sampling interval for 82 studies which monitored temporal patterns of settlement and/or recruitment in four classes of echinoderms (Tables 1 to 4; studies using larval distribution or genetic analysis to predict patterns, or with insufficient data to assess sampling interval, are excluded).

(Fig. 2), and sampling frequencies < 2 weeks were achieved in only 12 (15%) of these studies. Because mortality and migration of settlers can occur within days to weeks after settlement (Hunt & Scheibling 1997), studies with longer sampling intervals may be unable to reliably detect patterns of settlement. 21 of the 82 studies (26%) sampled at yearly intervals providing information on interannual patterns of recruitment (Fig. 2), particularly when these studies extended over several years.

The use of artificial collectors promises an efficient and effective way of frequent monitoring which may enable reliable prediction of recruitment events. This is particularly important for species which can severely impact benthic community structure or for commercially important species. For example, artificial collectors have been used to sample settlers and predict catch rates of rock lobster 4 years in advance for the fishery in Western Australia (Phillips 1986, Pearce & Phillips 1994). Artificial collectors also may provide a means of rigorously examining spatial and temporal patterns of settlement and recruitment. However, collector results should be compared to measurements on natural substrata to identify artifactual effects. Also, different collector types should be standardised, or cross-calibrated (e.g., Balch et al. 1998), to allow comparisons across studies.

5 POST-SETTLEMENT PROCESSES

5.1 Predation

Juvenile echinoids are prey to various invertebrate and fish predators but the relative importance of these predators as agents of mortality is not well understood (reviewed by Scheibling 1996). Highsmith (1982) found that the tanaid crustacean *Leptochelia dubia* consumed *Dendraster excentricus* which settled outside of adult sand dollar beds in the San Juan Islands, Washington. Larvae of *D. excentricus* that settled amongst adult conspecifics had higher survival rates because tanaids were excluded from sand dollar beds by bioturbation. Keats et al. (1985) concluded that size-selective predation of juvenile *Strongylocentrotus droebachiensis* by cunner and winter flounder may play an important role in regulating recruitment to echinoid populations in Newfoundland. In a cobble bed in Nova Scotia, Scheibling and Hamm (1991) found that juvenile rock crabs, lobsters and sculpins had a significant effect on the survival of juvenile *S. droebachiensis* in predator inclusion cages. In a Californian kelp forest, Pearse and Hines (1987) observed a dramatic decline in the density of recently recruited *S. purpuratus*, which they suggested was due to disease and/or predation by asteroids. Also in California, Rowley (1990) attributed a higher rate of mortality of newly settled *S. purpuratus* in a kelp forest than in a nearby barren ground to a difference in predation between the 2 habitats, although he did not identify predators. In New Zealand, Andrew and Choat (1982) observed enhanced recruitment of *Evechinus chloroticus* in predator exclusion cages in barren grounds and suggested that fish are important predators of juveniles. They also found that survival of caged juveniles was much higher in sparse and dense kelp forests than in barren habitats and concluded that processes other than predation regulate echinoid abundance in algal covered habitats (Andrew & Choat 1985).

The importance of spatial refugia from predation has been demonstrated for several species of echinoids which find shelter in a variety of micro-habitats. Mussel beds were shown to provide a spatial refuge for juvenile *Strongylocentrotus droebachiensis* from predation by fish, crabs and lobster in New England (Witman 1985). In cage experiments in Nova Scotia, Scheibling and Hamm (1991) recorded a lower rate of predation on juveniles of *S. droebachiensis* which sheltered among cobbles compared to those without a spatial refuge. Juveniles of *S. franciscanus* are often observed under the spine canopies of conspecific adults which provide protection from predators (Tegner & Dayton 1977, 1981, Tegner & Levin 1983, Breen et al. 1985, Sloan et al. 1987, Rogers-Bennett et al. 1995). In contrast, Andrew and Choat (1982, 1985) found no effect of conspecific adults on juvenile survival of *Evechinus chloroticus*.

Several studies have documented bimodal size distributions of echinoids with prominent juvenile and adult modes but low numbers of intermediate size animals (e.g., Tegner & Dayton 1981, Ojeda & Dearborn 1991, Rodriguez & Ojeda 1993, reviewed by Scheibling 1996). This pattern has been attributed to an ontogenetic shift in microhabitat as juveniles outgrow spatial refugia and are subjected to increased predation until they reach a refuge in size as adults (but see also Botsford et al. 1994). For example, Scheibling and Raymond (1990) found that juveniles of *S. droebachiensis* in a cobble bed declined in abundance once they outgrew refuges in the interstices and undersides of cobbles. This was attributed to predation since juveniles survived under boulders that were experimentally transplanted to the bed, providing a more suitable spatial refuge for larger individuals.

Predation of juveniles of the asteroid *Acanthaster planci* by a variety of animals (including fish, crabs, lobster, shrimp, gastropods, corals and worms) has been observed in field and laboratory studies (reviewed by Moran 1986, Keesing & Halford 1992a). Keesing and Halford (1992b) placed laboratory-reared juveniles of *A. planci* of different ages in open and closed cages to examine survival rates in the field. They found that juveniles move little in the presence of adequate food supplies, suggesting that preferential settlement in spatial refugia would enhance survival. They recorded high but declining rates of mortality (attributed to predation) for newly settled juveniles over the first 4 months, suggesting that early post-settlement mortality could substantially affect population dynamics. McCallum (1992) developed models to examine predation on juveniles as a mechanism of controlling populations of *A. planci* and concluded that our understanding of the stock-recruitment relationship and degree of openness of populations was insufficient to reliably address this problem.

Post-settlement predation has been less well documented for other species of asteroids. Dayton et al. (1974) concluded that populations of *Acodontaster conspicuus* and *Austrodoris mcmurdensis* are at least partially regulated by predation of juveniles by the detritivorous asteroid *Odontaster validus*. Day and Osman (1981) observed predation on juveniles of *Patiria miniata* by adult conspecifics and by cancrid crabs. In laboratory experiments, Rumrill (1989) recorded high survival rates for juveniles of *Asterina miniata* offered to a fish, 2 species of predatory crabs and 5 species of predatory asteroids. These results, combined with observations of juvenile abundance in the field, led him to conclude that post-settlement mortality due to predation is low.

Duineveld and Van Noort (1986) attributed the rapid decline in density of the ophiuroid *Amphiura filiformis* off the Netherlands to predation of juveniles by shrimp, crabs and polychaetes. They also proposed that high adult densities (1330 m^{-2}) might result in competition or cannibalism, but reported similar rates of post-settlement mortality at a nearby site with lower adult densities (300 m^{-2}). Packer et al. (1994) found that small *Ophiura sarsi* (3 to 13 mm) were the most common component of the stomach contents of American plaice (*Hippoglossoides platessoides*) in the Gulf of Maine. They concluded that, despite their low caloric value, juveniles were selectively preyed upon because of their abundance and/or accessibility.

Several predators of juvenile holothuroids have been identified but the impact of predation on recruitment to adult populations is not clear. Rutherford (1973) showed that recruits of the brooding holothuroid *Cucumaria pseudocurata* were reduced by 61% after 1 month and by more than 96 % after 1 year, and suggested predation by the asteroid *Pycnopodia helianthoides* as a source of post-settlement mortality. In contrast, Sewell (1994) observed "no dramatic decrease in numbers of juvenile *Leptosynapta clarki* in the first month" and followed the cohort for 6 months until it merged with the adult population. This discrepancy could be a result of the different habitats sampled and different predator assemblages. In the St. Lawrence Estuary, Hamel and Mercier (1996) observed predation of juveniles of *Cucumaria frondosa* by the echinoid *Strongylocentrotus droebachiensis* and the asteroid *Solaster endeca*, but these occurrences were rare. In laboratory experiments, Cameron and Fankboner (1989) showed that *Solaster dawsoni* and a hermit crab *Pagarus hirsutiusculus* selectively preyed on juveniles of *Parastichopus californicus* and suggested that predation may limit recruitment to areas free of predators. In cage experiments off Okinawa Island,

Wiedemeyer (1994) showed low rates of natural mortality (0.6% mo^{-1}) of juvenile *Actinopyga echinites*. However, when the cage lids were removed and the juveniles exposed to potential predators (a gastropod and several species of fish), mortality rate increased 5 fold (to 3.3% mo^{-1}) due to predation. This predation rate is still low compared to that observed for other holothuroid species, leading Wiedemeyer (1994) to suggest that the juveniles of *A. echinites* are defended by toxins.

5.2 *Migration and dispersal*

Migration of juvenile echinoids between distinct habitats such as kelp forests and barren grounds is generally discounted due to the large distances relative to the size and rates of movement of the juveniles (Rowley 1989) or because of the presence of some physical barrier such as sand (Watanabe & Harrold 1991). With increasing age and size, however, the likelihood of migration increases. DeRidder et al. (1991) found no recruitment of *Echinocardium cordatum* in the littoral zone and suggested that 2 to 4 year old echinoids migrated there from deeper water.

Small scale migrations have been observed for juvenile asteroids that settle in nursery areas and then move to adult habitats (reviewed by Chia et al. 1984). Birkeland et al. (1971) commonly found several asteroids (*Mediaster aequalis, Luidia foliolata, Crossaster papposas, Henricia leviuscula, Solaster stimpsoni, S. dawsoni, Pteraster tesselatus*) on the tubes of the polychaete *Phyllochaetopterus prolifica* and suggested they settle there and feed on epizoites before migrating to sandy areas to feed on larger prey. Scheibling (1980a, b) suggested that *Oreaster reticulatus* individuals settle in seagrass beds and then migrate to adult populations in open sandy areas. Similarly, Jost and Rein (1985) proposed that migration of juvenile *Astropecten bispinosus* from *Zostera marina* beds compensates for losses due to predation on a sand bottom. In contrast, Rumrill (1989) measured low rates of movement of settlers and juveniles of *Asterina miniata* in the laboratory and concluded that migration was limited in natural populations. Extensive migrations of juveniles or adults of *Acanthaster planci* to form aggregations of different year classes have been proposed as a causal mechanism for population outbreaks, although there is no obvious cue to trigger such behaviour (reviewed by Ebert 1983, Moran 1986, Johnson 1992b).

Post-settlement migration of holothuroids has been observed at small and large spatial scales. In laboratory and field studies in the San Juan Islands, Washington, Young and Chia (1982) showed that post-settlement migration of *Psolus chitonoides* toward shaded areas such as cracks and overhangs occurs within the first week of settlement. After 1 month, juveniles in the field could no longer be located and were presumed to have dispersed. Young and Chia (1982) suggest that post-settlement processes such as migration are more important than substratum choice at settlement in determining the spatial distribution of *P. chitonoides*. Hamel and Mercier (1996) showed a 3-stage migration of *Cucumaria frondosa* in the St. Lawrence Estuary: 4 to 5 months after settling on the undersides of rocks, juveniles migrate to crevices; after 19 months, when they reach a size refuge from predation (~ 35 mm), they move to exposed rock surfaces; finally, once sexual maturity is reached after ~ 3 years, the adults migrate to deeper water. Similarly, Bulteel et al. (1992) concluded that settlement of *Holothuria tubulosa* occurs in shallow seagrass beds and that some individuals migrate to deeper water to reproduce.

For small fissiparous or brooding echinoderms, rafting on drift algae may be an effective mode of dispersal for both adults and juveniles. Highsmith (1985) observed rafting on drift algae for several brooding species of invertebrates on San Juan Island, Washington, including the ophiuroid *Amphipholus squamata*. He concluded that the small size of brooders, especially the juveniles, makes them well suited to this mode of dispersal. Mladenov and Emson (1988) proposed that rafting on floating clumps of algae such as *Halimeda* and *Amphiroa* may serve as a dispersal mechanism for strictly asexual populations of the ophiuroids *Ophiactis savignyi* and *Ophiocomella ophiactoides* in Jamaica. Sewell (1994) provides anecdotal evidence for dispersal of the brooding holothurian *Leptosynapta clarki* in Bamfield Inlet, British Columbia, by rafting, floating, swimming, and transport by waves or currents. In contrast, Hess et al. (1988) showed much greater gene flow between populations of *L. clarki* < 500 m apart than between those 11 to 24 km apart on San Juan Island, and concluded that dispersal between distant locations was limited due to barriers of unfavourable habitat.

5.3 *Disease, starvation and other mortality*

Disease can have catastrophic effects on echinoid populations, as evidenced by epizootics affecting *Strongylocentrotus droebachiensis* in Nova Scotia (Miller & Colodey 1983, Scheibling 1986) and *Diadema antillarum* in the Caribbean (Lessios et al. 1984, Hughes et al. 1985). Scheibling and Stephenson (1984) reported higher mortality of adults than juveniles of *S. droebachiensis* during a disease outbreak in Nova Scotia in 1983. They suggested that juveniles, which shelter beneath rocks and in crevices, are more likely to avoid exposure to a waterborne pathogen. Following mass mortality of *D. antillarum* in Barbados in 1983, recruitment in two successive years enabled populations to recover to within 57% of their pre-mortality levels by 1985 (Hunte & Younglao 1988). In other areas, however, recruitment occurred at relatively low levels and recruits tended not to persist (Bak et al. 1984, Lessios 1988, 1994, Carpenter 1990, Karlson & Levitan 1990).

Mass mortalities have also been observed for juveniles of *Acanthaster planci* in Fiji (Zann et al. 1987, 1990). In some areas, the 1984 cohort suffered complete mortality by 1987 due to an undescribed sporozoan pathogen, while juveniles from the same cohort survived in other areas. Necrotic lesions are typically observed in juveniles that have been nutritionally stressed and Keesing and Halford (1992a) concluded that density-dependant mortality, in the absence of sufficient coral food, could limit survival of juvenile *A. planci*.

Abiotic sources of mortality such as storms and temperature and salinity fluctuations also may affect recruitment (Lawrence 1996). In British Columbia, mass mortality of recently settled *Strongylocentrotus droebachiensis* and *S. franciscanus* was attributed to rainwater runoff during the winter (Cameron & Fankboner 1989). In laboratory and field experiments Himmelman et al. (1983, 1984) showed that juveniles of *S. droebachiensis* suffered greater rates of mortality than adults when exposed to low salinity conditions.

5.4 *Summary and conclusions*

Most species of benthic marine invertebrates suffer very high mortality within the first days to months of life after settlement (Gosselin & Qian 1997, Hunt & Scheibling 1997). Gosselin and Qian (1997) conclude that common processes may influence the early post-settlement survival of most species, which are at least as important as processes affecting larvae. However, the role of post-settlement processes in determining populations of echinoderms remains a poorly studied component of their early life history. High rates of settlement and recruitment are often recorded but the fate of recruits is usually unknown. Various factors such as migration, predation, disease, storms and reduced salinities have been implicated as sources of high post-settlement mortality or loss from a population. However, separating these factors and determining their relative importance has proven difficult, particularly in field studies. Although early survival can be tracked by using methods as simple as size or age distributions, the sampling resolution of most studies (Tables 1 to 4) is inadequate to make clear statements about meaningful time scales and possible causes of mortality. Many predators of juvenile echinoderms have been identified from laboratory studies and analysis of gut contents, but further field studies are required to quantify predation rates under natural conditions and to examine predator-prey interactions. Field enclosures and tethering are effective means of manipulating predators and prey, although they may introduce experimental artefacts which must be assessed using appropriate controls (Scheibling 1996, Hunt & Scheibling 1997).

6 GENERAL CONCLUSIONS AND RECOMMENDATIONS FOR FUTURE RESEARCH

In the fourteen years since Ebert's 1983 review, there has been a substantial increase in research on the early life-history stages of echinoderms, particularly in relation to settlement and recruitment patterns and processes. For example, of the 108 studies included in the tables, 80 were published after 1982. We have compartmentalised this research into three components: larval supply, settlement and recruitment. Most of the studies that we reviewed dealt with only one of these components, some in only a peripheral manner. In population studies, for example, it is often concluded that recruitment is highly variable and probably important in population regulation without strong empirical or experimental support for these conclusions. We contend that each of these components represents a critical stage of early life-history of echinoderms, and that studies which integrate all components for a given species will best enable us to understand the relative importance of each. Also, it is only through an integrated approach that we can reasonably parametrise population dynamics models (e.g., stage-based matrix or simultaneous differential equation models) to predict patterns of settlement and recruitment for echinoderm populations. To date, there has been little application of such models to benthic marine invertebrates in general (Eckman 1996). However, models will only be as good as the data used to construct them (Grant 1989) and, as indicated by this review, there remain many unanswered questions about the early life histories of echinoderms.

Many species of echinoderms produce vast quantities of planktonic larvae that are dispersed over great distances and suffer huge losses before settling. Larval supply is a crucial determinant of spatial and temporal patterns of recruitment in these species, and a potential bottleneck to population growth. However, the fate of larvae in the plankton remains the most poorly known aspect of echinoderm early life histories, a "black box" which we are only beginning to penetrate (McEdward 1995). The behaviour of echinoderm larvae has been studied almost exclusively in the laboratory (mostly under static conditions) and the relevance of observed behaviours to the situation in nature is difficult to predict. Experimental mesocosms may circumvent this problem by approximating natural conditions while enabling some degree of control over factors which regulate larval behaviour. Natural mesocosms such as tide pools also may prove useful in studies of larval behaviour (Metaxas & Scheibling 1993). Field studies may benefit from new methods of larval tracking (reviewed by Levin 1990) which, when combined with increasingly more sophisticated hydrodynamic models (e.g., Griffin & Thompson 1996), could yield more accurate predictions of advective transport of larvae from spawning sources, and elucidate physical factors which influence larval supply (e.g., Taggart et al. 1996).

Studies of settlement of echinoderm larvae also have been largely restricted to static conditions in the laboratory. These studies have shown that many species can be induced to settle on a variety of substrata, including microbial films, suggesting that settlement is not substratum specific (Pearce 1997). Settlement induction, therefore, may be less important than larval supply in determining rates and patterns of settlement. Once competent larvae are delivered to an area by large-scale hydrodynamic processes, specific settlement sites may be determined by variations in boundary-layer flow, either by passive deposition of larvae or active selection of particular substrates when flow conditions permit (Butman 1987). Field experiments which track competent larvae through to settlement, particularly in areas where they will encounter a variety of substrata or microhabitats, could advance our understanding of the role of larval behaviour in determining settlement patterns in nature. A logical extension of such studies would be to monitor the fate of settlers in different microhabitats to determine whether larvae tend to select sites to maximise post-metamorphic survival.

Although the number of studies examining spatial and temporal patterns of settlement and recruitment in echinoderms has increased markedly over the past fourteen years, the distribution of these studies across taxa and geographic regions is highly skewed. Of the 108 studies on this topic that we reviewed (Tables 1 to 4), 77 were done in northern temperate waters and 36 of those were on echinoids, of which 27 were on strongylocentrotids. The work on *Strongylocentrotus* has revealed that recruitment patterns can be highly variable, both temporally and spatially, even within the geographic range of a species. However, this emphasis on a single northern temperate genus may bias our perception of recruitment variability among echinoderms in general. Only 24 of the studies on recruitment patterns have involved tropical echinoderms, and 14 of those have been on two species, *Diadema antillarum* and *Acanthaster planci*, which have undergone large population fluctuations, atypical of tropical species. Most studies of recruitment of *D. antillarum* have been aimed at investigating population recovery after a mass-mortality in the Caribbean, whereas studies *A. planci* have been prompted by attempts to understand population outbreaks on the Great Barrier Reef. Unfortunately, recruits rarely have been found in these cases, offering few insights for recruitment of even these most studied species. Furthermore, the paucity of studies from southern temperate (5) and polar (2, both Antarctic) regions provides little direct evidence to assess geographic variation in patterns and processes of recruitment of echinoderms. Because spatial and temporal variability in recruitment are common at all scales, among all echinoderm taxa, and throughout all geographic regions, data from one or a few species in a region will likely not be representative of other species in that area. Nevertheless, knowledge gained from intensively studied species can guide future research on the less studied species and geographic regions.

Although the importance of recruitment in determining the distribution and abundance of echinoderm populations is widely recognised, the inherent variability in the definition of recruitment often causes confusion when it fails to identify which stage of the early life-history is being considered (for a more general discussion of this problem, see Hunt & Scheibling 1997). Recruitment typically is defined operationally by the method used to sample juveniles which can vary greatly among studies, even of the same species (Tables 1 to 4). Because post-settlement processes such as predation and migration can alter the pattern of juvenile abundance, the measure of recruitment may depend largely on the time elapsed since settlement. Without isolating the effects of these processes, the researcher is left without a clear understanding of the factors responsible for the pattern observed or whether it reflects the pattern of settlement. Although sampling early juveniles presents significant challenges, the development of new techniques for tagging and monitoring (e.g., time-lapse video) recruits in the field can greatly extend our ability to track cohorts from settlement or at least shortly thereafter. The use of artificial collectors, coupled with continuous temperature and current records, is facilitating the detection of settlement patterns over ecologically relevant spatial and temporal scales, and in relation to local hydrographic conditions (e.g., Miller & Emlet 1997, Balch et al. 1998, 1999, Balch & Scheibling in press). Conventional procedures for tracking cohorts and detecting recruitment events, such as modal analysis of size-frequency distributions, also have benefited from advancements in analytical methods and improved techniques of aging individuals. By increasing the efficiency and precision of sampling, these technological and methodological advances will enable future researchers to better understand the mechanistic links between the early life-history stages of echinoderms which cause the patterns that we observe.

ACKNOWLEDGMENTS

We thank Bruce Hatcher and four anonymous reviewers for commenting on earlier versions of this manuscript. This research was supported by an NSERC Research Grant to RES and a Dalhousie University Graduate Scholarship and Patrick Lett Bursary to TB.

REFERENCES

Allison, G.W. 1994. Effects of temporary starvation on larvae of the sea star *Asterina miniata*. *Mar. Biol.* 118: 255-261.
Andrew, N.L. 1991. Changes in subtidal habitat following mass mortality of sea urchins in Botany Bay, New South Wales. *Aust. J. Ecol.* 16: 353-373.

Andrew, N.L. & Choat J.H. 1982. The influence of predation and conspecific adults on the abundance of juvenile *Evechinus chloroticus* (Echinodermata: Echinometridae). *Oecologia* 54: 80-87.

Andrew, N.L. & Choat J.H. 1985. Habitat related differences in the survivorship and growth of juvenile sea urchins. *Mar Ecol Prog Ser* 27: 155-161.

Andrew, N.L. & Underwood A.J. 1989. Patterns and abundance of the sea urchin *Centrostephanus rodgersii* (Agassiz) on the central coast of New South Wales, Australia. *J. Exp. Mar. Biol. Ecol.* 131: 61-80.

Bak, R.P.M. 1985. Recruitment patterns and mass mortalities in the sea urchin *Diadema antillarum*. In: Delesalle, B., Gabrie, C., Galzin, R., Harmelin-Vivien, M., Toffart, J-L. & Salvat, B. (eds), *Proc. 5th Int. Coral Reef Congr.* Antenne Muséum-EPHE, Moorea, 5: 267-272.

Bak, R.P.M., Carpay, M.J.E. & de Ruyter van Steveninck E.D. 1984. Densities of the sea urchin *Diadema antillarum* before and after mass mortalities on the coral reefs of Curacao. *Mar. Ecol. Prog. Ser.* 17: 105-108.

Balch, T., Hatcher B.G. & Scheibling R.E. 1999. A major settlement event associated with minor meteorologic and oceanographic fluctuations. *Can. J. Zool.* 77: 1657-1662.

Balch, T. & Scheibling R.E. 2000. Temporal and spatial variability in settlement and recruitment of echinoderms in kelp beds and barrens in Nova Scotia. *Mar. Ecol. Prog. Ser.* 205: 139-154.

Balch, T., Scheibling, R.E., Harris, L.G., Chester, C.M. & Robinson, S.M.C. 1998. Variation in settlement of *Strongylocentrotus droebachiensis* in the northwest Atlantic: Effects of spatial scale and sampling method. In: Mooi, R. & Telford, M. (eds), *Echinoderms San Francisco*: 555-560. A.A. Balkema, Rotterdam.

Balch, T. & Scheibling, R.E. 2000. Temporal and spatial variability in settlement and recruitment of echinoderms in kelp beds and barrens in Nova Scotia. *Mar. Ecol. Prog. Ser.* 205: 139-154.

Barker, M.F. 1977. Observations on the settlement of the brachiolaria larvae of *Stichaster australis* (Verrill) and *Coscinasterias calamaria* (Grey) (Echinodermata: Asteroidea) in the laboratory and on the shore. *J. Exp. Mar. Biol. Ecol.* 30: 95-108.

Barker, M.F. 1979. Breeding and recruitment in a population of the New Zealand starfish *Stichaster australis* (Verrill). *J. Exp. Mar. Biol. Ecol.* 41: 195-211.

Barker, M.F. & Nichols, D. 1983. Reproduction, recruitment and juvenile ecology of the starfish, *Asterias rubens* and *Marthasterias glacialis*. *J. Mar. Biol. Ass. UK.* 63: 745-765.

Basch, L.V. & Pearse, J.S. 1996. Consequences of larval feeding environment for settlement and metamorphosis of a temperate echinoderm. *Oceanol. Acta.* 19: 273-285

Basch, L. & Tegner, M.J. 1995. Sea urchin larval supply, settlement, recruitment and adult density in a large kelp forest: Pattern and process. In: *Abstracts of the 2nd Biennial Larval Biology Meetings*: 5. Harbour Branch Oceanographic Institution, Fort Pierce.

Benzie, J.A.H. 1992. Review of the genetics, dispersal and recruitment of crown-of-thorns starfish (*Acanthaster planci*). *Aust. J. Mar. Freshwater Res.* 43: 597-610.

Bernstein, B.B. & Jung, N.C. 1979. Selective pressures and coevolution in a kelp canopy community in southern California. *Ecol. Monogr.* 49: 335-355.

Beukema, J.J. 1985. Growth and dynamics in populations of *Echinocardium cordatum* living in the North Sea off the Dutch north coast. *Neth. J. Sea. Res.* 19: 129-134.

Birkeland, C. 1982. Terrestrial runoff as a cause of outbreaks of *Acanthaster planci* (Echinodermata: Asteroidea). *Mar. Biol.* 69: 175-185.

Birkeland, C., Chia, F.-S. & Strathmann, R.R. 1971. Development, substratum selection, delay of metamorphosis and growth in the seastar, *Mediaster aequalis* Stimpson. *Biol. Bull.* 141: 99-108.

Birkeland, C. & Lucas, J.S. 1990. *Acanthaster planci: Major Management Problem of Coral Reefs.* CRC Press, Boca Raton, 257 pp.

Black, K.P. & Moran, P.J. 1991. Influence of hydrodynamics on the passive dispersal and initial recruitment of *Acanthaster planci* (Echinodermata: Asteroidea) on the Great Barrier Reef. *Mar. Ecol. Prog. Ser.* 69: 55-65.

Boivin, Y., Larrivee, D. & Himmelman, J.H. 1986. Reproductive cycle of the subarctic brooding asteroid *Lepasterias polaris. Mar. Biol.* 92: 329-337.

Booth, D.J. & Brosnan, D.M. 1995. The role of recruitment dynamics in rocky shore and coral reef fish communities. *Adv. Ecol. Res.* 34: 309-385.

Bosch, I., Rivkin, R.B. & Alexander, S.P. 1989. Asexual reproduction by oceanic planktotrophic echinoderm larvae. *Nature* 337: 169-170.

Bosselmann, A. 1989. Larval plankton and recruitment of macrofauna in a subtidal area in the German Bight. In: Ryland, J.S. & Tyler, P.A. (eds), *Reproduction, genetics and distributions of marine organisms:* 43-54. Olsen and Olsen, Fredensborg.

Botsford, L.W., Smith, B.D. & Quinn, J.F. 1994. Bimodality in size distributions: The red sea urchin *Strongylocentrotus franciscanus* as an example. *Ecol. Appl.* 4: 42-50.

Bourgoin, A. & Guillou, M. 1988. Demographic study of *Amphiura filiformis* (Echinodermata: Ophiuroidea) in Concarneau Bay (Finistére, France). *Oceanol. Acta* 11: 79-87.

Bourgoin, A., Guillou, M. & Glémarec, M. 1990. Environmental instability and demographic variability in *Acrocnida brachiata* (Echinodermata: Ophiuroidea) in Douarnenez Bay (Brittany: France). *Mar. Ecol.* 12: 89-104.

Breen, P.A., Carolsfeld, W. & Yamanaka, K.L. 1985. Social behaviour of juvenile red sea urchins, *Strongylocentrotus franciscanus* (Agassiz). *J. Exp. Mar. Biol. Ecol.* 92: 45-61.

Brey, T., Pearse, J., Basch, L., McClintock, J. & Slattery, M. 1995. Growth and production of *Sterechinus neumayeri* (Echinoidea: Echinodermata) in McMurdo Sound, Antarctica. *Mar. Biol.* 124: 279-292.

Brodie, J.E. 1992. Enhancement of larval and juvenile survival and recruitment of *Acanthaster planci* from the effects of terrestrial runoff: a review. *Aust. J. Mar. Freshwater Res.* 43: 539-554.

Bulteel, P, Jangoux, M. & Coulon, P. 1992. Biometry, bathymetric distribution, and reproductive cycle of the holothuroid *Holothuria tubulosa* (Echinodermata) from Mediterranean seagrass beds. *Mar. Ecol.* 13: 53-62.

Burkenroad, M.D. 1957. Intensity of setting of starfish in Long Island Sound in relation to fluctuations of the stock of adult starfish and in the setting of oysters. *Ecology* 38: 164-165.

Butman, C.A. 1987. Larval settlement of soft-sediment invertebrates: the spatial scales of pattern explained by active habitat selection and the emerging role of hydrodynamical processes. *Oceanogr. Mar. Biol. Annu. Rev.* 25: 113-165.

Caley, M.J., Carr, M.H., Hixon, M.A., Hughes, T.P., Jones, G.P. & Menge, B.A. 1996. Recruitment and the local dynamics of open marine populations. *Ann. Rev. Ecol. Syst.* 27: 477-500.

Cameron, J.L. & Fankboner, P.V. 1989. Reproductive biology of the commercial sea cucumber *Parastichopus californicus* (Stimpson) (Echinodermata: Holothuroidea). II. Observations on the ecology of development, recruitment, and juvenile life stage. *J. Exp. Mar. Biol. Ecol.* 127: 43-67.

Cameron, R.A. & Hinegardner, R.T. 1974. Initiation of metamorphosis in laboratory cultured sea urchins. *Biol. Bull.* 146: 335-342.

Cameron, R.A. & Rumrill, S.S. 1982. Larval abundance and recruitment of the sand dollar *Dendraster excentricus* in Monterey Bay, California, USA. *Mar. Biol.* 71: 197-202.

Cameron, R.A. & Schroeter, S.C. 1980. Sea urchin recruitment: effect of substrate selection on juvenile distribution. *Mar. Ecol. Prog. Ser.* 2: 243-247.

Carpenter, R.C. 1990. Mass mortality of *Diadema antillarum* I. Long-term effects on sea urchin population-dynamics and coral reef algal communities. *Mar. Biol.* 104: 67-77.

Chao, S.-M., Chen, C.-P. & Alexander, P.S. 1994. Reproduction and growth of *Holothuria atra* (Echinodermata: Holothuroidea) at two contrasting sites in southern Taiwan. *Mar. Biol.* 119: 565-570.

Chao, S.-M. & Tsai, C.-C. 1995. Reproduction and population dynamics of the fissiparous brittle star *Ophiactis savignyi* (Echinodermata: Ophiuroidea). *Mar. Biol.* 124: 77-83.

Chapman, A.R.O. & Johnson, C.R. 1990. Disturbance and organization of macroalgal assemblages in the northwest Atlantic. *Hydrobiologia* 192: 77-121.

Chia, F.-S. 1989. Differential larval settlement of benthic marine invertebrates. In: Ryland, J.S., Tyler, P.A. (eds), *Reproduction, genetics and distributions of marine organisms*: 3-12. Olsen and Olsen, Fredensborg, Denmark.

Chia, F.-S. & Walker, C.W. 1991. Echinodermata: Asteroidea. In: Giese, A.C., Pearse, J.S., Pearse, V.B. (eds), *Reproduction of Marine Invertebrates: Echinoderms and Lophophorates:* 301-353. The Boxwood Press, Pacific Grove.

Chia, F.-S., Young, C.M. & McEuen, F.S. 1984. The role of larval settlement behaviour in controlling patterns of abundance in echinoderms. In: Engels, W., Clark, Jr W.H., Fisher, A., Olive, P.J.W., Wend, D.F. (eds), *Adv. Invert. Reprod.* 3: 409-424.

Chiu, S.T. 1986. Effects of the urchin fishery on the population structure of *Anthocidaris crassispina* (Echinodermata, Echinoidea) in Hong Kong. In: MacLean, J.L., Dizon, L.B., Hosillos, L.V. (eds), *The First Asian Fisheries Forum:* 361-366. Asian Fisheries Society, Manila.

Connell, J.H. 1985. The consequences of variation in initial settlement vs. postsettlement mortality in rocky intertidal communities. *J. Exp. Mar. Biol. Ecol.* 93: 11-45.

Cowden, C., Young, C.M. & Chia, F.-S. 1984. Differential predation on marine invertebrate larvae by two benthic predators. *Mar. Ecol. Prog. Ser.* 14: 145-149.

Day, R.W. & Osman, R.W. 1981. Predation by *Patiria miniata* (Asteroidea) on bryozoans: Prey diversity may depend on the mechanism of succession. *Oecologia* 51: 300-309.

Dayton, P.K., Robilliard, G.A., Paine, R.T. & Dayton, L.B. 1974. Biological accommodation in the benthic community at McMurdo Sound, Antarctica. *Ecol. Monogr.* 44: 105-128.

Dayton, P.K. & Tegner, M.J. 1984. The importance of scale in community ecology: a kelp forest example with terrestrial analogs. In: Price, P.W., Slobodchitkoff, C.N., Gaud, W.S. (eds), *A new ecology – novel approaches to interactive systems:* 457-481. John Wiley and Sons, New York.

DeRidder, C., David, B., Laurin, B. & LeGall, P. 1991. Population dynamics of the spatangoid echinoid *Echinocardium cordatum* (Pennant) in the Bay of Seine, Normandy. In: Yanagisawa, T., Yasumasu, I., Oguro, C., Suzuki, N. & Motokawa, T. (eds), *Biology of Echinodermata*: 153-158. A.A. Balkema, Rotterdam.

DeVantier, L.M. & Deacon, G. 1990. Distribution of *Acanthaster planci* at Lord Howe Island, the southern-most Indo-Pacific reef. *Coral Reefs* 9: 145-148.

Doherty, P.J. & Davidson, J. 1988. Monitoring the distribution and abundance of juvenile *Acanthaster planci* in the central Great Barrier Reef. In: Choat, J.H., Barnes, D., Borowitzka, M.A., Coll, J.C., Davies, P.J., Flood, P., Hatcher, B.G., Hopley, D., Hutchings, P.A., Kinsey, D., Orme, G.R., Pichon, M., Sale, P.F., Sammarco, P., Wallace, C.C., Wilkinson, C., Wolanski, E. & Bellwood, O. (eds), *Proc 6th Int Coral Reef Symp:* 131-136. The Sixth International Coral Reef Symposium Executive Committee, Townsville.

Duggins, D.O., Eckman, J.E. & Sewell, A.T. 1990. Ecology of understory kelp environments. II. Effects of kelps on recruitment of benthic invertebrates. *J. Exp. Mar. Biol. Ecol.* 143: 27-45.

Duineveld, G.C.A. & Van Noort, G.J. 1986. Observations on the population dynamics of *Amphiura filiformis* (Echinodermata: Ophiuroidea) in the southern North Sea and its exploitation by the dab, *Limanda limanda. Neth. J. Sea. Res.* 20: 85-94.

Ebert, T.A. 1983. Recruitment in echinoderms. In: Jangoux, M. & Lawrence, J. (eds), *Echinoderm Studies* 1: 169-203. A.A. Balkema, Rotterdam.

Ebert, T.A. & Russel, M.P. 1988. Latitudinal variation in size structure of the west coast purple sea urchin: A correlation with headlands. *Limnol. Oceanogr.* 33: 286-294.

Ebert, T.A., Schroeter, S.C., Dixon, J.D. & Kalvass, P. 1994. Settlement patterns of red and purple sea urchins (*Strongylocentrotus franciscanus* and *S. purpuratus*) in California, USA. *Mar. Ecol. Prog. Ser.* 111: 41-52.

Eckelbarger, K.J. 1994. Ultrastructural features of gonads and gametes in deep-sea invertebrates. In: Young, C.M. & Eckelbarger, K.J. (eds), *Reproduction, larval biology, and recruitment of the deep-sea benthos*: 137-157. Columbia University Press, New York.

Eckman, J.E. 1996. Closing the larval loop: linking larval ecology to the population dynamics of marine benthic invertebrates. *J. Exp. Mar. Biol. Ecol.* 200: 207-237.

Eckman, J.E., Duggins, D.O. & Sewell, A.T. 1989. Ecology of understory kelp environments. I. Effects of kelps on flow and particle transport near the bottom. *J. Exp. Mar. Biol. Ecol.* 129: 173-187.

Emson, R.H., Jones, M.B. & Whitfield, P.J. 1989. Habitat and latitude differences in reproductive pattern and life-history in the cosmopolitan brittle-star *Amphipholis squamata* (Echinodermata). In: Ryland, J.S. & Tyler, P.A. (eds), *Reproduction, genetics and distributions of marine organisms*: 75-82. Olsen and Olsen, Fredensborg.

Estes, J.A. & Duggins, D.O. 1995. Sea otters and kelp forests in Alaska: Generality and variation in a community ecological paradigm. *Ecol. Monogr.* 65: 75-100.

Fisk, D.A. 1992. Recruitment of *Acanthaster planci* over a five-year period at Green Island reef. *Aust. J. Mar Freshwater Res.* 43: 629-633.

Fisk, D.A., Harriot, V.J., Pearson, R.G. 1988. The history and status of Crown-of-thorns starfish and corals at Green Island Reef, Great Barrier Reef. In: Choat, J.H., Barnes, D., Borowitzka, M.A., Coll, J.C., Davies, P.J., Flood, P., Hatcher,

B.G., Hopley, D., Hutchings, P.A., Kinsey, D., Orme, G.R., Pichon, M., Sale, P.F., Sammarco, P., Wallace, C.C., Wilkinson, C., Wolanski, E. & Bellwood, O. (eds), *Proc 6th Int Coral Reef Symp*: 149-155. The Sixth International Coral Reef Symposium Executive Committee, Townsville.

Forcucci, D. 1994. Population density, recruitment and 1991 mortality event of *Diadema antillarum* in the Florida Keys. *Bull. Mar. Sci.* 54: 917-928.

Foreman, R.E. 1977. Benthic community modification and recovery following intensive grazing by *Strongylocentrotus droebachiensis. Helgoländer wiss Meeresunters* 30: 468-484.

Fujita, T. & Ohta, S. 1990. Size structure of dense populations of the brittle star *Ophiura sarsi* (Ophiuroidea: Echinodermata) in the bathyl zone around Japan. *Mar. Ecol. Prog. Ser.* 64: 113-122.

Gage, J.D. 1982. Age structure in populations of the deep-sea brittle-star *Ophiomusium lymani*: a regional comparison. *Deep-Sea Res.* 29: 1565-1586.

Gage, J.D. 1985. The analysis of population dynamics in deep-sea benthos. In: Gibbs PE (ed.), *Proceedings of the Nineteenth European Marine Biology Symposium*: 201-212. Cambridge University Press, Plymouth.

Gage, J.D. 1994. Recruitment ecology and age structure of deep-sea invertebrate populations. In: Young, C.M., & Eckelbarger, K.J. (eds), *Reproduction, larval biology, and recruitment of the deep-sea benthos*: 223-242. Columbia University Press, New York.

Gage, J.D., Lightfoot, R.H., Pearson, M. & Tyler, P.A. 1980. An introduction to a sample time-series of abyssal macrobenthos: methods and principle sources of variability. *Oceanol. Acta* 3: 169-176.

Gage, J.D. & Tyler, P.A. 1981a. Non-viable seasonal settlement of larvae of the upper bathyal brittle star *Ophiocten gracilis* in the Rockall Trough abyssal. *Mar. Biol.* 64: 153-161.

Gage, J.D. & Tyler, P.A. 1981b. Re-appraisal of age composition, growth and survivorship of the deep-sea brittle star *Ophiura ljungmani* from size structure in a sample time series from the Rockall Trough. *Mar. Biol.* 64: 163-172.

Gage, J.D. & Tyler, P.A. 1982a. Growth and reproduction of the deep-sea brittlestar *Ophiomusium lymani* Wyville Thomson. *Oceanol. Acta* 5: 73-83.

Gage, J.D. & Tyler, P.A. 1982b. Growth strategies of deep-sea ophiuroids. In: Lawrence J.M. (ed), *Echinoderms: Proceedings of the International Conference, Tampa Bay*: 305-311. A.A. Balkema, Rotterdam.

Gage, J.D. & Tyler, P.A. 1982c. Depth-related gradients in size structure and bathymetric zonation of deep-sea brittle stars. *Mar. Biol.* 71: 299-308.

Gage, J.D. & Tyler, P.A. 1985 Growth and recruitment of the deep-sea urchin *Echinus affinis. Mar. Biol.* 90: 41-53.

Gaines, S.D. & Roughgarden, J. 1985. Larval settlement rate: a leading determinant of structure in an ecological community of the marine intertidal zone. *Proc. Nat. Acad. Sci. USA* 82: 3707-3711.

Gaines, S.D. & Roughgarden, J. 1987. Fish in offshore kelp forests affect recruitment to intertidal barnacle populations. *Science* 235: 479-481.

Glémarec, M. & Guillou, M. 1996. Recruitment and year-class segregation in response to abiotic and biotic factors. *Oceanol. Acta* 19: 409-414.

Gosselin, L.A. & Qian, P.-Y. 1997. Juvenile mortality in benthic marine invertebrates. *Mar. Ecol. Prog. Ser.* 146: 265-282.

Grant, A. 1989. Marine invertebrate life histories – is there any value in mathematical modelling? In: Ryland, J.S. & Tyler, P.A. (eds), *Reproduction, genetics and distributions of marine organisms*: 91-94. Olsen and Olsen, Fredensborg.

Grassle, J.F. 1994. Ecological patterns in the deep-sea: How are they related to repro-
duction, larval biolgy, and recruitment? In: Young, C.M., Eckelbarger, K.J. (eds).
Reproduction, larval biology, and recruitment of the deep-sea benthos: 306-314.
Columbia University Press, New York.

Griffin, D.A. & Thompson, K.R. 1996. The adjoint method of data assimilation used
operationally for shelf circulation. *J. Geophys. Res.* 101: 3457-3477.

Guillou, M. & Diop, M. 1988. Ecology and demography of a population of *Anseropoda placenta* (Echinodermata: Asteroidea) in the Bay of Brest, Brittany. *J. Mar.
Biol. Ass. UK* 68: 41-54.

Guillou, M. & Michel, C. 1993a. Reproduction and growth of *Sphaerechinus granularis* (Echinodermata: Echinoidea) in southern Brittany. *J. Mar. Biol. Ass. UK*
73: 179-192.

Guillou, M. & Michel, C. 1993b. Impact de la variabilité du recrutement sur le stock
d'oursins commercialisables de l'archipel de Glénan (Sud-Bretagne). *Oecea-
nolgica Acta* 16: 423-430.

Haedrich, R.L., Rowe, G.T. & Polloni, P.T. 1980. The megabenthic fauna in the deep
sea south of New England, USA. *Mar. Biol.* 57: 165-179.

Hamel, J.-F. & Mercier, A. 1996. Early development, settlement, growth, and spatial distribution of the sea cucumber *Cucumaria frondosa* (Echinodermata: Holothuroidea). *Can. J. Fish. Aquat. Sci.* 53: 253-271.

Harris, L.G. & Chester, C.M. 1996. Effects of location, exposure and physical structure on juvenile recruitment of the sea urchin *Strongylocentrotus droebachiensis*
in the Gulf of Maine. *J. Invert. Repro. Dev.* 30: 207-215.

Harris, L.G., Rice, B. & Nestler, E.C. 1994. Settlement, early survival and growth
in a southern Gulf of Maine population of *Strongylocentrotus droebachiensis*
(Muller). In: David, B., Guille, A., Féral, J.-P. & Roux, M. (eds), *Echinoderms
Through Time*: 701-706. A.A. Balkema, Rotterdam.

Harris, L.G., Witman, J.D., Rowley, R. 1985) A comparison of sea urchin recruitment at sites on the Atlantic and Pacific coasts of North America. In: Keegan,
B.F., O'Connor, B.D.S. (eds), *Echinodermata*: 389. A.A. Balkema, Rotterdam.

Harrold, C., Lisin, S., Light, K.H. & Tudor, S. 1991. Isolating settlement from
recruitment of sea urchins. *J. Exp. Mar. Biol. Ecol.* 147: 81-94.

Harrold, C., Pearse, J.S. 1987. The ecological role of echinoderms in kelp forests. In:
Jangoux M, Lawrence JM (eds), *Echinoderm Studies* 2: 137-233. A.A. Balkema,
Rotterdam.

Hart, M.W. & Scheibling, R.E. 1988. Heat waves, baby booms, and the destruction
of kelp beds by sea urchins. *Mar. Biol.* 99: 167-176.

Hendler, G. 1991. Echinodermata: Ophiuroidea. In: Giese, A.C., Pearse, J.S. &
Pearse, V.B. (eds), *Reproduction of Marine Invertebrates: Echinoderms and
Lophophorates:* 356-511. The Boxwood Press, Pacific Grove.

Hess, H., Bingam, B., Cohen, S., Grosberg, R.K., Jefferson, W. & Walters, L. 1988.
The scale of genetic differentiation in *Leptosynapta clarki* (Heding), an infaunal
brooding holothuroid. *J. Exp. Mar. Biol. Ecol.* 122: 187-194.

Highsmith, R.C. 1982. Induced settlement and metamorphosis of sand dollar (*Dendraster excentricus*) larvae in predator-free sites: Adult sand dollar beds. *Ecology*
63: 329-337.

Highsmith, R.C. 1985. Floating and algal rafting as potential dispersal mechanisms
in brooding invertebrates. *Mar. Ecol. Prog. Ser.* 25: 169-179.

Himmelman, J.H. 1986. Population biology of green sea urchins on rocky barrens.
Mar. Ecol. Prog. Ser. 33: 295-306.

Himmelman, J.H. & Dutil, C. 1991. Distribution, population structure and feeding

74

of subtidal seastars in the northern Gulf of St. Lawrence. *Mar. Ecol. Prog. Ser.* 76: 61-72.

Himmelman, J.H., Guderley, H., Vignault, G., Drouin, G. & Wells, P.G. 1984. Response of the sea urchin, *Strongylocentrotus droebachiensis,* to reduced salinities: importance of size, acclimation, and interpopulation differences. *Can. J. Zool.* 62: 1015-1021.

Himmelman, J.H., Lavergne, Y., Axelsen, F., Cardinal, A. & Bourget, E. 1983. Sea urchins in the St. Lawrence Estuary: their abundance size-structure, and suitability for commercial exploitation. *Can. J. Fish. Aquat. Sci.* 40: 474-486.

Himmelman, J.H., Leveigne, Y., Cardinal, A., Martel, G. & Jalbert, P. 1982. Brooding behaviour in the northern sea star *Leptasterias polaris. Mar. Biol.* 68: 235-240.

Holland, N.D. 1991. Echinodermata: Crinoidea. In: Giese, A.C., Pearse, J.S. & Pearse, V.B. (eds), *Reproduction of Marine Invertebrates: Echinoderms and Lophophorates:* 247-299. The Boxwood Press, Pacific Grove.

Hooper, R. 1980. Observations on algal-grazer interactions in Newfoundland and Labrador. In Pringle, J.D., Sharp, G.J., Caddy, J.F. (eds), Proceedings of the workshop on the relationship between sea urchin grazing and commercial plant/animal harvesting. *Can. Tech. Rep. Fish. Aquat. Sci. No.* 954, 120-124.

Hughes, T.P., Keller, B.D., Jackson, J.B.C. & Boyle, M.J. 1985. Mass mortality of the echinoid *Diadema antillarum* Phillippi in Jamaica. *Bull. Mar. Sci.* 36: 377-384.

Hunt, H.L. & Scheibling, R.E. 1997. Role of early post-settlement mortality in the recruitment of benthic marine invertebrates. *Mar. Ecol. Prog. Ser.* 155: 269-301.

Hunte, W. & Younglao, D. 1988. Recruitment and population recovery of *Diadema antillarum* (Echinodermata, Echinoidea) in Barbados. *Mar. Ecol. Prog. Ser.* 45: 109-119.

Jackson, G.A. & Strathmann, R.R. 1981. Larval mortality from offshore mixing as a link between precompetent periods of development. *Am. Nat.* 118: 16-26.

Jackson, G.A. & Winant, C.D. 1983. Effect of a kelp forest on coastal currents. *Cont. Shelf. Res.* 2: 75-80.

James, M.K. & Scandol, J.P. 1992. Larval dispersal simulations: correlation with the Crown-of-thorns starfish outbreaks database. *Aust. J. Mar. Freshwater Res.* 43: 569-582.

Johnson, C. 1992a. Reproduction, recruitment and hydrodynamics in the Crown-of-thorns phenomenon on the Great Barrier Reef: Introduction and synthesis. *Aust. J. Mar. Freshwater Res.* 43: 517-523.

Johnson, C. 1992b. Settlement and recruitment of *Acanthaster planci* on the Great Barrier Reef: Questions of process and scale. *Aust. J. Mar. Freshwater Res.* 43: 611-627.

Johnson, D.B., Sutton, D.C., Olson, R.R. & Giddins, R. 1991. Settlement of crown-of-thorns starfish: role of bacteria on surfaces of coralline algae and a hypothesis for deepwater recruitment. *Mar. Ecol. Prog. Ser.* 71: 143-162.

Jones, M.B. & Smaldon, G. 1989. Aspects of the biology of a population of the cosmopolitan brittlestar *Amphipholis squamata* (Echinodermata) from the Firth of Forth, Scotland. *J. Nat. Hist.* 23: 613-625.

Jost, P. & Rein, K. 1985. Migration from refuges: a stabilising factor for a sea-star community. In: Keegan, B.F., O'Connor, B.D.S. (eds), *Echinodermata*: 523-528. A.A. Balkema, Rotterdam.

Karlson, R.H. & Levitan, D.R. 1990. Recruitment limitation in open populations of *Diadema antillarum* an evaluation. *Oecologia* 82: 40-44.

Keats, D.W., South, G.R. & Steele, D.H. 1985. Ecology of juvenile green sea urchins (*Strongylocentrotus droebachiensis*) at an urchin dominated subtidal site in eastern Newfoundland. In: Keegan, B.F., O'Connor, B.D.S. (eds), *Echinodermata:* 295-302. A.A. Balkema, Rotterdam.

Keesing, J.K., Cartwright, C.M. & Hall, K.C. 1993. Measuring settlement intensity of echinoderms on coral reefs. *Mar. Biol.* 117: 399-407.

Keesing, J.K. & Halford, A.R. 1992a. Importance of postsettlement processes for the population dynamics of *Acanthaster planci* (L.). *Aust. J. Mar. Freshwater Res.* 43: 635-651.

Keesing, J.K. & Halford, A.R. 1992b. Field measurement of survival rates of juvenile *Acanthaster planci*: techniques and preliminary results. *Mar. Ecol. Prog. Ser.* 85: 107-114.

Keough, M.J. & Downes, B.J. 1982. Recruitment of marine invertebrates: the role of active larval choices and early mortality. *Oecologia* 54: 348-352.

Kitamura, H., Kitahara, S. & Koh, H.B. 1993. The induction of larval settlement and metamorphosis of two sea urchins, *Pseudocentrotus depressus* and *Anthocidaris crassispina*, by free fatty acids extracted from the coralline red alga *Corallina pilulifera*. *Mar. Biol.* 115: 387-392.

Künitzer, A. 1989. Factors affecting population dynamics of *Amphiura filiformis* (Echinodermata: Ophiuroidea) and *Mysella bidentata* (Bivalvia: Galeommatacea) in the North Sea. In: Ryland, J.S., Tyler, P.A. (eds), *Reproduction, genetics and distributions of marine organisms*: 395-406. Olsen and Olsen, Fredensborg, Denmark.

Lang, C., Mann, K.H. 1976 Changes in sea urchin populations after the destruction of kelp beds. *Mar. Biol.* 36: 321-326.

Lawrence, J.M. 1975. On the relationships between marine plants and sea urchins. *Oceanogr. Mar. Biol. Annu. Rev.* 13: 213-286.

Lawrence, J.M. 1996. Mass mortality of echinoderms from abiotic factors. In: Jangoux, M. & Lawrence, J.M. (eds), *Echinoderm Studies* 5: 103-137. A.A. Balkema, Rotterdam.

Leinaas, H.P. & Christie, H. 1996. Effects of removing sea urchins (*Strongylocentrotus droebachiensis*): Stability of the barren state and succession of kelp forest recovery in the east Atlantic. *Oecologia* 105: 524-536.

Lessios, H.A. 1988. Population dynamics of *Diadema antillarum* (Echinodermata: Echinoidea) following mass mortality in Panama. *Mar. Biol.* 95: 515-526.

Lessios, H.A. 1994. Status of *Diadema* populations ten years after mass mortality. In: David, B., Guille, A., Féral, J.-P. & Roux, M. (eds), *Echinoderms through Time*: 748. A.A. Balkema, Rotterdam.

Lessios, H.A., Robertson, D.R., Cubit, J.D. 1984. Spread of *Diadema* mass mortality through the Caribbean. *Science* 226: 335-337.

Levin, L.A. 1990. A review of methods for labeling and tracking marine invertebrate larvae. *Ophelia* 32: 115-144.

Lewin, R. 1986. Supply-side ecology. *Science* 234: 25-27.

Loosanoff, V.L. 1964. Variations in time and intensity of settling of the starfish, *Asterias forbesi*, in Long Island Sound during a twenty-five-year period. *Biol. Bull.* 126: 423-439.

Lubchenco-Menge, J. & Menge, B. 1974. Role of resource allocation, aggression and spatial heterogeneity in coexistence of two competing intertidal starfish. *Ecol. Monogr.* 44: 189-209.

Lucas, J.S. 1982. Quantitative studies of feeding and nutrition during larval development of the coral reef asteroid *Acanthaster planci* (L.). *J. Exp. Mar. Biol. Ecol.* 65: 173-193.

Lucas, J.S., Hart, R.J., Howden, M.E. & Salathe, R. 1979. Saponins in eggs and larvae of *Acanthaster planci* (L.) (Asteroidea) as chemical defences against planktivorous fish. *J. Exp. Mar. Biol. Ecol.* 40: 155-165.

McCallum, H. 1992. Completing the circle: stock-recruitment relationships and *Acanthaster*. *Aust. J. Mar. Freshwater Res.* 43: 653-662.

McClintock, J.B., Hopkins, T.S., Marion, K.R., Watts, S.A. & Schinner, G.1993. Population structure, growth and reproductive biology of the Gorgonocephalid brittlestar *Asteroporpa annulata*. *Bull. Mar. Sci.* 52: 925-936.

McClintock, J.B., Pearse, J.S. & Bosch, I. 1988. Population structure and energetics of the shallow-water antarctic sea star *Odontaster validus* in contrasting habitats. *Mar. Biol.* 99: 235-246.

McClintock, J.B., Vernon, J.D., Watts, S.A., Marion, K.R. & Hopkins, T.S. 1994. Size frequency, recruitment and adult growth in the sea biscuit *Clypeaster ravenelii* in the northern Gulf of Mexico. In: David, B., Guille, A., Féral, J.-P., Roux, M. (eds), *Echinoderms through Time:* 777-781. A.A. Balkema, Rotterdam.

McEdward, L. (ed). 1995. *Ecology of marine invertebrate larvae*. CRC Press, Boca Raton, 464 pp.

Medeiros-Bergen, D.E. & Miles, E. 1997. Recruitment in the holothurian *Cucumaria frondosa* in the Gulf of Maine. *Invert. Repro. Dev.* 31: 123-133.

Medeiros-Bergen, D.E., Olson, R.R., Conroy, J.A. & Kocher, T.D. 1995. Distribution of holothurian larvae determined with species-specific genetic probes. *Limnol. Oceanogr* 40: 1225-1235.

Medeiros-Bergen, D.E., Perna, N.T., Conroy, J.A. & Kocher, T.D. 1998. Identification of ophiuroid post-larvae using mitochondrial DNA. In: Mooi, R. & Telford, M. (eds), *Echinoderms San Francisco*: 399-404. A.A. Balkema, Rotterdam.

Menge, B. 1975. Brood or broadcast? The adaptive significance of different reproductive strategies in the two intertidal sea stars *Leptasterias hexactis* and *Pisaster ochraceus*. *Mar. Biol.* 31: 87-100.

Menge, B. 1991. Relative importance of recruitment and other causes of variation in rocky intertidal community structure. *J. Exp. Mar. Biol. Ecol.* 146: 69-100.

Metaxas, A. & Scheibling, R.E. 1993. Community structure and organization of tidepools. *Mar. Ecol. Prog. Ser.* 98: 187-198.

Mileikovsky, S.A. 1968. Some common features in the drift of pelagic larvae and juvenile stages of bottom invertebrates with marine currents in temperate regions. *Sarsia* 34: 209-216.

Miller, B.A. & Emlet, R.B. 1997. Influence of nearshore hydrodynamics on larval abundance and settlement of sea urchins *Strongylocentrotus franciscanus* and *S. purpuratus* in the Oregon upwelling zone. *Mar. Ecol. Prog. Ser.* 148: 83-94.

Miller, R.J. & Colodey, A.G. 1983. Widespread mass mortalities of the green sea urchin in Nova Scotia, Canada. *Mar. Biol.* 73: 263-267.

Miller, R.J. & Mann, K.H. 1973. Ecological energetics of the seaweed zone in a marine bay on the Atlantic coast of Canada. III Energy transformations by sea urchins. *Mar. Biol.* 18: 99-114.

Minchinton, T.E. & Scheibling, R.E. 1991. The influence of larval supply and settlement on the population structure of barnacles. *Ecology* 75: 1867-1879

Minchinton, T.E. & Scheibling, R.E. 1993. Variations in sampling procedure and frequency affect estimates of recruitment of barnacles. *Mar. Ecol. Prog. Ser.* 99: 83-88.

Miron, G., Boudreau, B. & Bourget, E. 1995. Use of larval supply in benthic ecology: testing correlations between larval supply and larval settlement. *Mar. Ecol. Prog. Ser.* 124: 301-305.

Mladenov, P.V. & Chia, F.-S. 1983. Development, settling behaviour, metamorphosis and pentacrinoid feeding and growth of the feather star *Florometra serratissima*. *Mar. Biol.* 73: 309-323.

Mladenov, P.V. & Emson, R.H. 1988. Density, size structure and reproductive characteristics of fissiparous brittle stars in algae and sponges: evidence for interpopulational variation in levels of sexual and asexual reproduction. *Mar. Ecol. Prog. Ser.* 42: 181-194.

Mladenov, P.V. & Emson, R.H., Colpit, L.V. & Wilkie, I.C. 1983. Asexual reproduction in the West Indian brittle star *Ophiocomella ophiactoides* (H.L. Clark) (Echinodermata: Ophiuroidea). *J. Exp. Mar. Biol. Ecol.* 72: 1-23.

Moran, P.J. 1986. The *Acanthaster* phenomenon. *Oceanogr. Mar. Biol. Annu. Rev.* 24: 379-480.

Moran, P.J., De'ath, G., Baker, V.J., Bass, D.K., Christie, C.A., Miller, I.R., Miller-Smith, B.A. & Thompson, A.A. 1992. Pattern of outbreaks of crown-of-thorns starfish (*Acanthaster planci* L.) along the Great Barrier Reef since 1966. *Aust. J. Mar. Freshwater Res.* 43: 555-568.

Morse, A.N.C. 1992. Role of algae in the recruitment of marine invertebrate larvae. In: John, D.M., Hawkins, S.J., Price, J.H. (eds), *Plant-animal interactions in the marine benthos*: 385-403. Clarendon Press, Oxford.

Mullineaux, L.S. 1994. Implications of mesoscale flows for dispersal of deep-sea larvae. In: Young, C.M. & Eckelbarger, K.J. (eds), *Reproduction, larval biology, and recruitment of the deep-sea benthos*: 201-222. Columbia University Press, New York.

Munday, B.W. & Keegan, B.F. 1992. Population dynamics of *Amphiura chiajei* (Echinodermata: Ophiuroidea) in Killary Harbour, on the west coast of Ireland. *Mar. Biol.* 114: 595-605.

Muus, K. 1981. Density and growth of juvenile *Amphiura filiformis* (Ophiuroidea) in the Øresund. *Ophelia* 20: 153-168.

Nash, W.J., Goddard, M. & Lucas, J.S. 1988. Population genetic studies of the crown-of-thorns starfish, *Acanthaster planci* (L.) in the Great Barrier Reef region. *Coral Reefs* 7: 11-18.

Nichols, D. & Barker, M.F. 1984. Growth of juvenile *Asterias rubens* L. (Echinodermata: Asteroidea) on an intertidal reef in southwestern Britain. *J. Exp. Mar. Biol. Ecol.* 78: 157-165.

O'Connor, B., Bowmer, T. & Grehan, A. 1983. Long-term assessment of the population dynamics of *Amphiura filiformis* (Echinodermata: Ophiuroidea) in Galway Bay (west coast of Ireland). *Mar. Biol.* 75: 279-286.

O'Connor, B. & McGrath, D. 1980. The population dynamics of *Amphiura filiformis* (O.F. Muller) in Galway Bay, west coast of Ireland. In: Jangoux, M. (ed.), *Echinoderms: Present and Past*: 219-222. A.A. Balkema, Rotterdam.

Ojeda, F.P. & Dearborn, J.H. 1991. Feeding ecology of benthic mobile predators: experimental analyses of their influence in rocky subtidal communities of the Gulf of Maine. *J. Exp. Mar. Biol. Ecol.* 149: 13-44.

Ólafsson, E.B., Peterson, C.H. & Ambrose, W.G. 1994. Does recruitment limitation structure populations and communities of macro-invertebrates in marine soft sediments: The relative significance of pre- and post-settlement processes. *Oceanogr. Mar. Biol. Annu. Rev.* 32: 65-109.

Olson, R.R. 1985. *In situ* culturing of larvae of the crown-of-thorns starfish *Acanthaster planci*. *Bull. Mar. Sci.* 25: 207-210.

Packer, D.B., Watling, L. & Langton, R.W. 1994. The population of the brittle star *Ophiura sarsi* Lutken in the Gulf of Maine and its trophic relationship to American plaice (*Hippoglossoides platessoides* Fabricius). *J. Exp. Mar. Biol. Ecol.* 179: 207-222.

Paine, R.T. 1986. Benthic community-water column coupling during the 1982-1983 El Niño. Are community changes at high latitudes attributable to cause or coincidence? *Limnol. Oceanogr.* 31: 351-360.

Palumbi, S.R. 1995. Using genetics as an indirect estimator of larval dispersal. In: McEdward, L. (ed.), *Ecology of marine invertebrate larvae*: 369-387. CRC Press, Boca Raton.

Pawlik, J.R. 1992. Chemical ecology of the settlement of benthic marine invertebrates. *Oceanogr. Mar. Biol. Annu. Rev.* 30: 273-335.

Pearce, A.F. & Phillips, B.F. 1994. Oceanic processes, puerulus settlement and recruitment of the western rock lobster *Panulirus cygnus*. In: Sammarco, P.W., Heron, M.L. (eds), *The bio-physics of marine larval dispersal*: 279-303. American Geophysical Union, Washington.

Pearce, C.M. 1997. Induction of settlement and metamorphosis in echinoderms. In: Fingerman, M., Nagabhushanam, R. & Thompson, M.-F. (eds), *Recent advances in marine biotechnology:* 283-341.Oxford and IBH Publishing Co. Pvt. Ltd., New Delhi.

Pearce, C.M. & Scheibling, R.E. 1990a. Induction of metamorphosis of larvae of the green sea urchin, *Strongylocentrotus droebachiensis*, by coralline red algae. *Biol. Bull.* 179: 304-311.

Pearce, C.M. & Scheibling, R.E. 1990b. Induction of settlement and metamorphosis in the sand dollar *Echinarachnius parma*: evidence for an adult-associated factor. *Mar. Biol.* 107: 363-369.

Pearce, C.M. & Scheibling, R.E. 1991. Effect of macroalgae, microbial films, and conspecifics on the induction of metamorphosis of the green sea urchin *Strongylocentrotus droebachiensis* (Müller). *J. Exp. Mar. Biol. Ecol.* 147: 147-162.

Pearse, J.S. 1994. Cold-water echinoderms break "Thorson's Rule". In: Young, C.M. & Eckelbarger, K.J. (eds), *Reproduction, larval biology, and recruitment of the deep-sea benthos*: 26-43. Columbia University Press, New York.

Pearse, J.S. & Cameron, R.A. 1991. Echinodermata: Echinoidea. In: Giese, A.C., Pearse, J.S. & Pearse, V.B. (eds), *Reproduction of Marine Invertebrates: Echinoderms and Lophophorates:* 513-662.The Boxwood Press, Pacific Grove.

Pearse, J.S., Clark, M.E., Leighton, D.L., Mitchell, C.T. & North, W.J. 1970. Marine waste disposal and sea urchin ecology. In: Kelp habitat improvement project, Annual report 1 July 1969-30 June 1970. California Institute of Technology, Pasedena, Appendix 93 pp.

Pearse, J.S. & Hines, A.H. 1987. Long-term population dynamics of sea urchins in a central California kelp forest: rare recruitment and rapid decline. *Mar. Ecol. Prog. Ser.* 39: 275-283.

Pedrotti, M.L. 1993. Spatial and temporal distribution and recruitment of echinoderm larva in the Ligurian sea. *J. Mar. Biol. Ass. UK* 73: 513-530.

Pedrotti, M.L. & Fenaux, L. 1992. Dispersal of echinoderm larvae in a geographical area marked by upwelling (Ligurian Sea, NW Mediterranean). *Mar. Ecol. Prog. Ser.* 86: 217-227.

Pennington, J.T., Rumrill, S.S. & Chia, F.-S. 1986. Stage-specific predation upon embryos and larvae of the Pacific sand dollar, *Dendraster excentricus* by 11 species of common zooplanktonic predators. *Bull. Mar. Sci.* 39: 234-240.

Phillips, B.F. 1986. Prediction of commercial catches of the western rock lobster *Panulirus cygnus*. *Can. J. Fish. Aquat. Sci.* 43: 2126-2130.

Prince, J. 1995a. Limited effects of the sea urchin *Echinometra mathaei* (de Blainville) on the recruitment of benthic algae and macroinvertebrates into intertidal rock platforms at Rottnest Island, Western Australia. *J. Exp. Mar. Biol. Ecol.* 186: 237-258.

Prince, J. 1995b. Spatial and temporal patterns of recruitment of the sea urchin *Echinometra mathaei*: at Rottnest Island, Western Australia. *Third International Temperate Reef Symposium Abstracts:* p 28. University of Sydney, Sydney.

Raimondi, P.T. 1990. Patterns, mechanisms, consequences of variability in settlement and recruitment of an intertidal barnacle. *Ecol. Monogr.* 60: 283-309.

Raymond, B.G. & Scheibling, R.E. 1987. Recruitment and growth of the sea urchin *Strongylocentrotus droebachiensis* (Müller) following mass mortalities off Nova Scotia, Canada. *J. Exp. Mar. Biol. Ecol.* 108: 31-54.

Rodriguez, S.R. & Ojeda, F.P. 1993. Distribution patterns of *Tetrapygus niger* (Echinodermata, Echinoidea) off the central Chilean coast. *Mar. Ecol. Prog. Ser.* 101: 157-162.

Rodriguez, S.R., Ojeda, F.P. & Inestrosa, N.C. 1993. Settlement of benthic marine invertebrates. *Mar. Ecol. Prog. Ser.* 97: 193-207.

Rogers-Bennett, L., Bennett, W.A., Fastenau, H.C. & Dewees, C.M. 1995. Spatial variation in red sea urchin reproduction and morphology: Implications for harvest refugia. *Ecol. Appl.* 5: 1171-1180.

Rowley, R.J. 1989. Settlement and recruitment of sea urchins (*Strongylocentrotus* spp.) in a sea-urchin barren ground and a kelp bed: are populations regulated by settlement or post-settlement processes? *Mar. Biol.* 100: 485-494.

Rowley, R.J. 1990. Newly settled sea urchins in a kelp bed and urchin barren ground: a comparison of growth and mortality. *Mar. Ecol. Prog. Ser.* 62: 229-240.

Rumrill, S.S. 1988a. Larval populations of *Strongylocentrotus droebachiensis* in Barkley Sound, British Columbia. In: Burke, R.D., Mladenov, P.V., Lambert, P. & Parsley, R.L. (eds), *Echinoderm Biology:* p. 811. A.A. Balkema, Rotterdam.

Rumrill, S.S. 1988b. Temporal and spatial variability in the intensity of recruitment of a sea star: Frequent recruitment and demise. *Amer. Zool.* 28: 123.

Rumrill, S.S. 1989. Population size-structure, juvenile growth, and breeding periodicity of the sea star *Asterina miniata* in Barkley Sound, British Columbia. *Mar. Ecol. Prog. Ser.* 56: 37-47.

Rumrill, S.S. 1990. Natural mortality of marine invertebrate larvae. *Ophelia* 32: 163-198.

Rumrill, S.S. & Chia, F.-S. 1985. Differential mortality during the embryonic and larval lives of northeast Pacific echinoids. In: Keegan, B.F., O'Connor, B.D.S. (eds), *Echinodermata*: 333-338. A.A. Balkema, Rotterdam.

Rutherford, J.C. 1973. Reproduction, growth and mortality of the holothurian *Cucumaria pseudocurata*. *Mar. Biol.* 22: 167-176.

Scandol, J.P. & James, M.K. 1992. Hydrodynamics and larval dispersal: a population model of *Acanthaster planci* on the Great Barrier Reef. *Aust. J. Mar. Freshwater Res.* 43: 583-596.

Scheibling, R.E. 1980a. Abundance, spatial distribution, and size structure of populations of *Oreaster reticulatus* (Echinodermata, Asteroidea) in seagrass beds. *Mar. Biol.* 57: 95-105.

Scheibling, R.E. 1980b. Abundance, spatial distribution, and size structure of populations of *Oreaster reticulatus* (Echinodermata, Asteroidea) on sand bottoms. *Mar. Biol.* 57: 107-119.

Scheibling, R.E. 1986. Increased macroalgal abundance following mass mortalities of sea urchins (*Strongylocentrotus droebachiensis*) along the Atlantic coast of Nova Scotia. *Oecologia* 68: 186-198.

Scheibling, R.E. 1996. The role of predation in regulating sea urchin populations in eastern Canada. *Oceanol. Acta* 19: 421-430.

Scheibling, R.E. & Hamm, J. 1991. Interactions between sea urchins (*Strongylocentrotus droebachiensis*) and their predators in field and laboratory experiments. *Mar. Biol.* 110: 105-116.

Scheibling, R.E. & Raymond, B.G. 1990. Community dynamics on a subtidal cobble bed following mass mortalities of sea urchins. *Mar. Ecol. Prog. Ser.* 63: 127-145.

Scheibling, R.E. & Stephenson, R.L. 1984. Mass mortality of *Strongylocentrotus droebachiensis* (Echinodermata: Echinoidea) off Nova Scotia, Canada. *Mar. Biol.* 78: 153-164.

Scheltema, R.S. 1986. Long distance dispersal by planktonic larvae of shoal-water benthic invertebrates among central Pacific islands. *Bull. Mar. Sci.* 39: 241-256.

Schoener, A. 1968. Evidence for reproductive periodicity in the deep sea. *Ecology* 49: 81-87.

Schroeter, S.C., Dixon, J.D., Ebert, T.A. & Rankin, J.V. 1996. Effects of kelp forests *Macrocystis pyrifera* on the larval distribution and settlement of red and purple sea urchins *Strongylocentrotus franciscanus* and *S. purpuratus*. *Mar. Ecol. Prog. Ser.* 133: 125-134.

Sewell, M.A. 1994. Birth, recruitment and juvenile growth in the intraovarian brooding sea cucumber *Leptosynapta clarki*. *Mar. Ecol. Prog. Ser.* 114: 149-156.

Sewell, M.A. & Watson, J.C. 1993. A "source" for asteroid larvae?: recruitment of *Pisaster ochraceus, Pycnopodia helianthoides* and *Dermasterias imbricata* in Nootka Sound, British Columbia. *Mar. Biol.* 117: 387-398.

Shanks, A.L. 1995. Mechanisms of cross-shelf dispersal of larval invertebrates and fish. In: McEdward, L. (ed), *Ecology of marine invertebrate larvae:* 323-367. CRC Press, Boca Raton.

Sloan, N.A. 1984. Echinoderm fisheries of the world: a review. In: Keegan, B.F., O'Connor, B.D.S. (eds), *Echinodermata:* 109-124. A.A. Balkema, Rotterdam.

Sloan, N.A., Lauridsen, C.P. & Harbo, R.M. 1987. Recruitment characteristics of the commercially harvested red sea urchin *Strongylocentrotus franciscanus* in southern British Columbia, Canada. *Fish. Res.* 5: 55-69.

Smiley, S., McEuen, F.S., Chafee, C. & Krishnan, S. 1991. Echinodermata: Holothuroidea. In: Giese, A.C., Pearse, J.S. & Pearse, V.B. (eds), *Reproduction of Marine Invertebrates: Echinoderms and Lophophorates:* 663-750. The Boxwood Press, Pacific Grove.

Steimle, F.W. 1990. Population dynamics, growth, and production estimates for the sand dollar *Echinarachnius parma*. *Fish. Bull.* 88: 179-189.

Strathmann, R.R. 1978a. Length of pelagic period in echinoderms with feeding larvae from the northwest Pacific. *J. Exp. Mar. Biol. Ecol.* 34: 23-27.

Strathmann, R.R. 1978b. Larval settlement in echinoderms. In: Chia, F.-S., Rice, M.E. (eds), *Settlement and metamorphosis of marine invertebrate larvae:* 235-246. Elsevier North Holland, Amsterdam.

Strathmann, R.R. 1996. Are planktonic larvae of marine benthic invertebrates too scarce to compete within species? *Oceanol. Acta* 19: 399-407.

Sutherland, J.P. 1987. Recruitment limitation in a tropical intertidal barnacle: *Tetraclita panamensis* (Pilsbry) on the Pacific coast of Costa Rica. *J. Exp. Mar. Biol. Ecol.* 113: 267-282.

Sutherland, J.P. 1990. Recruitment regulates demographic variation in a tropical intertidal barnacle. *Ecology* 71: 995-972.

Taggart, C.T., Thompson, K.R., Maillet, G.L., Lochmann, S.E. & Griffin, D.A. 1996. Abundance distribution of larval cod (*Gadus morhua*) and zooplankton in a gyre-like water mass on the Scotian Shelf. In: Watanabe, Y., Yamashita, Y. & Oocki, Y. (eds), *Survival strategies in early life stages of marine resources*: 155-173. A.A. Balkema, Rotterdam.

Tegner, M.J. & Dayton, P.K. 1977. Sea urchin recruitment patterns and implications of commercial fishing. *Science* 196: 324-326.

Tegner, M.J. & Dayton, P.K. 1981. Population structure, recruitment and mortality of two sea urchins (*Strongylocentrotus franciscanus* and *S. purpuratus*) in a kelp forest. *Mar. Ecol. Prog. Ser.* 5: 255-268.

Tegner, M.J. & Dayton, P.K. 1991. Sea urchins, El Niños, and the long term stability of southern California kelp forest communities. *Mar. Ecol. Prog. Ser.* 77: 49-63.

Tegner, M.J. & Levin, L.A. 1983. Spiny lobsters and sea urchins: analysis of a predator-prey interaction. *J. Exp. Mar. Biol. Ecol.* 73: 125-150.

Thorson, G. 1950. Reproductive and larval ecology of marine bottom invertebrates. *Biol. Rev.* 25: 1-45.

Turner, R.L. & Miller, J.E. 1988. Post-metamorphic recruitment and morphology of two sympatric brittlestars. In: Burke, R.D., Mladenov, P.V., Lambert, P. & Parsley, R.L. (eds), *Echinoderm Biology*: 493-502. A.A. Balkema, Rotterdam.

Tyler, P.A. 1988. Seasonality in the deep sea. *Oceanogr. Mar. Biol. Annu. Rev.* 26: 227-258.

Tyler, P.A., Campos-Creasey, L.S. & Giles, L.A. 1994. Environmental control of quasi-continuous and seasonal reproduction in deep-sea benthic invertebrates. In: Young, C.M. & Eckelbarger, K.J. (eds), *Reproduction, larval biology, and recruitment of the deep-sea benthos*: 158-178. Columbia University Press, New York.

Tyler, P.A. & Gage, J.D. 1980. Reproduction and growth of the deep-sea brittlestar *Ophiura ljungmani* (Lyman). *Oceanol. Acta* 3: 177-185.

Tyler, P.A., Grant, A., Pain, S.L. & Gage, J.D. 1982. Is annual reproduction in deep-sea echinoderms a response to variability in their environment? *Nature* 300: 747-750.

Underwood, A.J. & Fairweather, P.G. 1989) Supply-side ecology and benthic marine assemblages. *Trend. Evol. Ecol.* 4: 16-20.

Vadas, R.L. & Elner, R.W. 1992. Plant-animal interactions in the north-west Atlantic. In: John, D.M., Hawkins, S.J. & Price, J.H. (eds), *Plant-animal interactions in the marine benthos*: 33-60. Clarendon Press, Oxford.

Valentine, J.F. 1991. Temporal variation in populations of the brittlestars *Hemipholis elongata* (Say, 1825) and *Microphiopholis atra* (Stimpson, 1852) (Echinodermata: Ophiuroidea) in eastern Mississippi Sound. *Bull. Mar. Sci.* 48: 597-605.

Watanabe, J.M. & Harrold, C. 1991. Destructive grazing by sea urchins *Strongylocentrotus* spp. in a central Caifornia kelp forest: potential roles of recruitment, depth, and predation. *Mar. Ecol. Prog. Ser.* 71: 125-141.

Watts, R.J., Johnson, M.S. & Black, R. 1990. Effects of recruitment on genetic patchiness in the urchin *Echinometra mathaei* in Western Australia. *Mar. Biol.* 105: 145-151.

Wiedemeyer, W.L. 1994. Biology of small juveniles of the tropical holothurian *Actinopyga echinites*: growth, mortality, and habitat preferences. *Mar. Biol.* 120: 81-93.

Wing, S.R., Largier, J.L., Botsford, L.W. & Quinn, J.F. 1995a. Settlement and transport of benthic invertebrates in an intermittent upwelling region. *Limnol. Oceanogr.* 40: 316-329.

Wing, S.R., Botsford, L.W., Largier, J.L. & Morgan, L.E. 1995b. Spatial structure of relaxtion events and crab settlement in the northern California upwelling system. *Mar. Ecol. Prog. Ser.* 128: 199-211.

Witman, J.D. 1985. Refuges, biological disturbance, and rocky subtidal community structure in New England. *Ecol. Monogr.* 55: 421-445.

Yamaguchi, M. 1973. Early life histories of coral reef asteroids, with special reference to *Acanthaster planci* (L.). In: Jones, O.A. & Endean, R. (eds), *Biology and geology of coral reefs*: 369-387. Academic Press, New York.

Yamaguchi, M. 1975. Coral reef asteroids of Guam. Biotropica 7: 12-23.

Yokochi, H. & Ogura, M. 1987. Spawning period and discovery of juvenile *Acanthaster planci* (L.) (Echinodermata: Asteroidea) at northwestern Iriomote-jima, Ryukyu Islands. *Bull. Mar. Sci.* 41: 611-616.

Young, C.M. 1992. Episodic recruitment and cohort dominance in echinoid populations at bathyal depths. In: Colombo, G., Ferrari, I., Ceccherelli, V.U. & Rossi, R. (eds), *Marine Eutrophication and Population Dynamics*: 239-246. Olsen and Olsen, Fredensborg.

Young, C.M. 1994a. A tale of two dogmas: the early history of deep-sea reproductive biology. In: Young, C.M. & Eckelbarger, K.J. (eds), *Reproduction, larval biology, and recruitment of the deep-sea benthos*: 1-25. Columbia University Press, New York.

Young, C.M. 1994b. The biology of external fertilization in deep-sea echinoderms. In: Young, C.M. & Eckelbarger, K.J. (eds), *Reproduction, larval biology, and recruitment of the deep-sea benthos*: 179-200. Columbia University Press, New York.

Young, C.M. & Chia, F.-S. 1982. Factors controlling spatial distribution of the sea cucmber *Psolus chitonoides*: Settling and post-settling behaviour. *Mar. Biol.* 69: 195-205.

Young, C.M. & Chia, F.-S. 1987. Abundance and distribution of pelagic larvae as influenced by predation, behaviour, and hydrographic factors. In: Giese, A.C., Pearse, J.S. & Pearse, V.B. (eds), *Reproduction of marine invertebrates. General aspects: seeking unity in diversity*: 385-463. Blackwell Scientific Publications, Palo Alto.

Zann, L., Brodie, J., Berryman, C. & Nagasima, M. 1987. Recruitment, ecology, growth and behaviour of juvenile *Acanthaster planci* (L.) (Echinodermata, Asteroidea). *Bull. Mar. Sci.* 41: 561-575.

Zann, L., Brodie, J. & Vuki, V. 1990. History and dynamics of the crown-of-thorns starfish *Acanthaster planci* (L.) in the Suva area, Fiji. *Coral Reefs* 9: 135-144.

Energy metabolism of sea urchin spermatozoa: the endogenous substrate and ultrastructural correlates

M. MITA[1,2] AND M. NAKAMURA[3]
[1]Department of Biology, School of Education, Waseda University, Shinjuku-ku, Tokyo 1698050, Japan
[2]Teikyo Junior College, Shibuya-ku, Tokyo 151-0071, Japan
[3]Department of Basic Life Science. Teikyo University, Hachioji, Tokyo 192-0352, Japan

KEYWORDS: ATP, respiration, phosphatidylcholine, triglyceride, energy metabolism, spermatozoa, sea urchin.

CONTENTS

1 INTRODUCTION

In higher animals, spermatozoa have the biologically important roles of egg activation and transmission of the male genetic message to offspring. For this purpose, spermatozoa must swim towards the egg. Sea urchin spermatozoa which are stored in the testis are immotile and their respiration is low (Gray 1928; Rothschild 1956). Upon spawning in seawater, these spermatozoa begin flagellar movement and the respiration is activated. Since the flagellar movement in spermatozoa occurs partly through reactions catalyzed by dynein ATPase (Gibbons & Gibbons 1972, Christen et al. 1982, 1983a, Evans & Gibbons 1986), energy metabolism for production of ATP by mitochondrial respiration is indispensable for swimming. In this review, we describe energy metabolism in sea urchin spermatozoa.

Generally, sugar is a good candidate for energy metabolism as a substrate in most cells and tissues. In mammalian spermatozoa, sugar in seminal plasma and the female reproductive tract is known to be responsible for the motility (Peterson & Freund 1976). Without seminal plasma, however, mammalian spermatozoa can maintain their motility under aerobic conditions (Lardy & Phillips 1941a). During incubation, the endogenous phospholipid content used for metabolism diminishes (Lardy & Phillips 1941b). Thus, mammalian spermatozoa may also be capable of using endogenous lipids, particularly phospholipids, for energy metabolism.

It is likely that sea urchin spermatozoa use an endogenous substrate for energy metabolism, because they swim in seawater where hardly any nutrients are present. In earlier studies by Rothschild and Cleland (1952) and Mohri (1957a, 1964), the endogenous phospholipid content of sea urchin spermatozoa was shown to decrease following the initiation of flagellar movement. Our previous work has also confirmed it in sea urchins of the order Echinoida (Mita & Yasumasu 1983a, Mita & Ueta 1988, Mita & Nakamura 1993b, Mita et al. 1994d). These findings strongly suggest that sea urchin spermatozoa obtain energy for swimming through the oxidation of endogenous phospholipids.

With regard to the taxonomy of the echinoids (Shigei 1974), in addition to the order Echinoida including *Hemicentrotus pulcherrimus* and *Strongylocentrotus purpuratus*, the orders Arbacioida, Clypeasteroida, Spatangoida, Diadematoida, Echinothurioida and Cidaroida are included in the class Echinoidea (Fig. 1). We have found that sea urchin spermatozoa of the orders Arbacioida (Mita 1991, Mita et al. 1994d), Clypeasteroida (Mita et al. 1994e) and Diadematoida (Mita et al. 1995) use endogenous triglyceride (TG) instead of phospholipid for energy metabolism. This suggests that phospholipid is not a common substrate in the class Echinoidea.

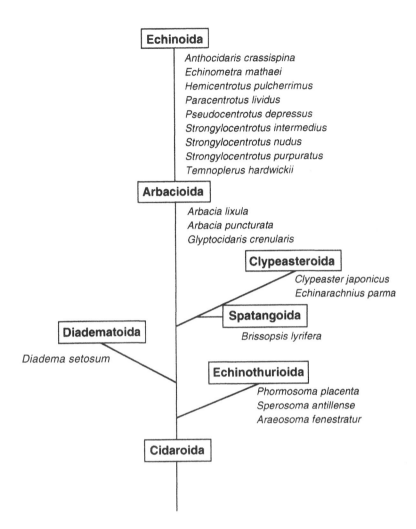

Echinoida

Anthocidaris crassispina
Echinometra mathaei
Hemicentrotus pulcherrimus
Paracentrotus lividus
Pseudocentrotus depressus
Strongylocentrotus intermedius
Strongylocentrotus nudus
Strongylocentrotus purpuratus
Temnoplerus hardwickii

Arbacioida

Arbacia lixula
Arbacia puncturata
Glyptocidaris crenularis

Clypeasteroida

Clypeaster japonicus
Echinarachnius parma

Spatangoida

Brissopsis lyrifera

Diadematoida

Diadema setosum

Echinothurioida

Phormosoma placenta
Sperosoma antillense
Araeosoma fenestratur

Cidaroida

Figure 1. Phylogeny of echinoids and species of sea urchins discussed in this review.

2 RESPIRATION

2.1 *Oxygen consumption*

It is known that spermatozoa are stored for months as immotile cells in male sea urchins. In semen, the rate of respiration in sea urchin spermatozoa is extremely low. This appears to be reasonable for saving energy in these mature spermatozoa stored in testis. When dry sperm (semen) of *H. pulcher-*

88

Table 1. Effect of pHi on motility, respiration and energy metabolism on *H. pulcherrimus* spermatozoa.

Conditions	pHi	Motility	PC consumption ($\mu g/h/10^9$ sperm)	O_2 consumption (nmol O_2/min/10^9 sperm)	$^{14}CO_2$ production dpm $10^{-3}/h/10^9$ sperm)
Seawater pH 8.2	7.5		4 ± 1	36 ± 4	29.5 ± 3.5
Seawater pH 6.6	7.0		1	15 ± 1	16.5 ± 1.7
Seawater + SAP pH 6.6	7.3		2 ± 1	28 ± 2	30.6 ± 2.8
Na$^+$-free seawater pH 8.2	6.8		not detectable	< 1	< 0.1

$^{14}CO_2$ was obtained from oxidation of [1-^{14}C]oleic acid. Each value is the mean ± SEM obtained in three separate experiments.

rimus are diluted in artificial seawater (ASW) at pH 8.2, spermatozoa begin to swim and their respiratory rate is activated (Table 1, Fig. 2a). The oxygen consumption in sea urchin spermatozoa is accelerated by increasing the dilution titer against seawater; that is called 'dilution effect' (Gray 1928, Rothschild 1951). The dilution effect is due to the accumulation of CO_2 produced by sperm respiration (Mohri 1956).

It has been demonstrated that sperm respiration is sensitive to rotenone, antimycin A and cyanide (Hino et al. 1980), suggesting that the oxygen consumption in sea urchin spermatozoa results from electron transport through the whole span of the mitochondrial respiratory chain. Also, oligomycin inhibits sperm respiration and the inhibition is released by 2,4-dinitrophenol which is an uncoupling agent of oxidative phosphorylation (Hino et al. 1980). This suggests that the electron transport in mitochondria of sea urchin spermatozoa couples with oxidative phosphorylation. ATP available for flagellar movement is produced by mitochondrial respiration.

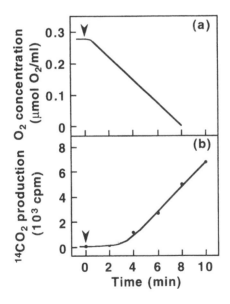

Figure 2. Oxygen consumption (a) and fatty acid oxidation (b) in *H. pulcherrimus* spermatozoa. (a) Dry sperm were diluted l00-fold and incubated in ASW at 20°C. Trace of change in the oxygen concentration in sperm suspension was recorded. (b) Dry sperm were diluted 100- fold and incubated in ASW containing [1-^{14}C]oleic acid and carbonic anhydrolase at 20°C. [14]CO_2 produced was counted by the thin-window gas flow counter. Arrowheads show the time of sperm addition.

2.2 *Fatty acid oxidation*

The respiratory quotient (R.Q.) in sea urchin spermatozoa is found to be about 0.7-0.8 (Mohri & Horiuchi 1961), suggesting that a substrate for energy metabolism in sea urchin spermatozoa is lipid. Indeed, following dilution of *H. pulcherrimus* spermatozoa in ASW containing [1-^{14}C]oleic acid and carbonic anhydrase, $^{14}CO_2$ is released within 2 min (Table 1, Fig. 2b). This is in accord with earlier work (Mohri 1957a, 1957b) indicating that a fatty acid oxidizing system is present in sea urchin spermatozoa. Thus, sea urchin spermatozoa obtain energy for movement through oxidation of fatty acids.

3 HIGH-ENERGY PHOSPHATE COMPOUNDS

3.1 *ATP*

The relation of ATP to sea urchin sperm motility is discussed in a number of reports (Gibbons & Gibbons 1972, Christen et al., 1982, 1983a, Evans & Gibbons 1986). ATP production is important for their movement. It has been shown that ATP is present in high concentration in the dry sperm (semen) but contents of ADP and AMP to be low (Mohri 1958, Moriwaki 1958, Taguchi et al. 1963, Yanagisawa 1967, Christen et al. 1983b, Mita & Yasumasu 1983a, Mita et al. 1994a). When *H. pulcherrimus* spermatozoa are diluted and incubated in ASW, a rapid decline in the level of ATP occurs (Fig. 3a). This strongly suggests that the ATP is used for motility. On the other hand, the level of ADP hardly changes after dilution and incubation in ASW (Fig. 3b), whereas the AMP level increases remarkably (Fig. 3c).

Based on levels of ATP, ADP and AMP in the sea urchin spermatozoa, the energy charge is calculated by using the equation (Atkinson, 1968):

$$\text{Energy charge} = \frac{1}{2} \left(\frac{2[\text{ATP}] + [\text{ADP}]}{[\text{ATP}] + [\text{ADP}] + [\text{AMP}]} \right)$$

The value of the energy charge is used for the index of energy metabolism activity. In semen, the value of energy charge is about 0.8 (Mita & Yasumasu 1983a, Mita et al. 1994a). After dilution and incubation of spermatozoa in ASW, the value decreases rapidly to become 0.3-0.5. These findings suggest that the amount of ATP hydrolyzed by dynein ATPase is much greater than that of ATP produced by mitochondrial respiration.

90

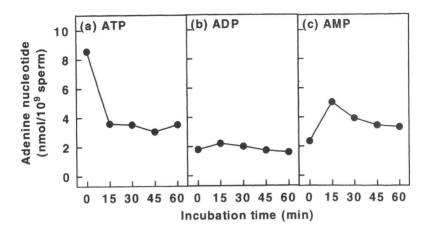

Figure 3. Changes in the levels of ATP (a), ADP (b) and AMP (c) following incubation of *H. pulcherrimus* spermatozoa in seawater. Dry sperm were diluted 100-fold
and incubated in ASW at 20°C.

3.2 *Phosphagen*

It is well known that creatine phosphate and/or arginine phosphate play
important roles in the ATP generate system. Creatine phosphate and creatine
kinase are observed in *Arbacia lixula* and *Paracentrotus lividus* spermatozoa, whereas neither arginine phosphate nor arginine kinase is detectable
(Table 2). Similar results are obtained in spermatozoa of other species of sea
urchins (Yanagisawa 1967, Christen et al. 1983b, Mita & Yasumasu 1983a,
Mita et al. 1994b). These suggest that phosphagen in sea urchin spermatozoa
is creatine phosphate alone. Though high concentration of creatine phosphate is observed in semen of *H. pulcherrimus*, the level decreases rapidly
following incubation in ASW (Fig. 4), suggesting that the creatine phosphate
is used for ATP generation.

It has been demonstrated that two isoforms of creatine kinase exist in
sperm mitochondria and tail (Christen et al. 1983a, Tombes & Shapiro 1985,
1987, Ratto et al. 1989). These creatine kinases play important roles for the
creatine-phosphate shuttle which is associated with transportation at sites
of energy synthesis and utilization (Tombes & Shapiro 1985, 1987, Tombes
et al. 1987). Furthermore, high adenylate kinase (myokinase) activities are
observed in *A. lixula* and *P. lividus* spermatozoa (Table 2). Since adenylate
kinase catalyzes the reaction as follows: ATP + AMP ⇔ 2ATP, the kinase
would attend the ATP generation system in sea urchin spermatozoa.

Table 2. Phosphagen contents and related kinase activities in *A. lixula and P. lividus* spermatozoa.

Sea urchin	Phosphagen (nmol/10⁹ sperm)		Enzyme activity (units/g protein)		
	Creatine phosphate	Arginine phosphate	Creatine kinase	Arginine kinase	Adenylate kinase
A. lixula	6.8 ± 1.1	> 0.1	117 ± 8	ND	33 ± 2
P. lividus	6.3 ± 0.7	> 0.1	45 ± 5	ND	46 ± 1

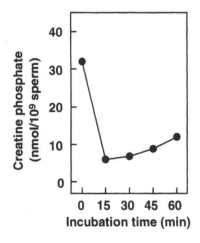

Figure 4. Changes in the level of creatine phosphate following incubation of *H. pulcherrimus* spermatozoa in seawater. Dry sperm were diluted 100-fold and incubated in ASW at 20°C.

4 ENDOGENOUS SUBSTRATE FOR ENERGY METABOLISM

4.1 *Glycogen*

In vertebrate cells, glucose is well used for the energy metabolism substrate because it is easy to obtain from blood fluid. Indeed, it has been demonstrated that [^{14}C]glucose is converted to $^{14}CO_2$ in sea urchin spermatozoa under hypotonic conditions (Mohri & Yasumasu 1966).

It might be possible that the sea urchin spermatozoa use glucose for energy metabolism. However, endogenous glycogen and glucose are present in very small quantities in spermatozoa of most species of sea urchins belonging to the orders Echinoida (Mita & Yasumasu 1983a, Mita & Naka-

92

Table 3. Concentrations of glycolysis intermediates in *H. pulcherrimus* spermatozoa.

	Concentration (nmol/10^9 sperm)	
	Dry sperm	Incubation for 30 min
Glycogen	1.4 ± 0.1	0.9 ± 0.1
Glucose	0.05 ±0.01	0.02 ± 0.01
G6P	0.05 ± 0.01	0.07 ± 0.01
F6P	0.02 ± 0.01	0.03 ± 0.01
FBP	0.02 ± 0. 01	0.03 ± 0.01
DHAP	0.05 ± 0.01	0.08 ± 0.01
GA3P	0.02 ± 0.01	0.02 ± 0.01
1.3PG	0.01	0.01
3PG	0.01	0.01
2PG	0.01	0.01
PEP	0.02 ± 0.01	0.03 ± 0 01
Pyruvate	0.07 ± 0.01	0.11 ± 0.01
Lactate	1.25 ± 0.29	1.08 ± 0.38
αGP	0.15 ± 0.02	1.80 ± 0.32
6PG	0.02 0.01	0.02 ± 0.01
Pi	25.2 ± 7.9	35.4 ± 5.4

The values are mean ± SEM obtained in three separate experiments.
Abbreviations: G6P, glucose-6-phosphate; F6P, fructose-1.6-bisphosphate; 1.3PG, 1.3-diphosphoglycerate; 3PG, 3-phosphoglycerate; 2PG, 2-phosphoglycerate; PEP, phosphoenolpyruvate; αPG, α-glycerophosphate; 6PG, 6-phosphogluconate; Pi, orthophosphate.

mura 1993b), Arbacioida (Mita 1991, Mita et al. 1994d), Clypeasteroida (Mita et al. 1994e) and Diadematoida (Mita et al. 1995). There is a small change in the levels of glucose and glycogen in *H. pulcherrimus* spermatozoa before and after swimming (Table 3). These findings suggest that sea urchin spermatozoa do not use sugar as a principal substrate for energy metabolism.

Though sugar is not the main energy source in sea urchin spermatozoa, glycolytic pathway is active in sea urchin spermatozoa (Mita & Yasumasu 1983a). Table 3 shows the concentrations of glycolytic intermediates in *H. pulcherrimus* spermatozoa. These data also reveal that there are four rate-Iimiting steps which are α-glucan phosphorylase, phosphofructokinase, glyceraldehyde-3-phosphate dehydrogenase and pyruvate kinase, in the glycolysis of sea urchin spermatozoa (Mita & Yasumasu 1983a). After dilution and incubation of spermatozoa in ASW, the level of α-glycerophosphate (αGP) increases significantly. Since αGP is produced from phospholipid, the lipid metabolism in sea urchin spermatozoa is activated following dilution in seawater. However, the levels of glycolytic intermediates and of 6-phosphogluconate (6PG) in the pentose monophosphate shunt remain essentially constant. These suggest that the glycolysis in sea urchin spermatozoa also plays an important for lipid metabolism.

4.2 *Phosphatidylcholine*

Lipids extracted from dry sperm of *H. pulcherrimus* consist primarily of phospholipids and cholesterol (CH) (Table 4). TG and cholesterol-ester (CE) are also present, but at extremely low levels. Similar results have been reported for other species of Echinoida spermatozoa (Mita & Ueta 1988, 1989, Mita & Nakamura 1993b, Mita et al. 1994c, 1994d). When dry sperm is diluted and incubated in seawater, the phospholipid content decreases rapidly (Mita & Ueta 1988, Mita & Nakamura 1993b, Mita et al. 1994c, 1994d). These findings confirm the observations made by Rothschild and Cleand (1952) and Mohri (1957a), suggesting that sea urchin spermatozoa use phospholipids as the source of energy for metabolism. Among phospholipids, phosphatidylcholine (PC), phosphatidylserine (PS), phosphatidylethanolamine (PE) and cardiolipin (CL) are found in spermatozoa of *H. pulcherrimus*, but not phosphatidic acid (PA) or sphingophospholipid (Table 4). It is notable that during incubation in seawater, only the level of PC decreases rapidly (Mita & Ueta 1988, Mita & Nakamura 1993b, Mita

Table 4. Change in lipid levels following incubation of *H. pulcherrimus* spermatozoa with seawater.

Lipids	Concentration (μg/10^9 spermatozoa)	
	dry sperm	incubation for 1 h
Phosphatidylcholine	27 ± 1	23 ± 1
Phosphatidylinositol	11	
Phosphatidylserine	13 ± 2	11 ± 1
Phosphatidylethanolamine	22 ± 1	22 ± 1
Cardiolipin	4 ± 1	4 ± 1
Triglyceride	tr.	tr.
Free fatty acid	12 ± 1	
Cholesterol	11 ± 1	10 ± 1
Cholesterol ester	tr.	tr.

Each value is the mean ± SEM of four separate experiments. Tr., trace amount (less than 1μg/109 spermatozoa).

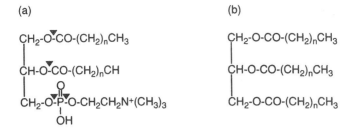

(a)   (b)

CH_2-O-CO-$(CH_2)_n CH_3$ CH_2-O-CO-$(CH_2)_n CH_3$

CH-O-CO-$(CH_2)_n CH$ CH-O-CO-$(CH_2)_n CH_3$

CH_2-O-P-O-$CH_2 CH_2 N^+(CH_3)_3$ CH_2-O-CO-$(CH_2)_n CH_3$

Figure 5. Chemical structures of phosphatidylcholine (a) and triglyceride (b). Arrowheads show sites of hydrolysis by phospholipases.

et al. 1994c, 1994d). In contrast to PC, the levels of other phospholipids remain almost constant (Table 4).

In addition to diacyl phospholipids, alkenyl and alkyl derivatives are known to be present in sea urchin spermatozoa (Mohri 1959a, 1961). In *H. pulcherrimus* spermatozoa, PC contains alkylacyl (19%) and diacyl (81%) components (Mita 1992; Mita & Ueta 1992). After incubation of these spermatozoa in seawater, only the diacyl PC content decreases significantly, and no changes are detectable in the other phospholipids (Mita 1992). Palmitic (16:0), stearic (18:0), eicosamonoenoic (20:1), arachidonic (20:4) and eicosapentaenoic (20:5) acids at the 1-position and arachidonic (20:4) and eicosapentaenoic (20:5) acids at the 2-position of diacyl PC are also found to decrease during incubation (Mita 1992). These observations suggest that the diacyl PC composed of unsaturated fatty acid is available for utilization in energy production by sea urchin spermatozoa of the order Echinoida. In contrast, during incubation of *H. pulcherrimus* spermatozoa in Na^+ free seawater, in which spermatozoa are immotile, there is no decrease in the level of PC (Mita & Yasumasu 1983b, 1984), suggesting that activation of phospholipid hydrolysis is closely related to the initiation of sperm motility (Table 1).

4.3 *Phospholipid metabolism*

Previous studies have shown that phospholipase activity is present in sea urchin spermatozoa (Mohri 1959b, 1964, Mita & Yasumasu 1983a). The hydrolysis of phospholipid is generally catalyzed by phospholipases, which are classified into A_1, A_2, C and D based on the site of phospholipid hydrolysis (Fig. 6). To determine the phospholipase associated with energy metabolism in sea urchin spermatozoa, the conversion of radioactivity from 1-palmitoyl-2[1-^{14}C]linoleoyl-PC has been examined. After incubation with

95

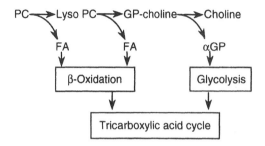

Figure 6. Phosphatidylcholine metabolism in sea urchin spermatozoa of the order Echinoida.

the homogenate of *H. pulcherrimus* spermatozoa, about 80% of the radioactivity is recovered in the free fatty acid (FA) fraction (Mita & Ueta 1988, 1990). Only 4% and 7% of the radioactivity is observed in lysophosphatidylcholine (LysoPC) and PS, respectively. There is only a trace of radioactivity in PE, PA, CL and diacylglycerol (DG). Thus it is concluded that the hydrolysis of PC takes place through the action of phospholipase A_2. On the other hand, incubation in the presence of 1-palmitoyl-2-[1-^{14}C]linoleoyl-PE shows that only 20% of the radioactivity is distributed in FA (Mita & Ueta 1990). More than 50% of the radioactivity from PE is converted to PS. This suggests that phospholipase A_2 in *H. pulcherrimus* spermatozoa has strict substrate specificity for PC, thus explaining the selective hydrolysis of PC. Similar results have been obtained from other Echinoida spermatozoa, such as *Anthocidaris crassispina* (Mita & Nakamura 1993b), *Echinometra mathaei* (Mita & Nakamura 1993b), *Paracentrotus lividus* (Mita et al. 1994d), *Pseudocentrotus depressus* (Mita & Nakamura 1993), *Strogylocentrotus intermedius* (Mita & Nakamura 1993b), *Strongylocentrotus nudus* (Mita & Nakamura 1993b), and *Temnoplerus hardwickii* (Mita & Nakamura 1993b). It is concluded that sea urchin spermatozoa of the order Echinoida commonly use PC as an endogenous substrate for energy metabolism. Furthermore, accumulation of choline has been observed in *H. pulcherrimus* spermatozoa following incubation in seawater (Mita & Ueta 1990). Since the amount of PC consumed is essentially the same as that of choline accumulated (Mita & Ueta 1990), it is suggested that PC is finally metabolized to choline in sea urchin spermatozoa. The major degradative pathway for PC in Echinoida spermatozoa is considered to be: PC → LysoPC → glycerophosphocholine → choline (Fig. 6). It has also been shown that the content of long chain fatty acyl-CoA increases slightly following dilution of *H. pulcherrimus* spermatozoa in seawater (Mita & Yasumasu 1983b). Finally, fatty acids obtained by hydrolysis of PC are metabolized through β-oxidation and the tricarboxylic acid (TCA) cycle to CO_2 and H_2O (Fig. 6). α-Glycerophosphate derived from PC is also metabolized by glycolysis and the TCA cycle.

Table 5. Concentrations of tricarboxylic acid cycle intermediates in *H. pulcherrimus* spermatozoa.

	Concentration (nmol/10⁹ sperm)	
	Dry sperm	Incubation for 30 min
Citrate	1.10 ± 0.23	1.36 ± 0.21
Isocitrate	0.04 ± 0.02	0.08 ± 0.01
α-Ketoglutarate	0.09 ± 0.02	0.10 ± 0.03
Fumarate	0.18 ± 0.07	0.46 ± 0.25
Malate	0.61 ± 0.11	5.15 ± 1.23
Oxalacetate	0.02 ± 0.01	0.02 ± 0.01

The values are mean ± SEM obtained in three separate experiments.

Table 5 shows the concentrations of the intermediates of the TCA cycle in *H. pulcherrimus* spermatozoa. The concentrations of citrate and malate are relatively higher than those other intermediates in dry sperm. After dilution and incubation in ASW, the malate level strongly increases. The levels of other intermediates remains essentially constant, although the fumarate level increases slightly. These suggest that the TCA cycle in sea urchin spermatozoa is activated following dilution in sea water, with concomitant initiation of motility.

Recently, it was reported that *H. pulcherrimus* spermatozoa can maintain their motility in seawater for half a day (Ohtake et al. 1996). It is likely that the endogenous PC is enough to support the motility through the oxidation of fatty acid to produce ATP.

4.4 Triglyceride

Similar to the Echinoida, sea urchin spermatozoa of the orders Arbacioida, Clypeasteroida and Diadematoida are quiescent in undiluted semen. After dilution and incubation in seawater, spermatozoa begin flagellar movement and respiration is activated.

Simultaneously, a rapid decline of ATP and creatine phosphate occurs in these spermatozoa (Mita et al. 1994c). Thus, spermatozoa of the orders Arbacioida, Clypeasteroida and Diadematoida also obtain energy for swimming through oxidation of an endogenous substrate (Mita & Nakamura 1998). As described already, however, it is unlikely that a carbohydrate is the substrate for energy metabolism, since the endogenous glucose and glycogen contents are quite low in spermatozoa of the sea urchins *A. lixula* (Mita et al. 1994d), *Glyptocidaris crenularis* (Mita 1991), *Clypeaster japon-*

icus (Mita et al. 1994e) and *Diadema setosum* (Mita et al. 1995). It is notable that the lipid composition of these spermatozoa is different from that in the Echinoida. The lipids in spermatozoa of *G. crenularis*, *A. lixula* and *D. setosum* contain CH, TG and several kinds of phospholipids including PC (Mita 1991, Mita et al. 1994d, 1995). CE in addition to similar phospholipids, CH and TG, is present in *C. japonicus* spermatozoa (Mita et al. 1994e). After incubation of these spermatozoa in seawater, the level of TG decreases without any change in the levels of phospholipids (Mita 1991, Mita et al. 1994d, 1994e, 1995). This suggests that spermatozoa of sea urchins in the orders Arbacioida, Diadematoida and Clypeaseteroida use TG as a substrate for energy metabolism (Fig. 5). Although PC is present in these spermatozoa, it is unclear why they do not use it for energy metabolism. In the spermatozoa of *A. lixula* (Mita et al. 1994d), *G. crenularis* (Mita 1991), *C. japonicus* (Mita et al. 1994e) and *D. setosum* (Mita et al. 1995), the activity of lipase is fairly high, but that of phospholipase A_2 is extremely low. Since Echinoida spermatozoa have high phospholipase A_2 activity, the lack of change in the level of PC appears to be due to the low activity of phospholipase in Arbacioida, Clypeasteroida and Diadematoida spermatozoa. Thus, it is considered that the preferential hydrolysis of PC and TG is related to the properties of phospholipase A_2 with respect to the Echinoida and of lipase with respect to the Arbacioida, Clypeasteroida and Diadematoida respectively.

5 REGULATION OF ENERGY METABOLISM

5.1 *Intracellular pH (pHi)*

Since it is known that a specific pHi value is essential for the activation of sea urchin sperm respiration and motility (Christen et al. 1982, 1983a; Lee et al. 1983; Bibring et al.1984), sperm metabolism is likely to be regulated through the alteration of pHi. The mechanism responsible for their flagellar movement has been discussed in several reviews (Mohri et al. 1979, Brokaw 1987, 1990, Shapiro 1990). The initiation of sea urchin sperm motility requires external Na^+ and is associated with Na^+ dependent acid release (Nishioka & Cross 1978). Following dilution in seawater, the intracellular pH (pHi) of sea urchin spermatozoa rises from 6.8 to 7.4 (Christen et al. 1982, 1983a, Lee et al. 1983, Bibring et al. 1984). CO_2 is responsible for the immobility of sea urchin spermatozoa in semen, because CO_2 lowers the pHi of spermatozoa by entering the cells as a neutral weak acid (Mohri & Yasumasu 1963, Johnson et al. 1983). Internal alkalization following dilution in seawater leads to activation of dynein ATPase and to initiation of motility (Christen et al. 1982, 1983a). The ADP molecule derived from ATP by hydrolysis is used as a substrate for mitochondrial respiration. As a result, the respiration is also activated, accompanied by initiation of motility (Table 1).

The pHi is also important to regulate activities of phospholipase A_2 and fatty acid oxidation in sea urchin spermatozoa. The activity of phospholipase A_2 in *H. pulcherrimus* spermatozoa is extremely low at pH 6.5, and it accelerates gradually as the pH increases, reaching maximal activity near pH 8 (Fig. 7a). In addition, the oxidation of [1-^{14}C]oleic acid to $^{14}CO_2$ in a cell-free system of *H. pulcherrimus* spermatozoa is also pH-dependent (Fig. 7b). It is relatively low under acidic (pH 6.5) and alkaline (pH 8) conditions and is maximal at about pH 7.5. The activation of PC metabolism would thus appear to be mediated by increase in pHi. On increasing in seawater, PC metabolism in sea urchin spermatozoa is assumed to be enhanced, with concomitant activation of respiration and motility (Table 1).

It has been demonstrated that a sperm-activating peptide (SAP) isolated from the egg jelly (Hansbrough & Garbers 1981, Suzuki et al. 1981, Garbers et al. 1982, Suzuki & Garbers 1984, Suzuki 1990) stimulates sperm respiration and motility in slightly acidic seawater (pH 6.6). When *H. pulcherrimus* spermatozoa are diluted in seawater at pH 6.6, PC metabolism is sluggish (Table 1). The pHi of spermatozoa diluted in seawater at pH 6.6 remains 7.0 (Lee & Garbers 1986). Since SAP increases the pHi of spermatozoa (Repaske & Garbers 1983, Lee & Garbers 1986), the SAP causes an increase of PC consumption and fatty acid oxidation in *H. pulcherrimus* spermatozoa at pH 6.6 (Table 1) (Mita et al. 1990).

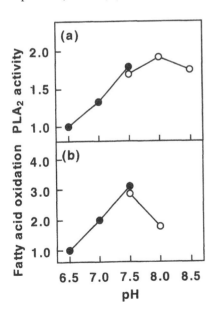

Figure 7. Effect of pH on phospholipase A_2 activities (a) and fatty acid oxidation (b) in *H. pulcherrimus* spermatozoa. The enzyme activities and fatty acid oxidation were measured at the pH indicated by Aces (pH 6.5-7.5)(O) and Tris (pH 7.5-8.5)(●).

The lipase activity in *G. crenularis* (the order Arbacioida) is also influenced by pH (Mita 1991). The activity is relatively low under acidic (pH 6.0) and alkaline (pH 9.0) conditions, and maximal activity is attained at pH 7.5. Thus, these findings suggest that the activity of phospholipase A_2 or lipase and fatty acid oxidation are enhanced by a rise in pHi from 6.5 to 7.5 following sperm dilution in seawater, accompanied by initiation of motility and activation of respiration.

5.2 Creatine-phosphate shuttle and adenylate kinase

The sea urchin spermatozoon is an extremely polarized cell. The tip of the sperm head responds to contact with an egg surface by undergoing an acrosome reaction. The sperm tail is the agent of flagellar motility. The single mitochondrion is located in the midpiece region. The energy for swimming is produced by mitochondrial respiration, and ATP is utilized almost exclusively by dynein ATPase of the flagellar axoneme. In sea urchin spermatozoa, a creatine-phosphate shuttle apparently establishes a pool of ATP used primarily for motility (Tombes & Shapiro 1985, 1987, Tombes et al. 1987). At the junction of the mitochondrion and tail, a creatine kinase isozyme forms creatine phosphate, which then diffuses down the tail, where ATP is formed by another creatine kinase isozyme, presumably in close proximity to dynein ATPases (Tombes & Shapiro 1985, 1987, Tombes et al. 1987). An intracellular alkalization brings about the activation of the dynein ATPase (Christen et al. 1983a) responsible for motility (Gibbons & Gibbons 1972). Motility activation leads to the consumption of ~P (Christen et al. 1982, 1983a) and the initiation of respiration (Christen et al. 1982, Lee et al. 1983, Bibring et al. 1984). Two creatine kinase isozymes would mediate transphosphorylations that drive a creatine-phosphate shuttle. This shows that only creatine phosphate, creatine and Pi diffuse between the mitochondrion and tail, where two distinct intracellular compartments of adenine nucleotides residue and are used locally for respiration and motility (Fig. 8a) (Tombes & Shapiro 1985).

However, following incubation in seawater, creatine phosphate levels decrease rapidly (Fig. 4) with concomitant decline of ATP (Fig. 3a), suggesting that the creatine phosphate pool is discharged immediately upon initiation of swimming. It is unlikely that the creatine phosphate shuttle can sustain sperm motility. In contrast to the decrease in ATP, the level of AMP increases markedly following dilution and incubation of *H. pulcherrimus* spermatozoa in ASW (Fig. 3c), whereas the ADP level shows only a slight change (Fig. 3b). In addition to creatine kinase, adenylate kinase activity is present in sea urchin spermatozoa (Table 2). Since adenylate kinase is distributed in sea urchin flagella (Yanagisawa et al. 1968, Brokaw & Gibbons,

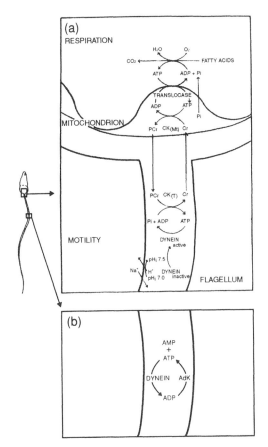

Figure 8. Possible regulation of sea urchin sperm motility and respiration. (a) The activation pathway based on two isozymes of creatine kinase, CrK (T) at the sperm tail and CrK (Mi) at the mitochondrion. A data from Tombes and Shapiro (1985). (b) Adenylate kinase (AdK) dependent ATP regeneration system.

1973, Tombes & Shapiro 1985, Mita et al. 1994a), the enzyme is considered to catalyze the conversion of ATP and AMP, as soon as ADP is produced from ATP, suggesting that ATP is directly regenerated from ADP by the reaction of adenylate kinase in the tail. It is also possible that not only the creatine phosphate shuttle but also adenylate kinase plays an important role in the supply of ATP for flagellar motility (Fig. 8b).

6 LOCATION OF ENDOGENOUS SUBSTRATE

6.1 Distribution of phosphatidylcholine

In general, phospholipids including PC are components of the plasma membrane. If PC located in the plasma membrane is digested during swimming,

Table 6. Distribution of total lipid, PC and phospholipase A_2 activity in *H. pulcherrimus* spermatozoa.

Fraction	Total lipid (mg/10^9 sperm equiv.)		PC (mg/10^9 sperm equiv.)		Phospholipase A_2 activity (nmol/h/mg protein)	
Whole sperm	77 ± 9		24 ± 1		10.7 ±1.0	
Head + Midpiece	48 ± 3	(62%)	22 ± 2	(92%)	12.2 ± 1.7	(96%)
Tail	20 ± 1	(26%)	1	(4%)	2.6 ± 0.4	(4%)
Midpiece	31 ± 3	(40%)	18 ± 3	(75%)	43.2 ± 1.6	(65%)

Values are the mean ± SEM of three separate experiments. Values in parentheses are percentage in the total.

the membrane will be destroyed, causing the spermatozoa to die. Thus, it seems that a special kind of PC reservoir is available for utilization in energy metabolism of Echinoida spermatozoa. Analysis of the distribution of PC in *H. pulcherrimus* sperm heads, tails and midpieces shows that about 75% of the PC is distributed in the midpiece (Table 6), whereas the head and tail contain only 17% and 4%, respectively. It is notable that the decrease in PC content during incubation in seawater is due to a change in the level of PC in the head and midpiece, without any change in the level of tail PC (Mita et al. 1991). In contrast, the levels of other lipids in the head and midpiece and the tail remain almost constant. It is also important that about 65% of phospholipase A_2 activity are distributed in the midpieces (Table 2). Since PC and phospholipase A_2 activity are abundant in the midpiece, these findings suggest that midpiece PC is available for utilization for energy metabolism in sea urchin spermatozoa.

6.2 *Lipid bodies*

Previously, Afzelius & Mohri (1966) have reported that sperm mitochondrial matrix of *Brissopsis lyrifera* is damaged after prolonged incubation in seawater. To ascertain this, we have ultrastructurally studied the endogenous substrate in spermatozoa. However, various structural features of the mitochondrion, such as the number of cristae and the thickness of the membranes, did not change during incubation of spermatozoa in seawater (Mita & Nakamura 1992). Observing a thin-section of spermatozoa of *H. pulcherrimus*, it is evident that the intramembrane space, i.e. the region between the mitochondrial outer and inner membranes, has a band-like dilation nearest the flagellum and contains lipid bodies (Fig. 9). These lipid bodies of

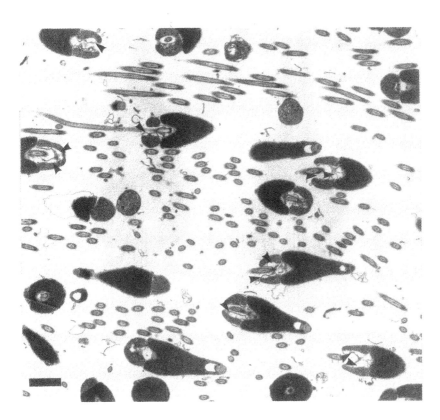

Figure 9. Electron micrograph of *H. pulcherrimus* spermatozoa. Arrows show lipid bodies. Bar shows 1 μm.

low electron density are irregular in profile and about 0.1-0.2 μm in diameter. The schematic structure of *H. pulcherrimus* spermatozoa is shown in Fig. l0a. When spermatozoa are diluted and incubated in seawater, the lipid bodies and the inner ring of the mitochondrion become small and finally disappears (Mita & Nakamura 1992). Thus, the disappearance of the lipid bodies is correlated with the decrease in the level of PC. Similar morphological changes in lipid bodies have been observed in sperm midpieces in other species of Echinoida (Mita & Nakamura 1993b. Mita et al. 1994b). These observations strongly suggest that PC available for use in Echinoida sperm energy metabolism is related to the lipid bodies within the mitochondria of the midpiece.

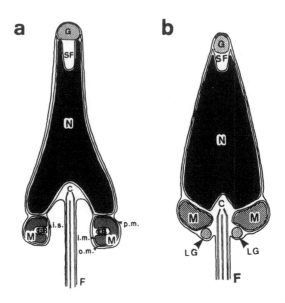

Figure 10. Schematic representation of a spermatozoon of *H. pulcherrimus* (the order Echinoida) (a) and *G. crenularis* (the order Arbacioida) (b). C: proximal centriole, F: flagellum, G: acrosomal granule, im: mitochondrial inner membrane, LB: lipid bodies, LG: lipid globules, is: intramembrane space, N: nucleus, om: mitochondrial outer membrane, pm: plasma membrane. SF: subacrosomal fossa.

6.3 *Lipid globules*

In contrast, the spermatozoa of *G. crenularis* (the order Arbacia) contain several lipid globules in the midpiece, although there are no lipid bodies (Fig. 11). The lipid globules differ from the lipid bodies in that the former are spherical and located in the posterior region between the basis of the mitochondrion and the plasma membrane (Fig. 10b). The long and short axes of the globules measure 0.23 and 0.21 μm, respectively. The lipid globules have been observed in the midpieces of *Arbacia punctulata* (Longo & Anderson 1969), *A. lixula* (Mita et al. 1994d), *C. japonicus* (Mita et al. 1994e), *Echinarachnius parma* (Summers & Hylander 1974), *B. lyrifera* (Afzelius & Mohri 1966), *Phormosoma placenta* (Eckelbarger et al. 1989), *Sperosoma antillense* (Eckelbarger et al. 1989) and *Araeosoma fenestratur* (Eckelbarger et al. 1989). After incubation of *G. crenularis* spermatozoa in seawater, morphological changes in the lipid globules occur: Some lipid globules lose their smooth spherical shape and their surface becomes irregular and uneven (Mita & Nakamura 1993a). In other cases, the lipid globules

104

Figure 11. Electron micrograph of *G. crenularis* spermatozoa. Arrows show lipid globules. Bar shows 1 µm.

have a bilobed appearance. Vacuoles of various sizes and forms also appear near the lipid globules. The volume of lipid globules in the spermatozoa decreases as the TG content declines following incubation in seawater (Mita & Nakamura 1993a). Similar results are obtained in *A. lixula* (order Arbacioida) (Mita et al. 1994b), *C. japonicus* (order Clypeasteroida) (Mita et al. 1994e) and *D. setosum* (order Diadematoida) (Mita et at. 1995). Thus, these findings strongly suggest that the spermatozoa of Arbacioida, Clypeasteroida and Diadematoida obtain energy through oxidation of fatty acid from TG stored in the lipid globules within their midpieces.

7 CONCLUSIONS

Sea urchin spermatozoa obtain ATP as an energy sourse for flagellar movement through oxidation of fatty acids. Although Echinoida spermatozoa use endogenous phospholipids, particularly PC, as a substrate for energy metabolism, the preferred substrate for spermatozoa of Arbacioida, Clypeasteroida and Diadematoida is TG. PC available for utilization in energy metabolism is contained in lipid bodies in the intramembrane space of sperm mitochondria. In contrast, TG is stored in lipid globules in sperm midpieces. Since both types of energy metabolism are enhanced by rise in pH from 6.8 to 7.4, not only sperm motility and respiration but also energy metabolism in sea urchin spermatozoa are activated by increase in pHi following dilution in seawater .

ACKNOWLEDGMENTS

The authors are grateful to Dr. Ikuo Yasumasu (Waseda University), Dr. Sakae Kikuyama (Waseda University), Dr. Tsuyoshi Uehara (University of the Ryukyus), and to Dr. Yoshitaka Nagahama (National Institute for Basic Biology) for their encouragement and valuable advice. Thanks are also extended to the staff of Asamushi Marine Biological Station, University of Tohoku, Misaki Marine Biological Station, University of Tokyo, Tateyama Marine Laboratory, Ochanomizu University, Sesoko Marine Station, University of the Ryukyus, Stazione Zoologica 'Anton Dohrn' di Napoli, and to Dr. Hiroaki Tosuji (Kagoshima University) for their affording us the opportunity to utilize their facilities and for kind assistance with the collection of sea urchins.

REFERENCES

Afzelius, B.A. & Mohri, H. 1966. Mitochondria respiring without exogenous substrate. A study of aged sea urchin spermatozoa. *Exp. Cell Res.* 42: 10-17.

Atkinson, D.E. 1968. The energy charge of the adenylate pool as a regulartory parameter: Interaction with feedback modifiers. *Biochemistry* 7: 4030-4034.

Bibring, T.J., Baxandall. J. & Harter, C.C. 1984. Sodium-dependent pH regulation in active sea urchin sperm. *Dev. Biol.* 101: 425-435.

Brokaw, C.J. & Gibbons, I.R. 1973. Localized activation of bending in proximal, medial and distal regions of sea-urchin sperm flagella. *J. Cell Sci.* 13: 1-10.

Brokaw, C.J. 1987. Regulation of sperm flagellar motility by calcium and cAMP-dependent phosphorylation. *J. Cell Biochem.* 35: 175-184.

Brokaw, C.J. 1990. The sea urchin spermatozoon. *Bioessays* 12: 449-452.

Christen, R., Schackmann, R.W. & Shapiro, B.M. 1982. Elevation of intracellular pH activated sperm respiration and motility of sperm of the sea urchin *Strongylocentrotus purpuratus. J. Biol. Chem.* 257: 14881-14890.

106

Chiristen, R., Schackmann, R.W. & Shapiro, B.M. 1983a. Metabolism of sea urchin sperm. Interrelationships between intracellular pH, ATPase activity, and mitochondrial respiration. *J. Biol. Chem.* 258: 5392-5399.

Chiristen, R., Schackmann, R.W. Dahlquist, F.W. & Shapiro, B.M. 1983b. ^{31}P-NMR analysis of sea urchin sperm activation: Reversible formation of high energy phosphate compounds by changes in intracellular pH. *Exp. Cell Res.* 149: 289-294.

Eckelbarger, K.L., Young, C.M. & Cameron, J.L. 1989. Modified sperm ultrastructure in four species of soft-bodies echinoids (Echinodermata: Echinothuridae) from the bathyal zone of the deep sea. *Biol. Bull.* 177: 230-236.

Evans, J.A. & Gibbons, I.R. 1986. Activation of dynein 1 adenosine triphosphate by organic solvents and by Triton X-l00. *J. Biol. Chem.* 261: 14044-14048.

Garbers, D.L., Watkins, H.D., Hansbrough, J.R., Smith, A. & Misono, K.S. 1982. The amino acid sequence and chemical synthesis of speract and speract analogues. *J. Biol. Chem.* 254: 2734-2737.

Gibbons, B.H. & Gibbons, I.R. 1972. Flagellar movement and adenosine triphosphatase activity in sea urchin sperm extracted with triton X-l00. *J. Cell. Biol.* 54: 75-97.

Gray, J. 1928.The effect of dilution of spermatozoa. *Brit. J. Exp. Biol.* 5: 337-344.

Hansbrough, J.R. & Garbers, D.L. 1981. Speract: Purification and characterization of a peptide associated with egg that activates spermatozoa. *J. Biol. Chem.* 256: 14471452.

Hino, A., Hiruma, T., Fujiwara, A. & Yasumasu, I. 1980. Inhibition of respiration in sea urchin spermatozoa following interaction with fixed unfertilized eggs. IV. State 4 respiration in spermatozoa of *Hemicentrotus pulcherrimus* after their interaction with fixed unfertilized eggs. *Develop. Growth Differ.* 22: 813-820.

Johnson, C.H., Clapper, D.L., Winkler, M.M., Lee, H.C. & Epel, D. 1983. A volatile inhibitor immobilizes sea urchin sperm in semen by depressing the intracellular pH. *Dev. Biol.* 98: 493-501.

Lardy, H.A. & Phillips, P.H. 1941a. The interrelation of oxidative and glycolytic processes as sources of energy for bull spermatozoa. *Am. J. Physiol.* 133: 602-609.

Lardy, H.A. & Phillips, P.H. 1941b. Phospholipids as a source of energy for motility of bull spermatozoa. *Am. J. Physiol.* 134: 542-548.

Lee, H.C., Johnson, C. & Epel, D. 1983. Changes in internal pH associated with initiation of motility and acrosome reaction of sea urchin sperm. *Dev. Biol.* 95: 31-45.

Lee, H.C. & Garbers, D.L. 1986. Modulation of the voltage-sensitive Na^+/H^+ exchange in sea urchin spermatozoa through membrane potential changes induced by the egg peptide speract. *J. Biol. Chem.* 261: 16026-16032.

Longo, F.J., Anderson, E. 1969. Sperm differentiation in the sea urchins *Arbacia punctulata* and *Strongylocentrotus purpuratus. J. Ultrastruct. Res.* 27: 486-507.

Mita, M. & Yasumasu, I. 1983a. Metabolism of lipid and carbohydrate in sea urchin spermatozoa. *Gamete. Res.* 7:133-144.

Mita, M. & Yasumasu, I. 1983b. Effect of Na^+-free seawater on energy metabolism in sea urchin spermatozoa with special reference to coenzyme A and carnitine derivatives. *Gamete. Res.* 7: 259-267.

Mita, M. & Yasumasu, I. 1984. The role of external potassium ion in activation of sea urchin spermatozoa. *Dev. Growth Differ.* 26: 489-495.

Mita, M. & Ueta, N. 1988. Energy metabolism of sea urchin spermatozoa, with phosphatidylcholine as the preferred substrate. *Biochim. Biophys. Acta* 959: 361-369.

Mita, M. & Ueta, N. 1989. Fatty chain composition of phospholipids in sea urchin spermatozoa. *Comp. Biochem. Physiol.* 92B: 319-322.

Mita, M. & Ueta, N. 1990. Phosphatidylcholine metabolism for energy production in sea urchin spermatozoa. *Biochim. Biophys. Acta* 1047: 175-179.

Mita, M., Ueta, N., Harumi, T. & Suzuki, N. 1990. The influence of an egg-associated peptide on energy metabolism in sea-urchin spermatozoa: the peptide stimulates preferential hydrolysis of phosphatidylcholine and oxidation of fatty acid. *Biochim. Biophys. Acta* 1035: 175-181.

Mita, M. 1991. Energy metabolism of spermatozoa of the sea urchin *Glyptocidaris crenularis. Mol. Reprod. Develop.* 28: 280-285.

Mita, M., Harumi, T., Suzuki, N. & Ueta, N. 1991. Localization and characterization of phosphatidylcholine in sea urchin spermatozoa. *J. Biochem.* 109: 238-242.

Mita, M. 1992. Diacyl choline phosphoglycerides: The endogenous substrate for energy metabolism in sea urchin spermatozoa. *Zool. Sci.* 9: 563-568.

Mita, M. & Nakamura, M. 1992. Ultrastructural study of an endogenous energy substrate in spermatozoa of the sea urchin *Hemicentrotus pulcherrimus. Biol. Bull.* 182: 298-304.

Mita, M. & Ueta, N. 1992. Fatty chains of alkenylacyl, alkylacyl and diacyl phospholipids in sea urchin spermatozoa. *Comp. Biochem. Physiol.* 102B: 15-18.

Mita, M. & Nakamura, M. 1993a. Lipid globules at the midpieces of *Glyptocidaris crenularis* spermatozoa and their relation with energy metabolism. *Mol. Reprod. Develop.* 34: 158-163.

Mita, M. & Nakamura, M. 1993b. Phosphatidylcholine is an endogenous substrate for energy metabolism in spermatozoa of sea urchins of the order Echinoidea. *Zool. Sci.* 10: 73-83.

Mita, M., Fujiwara, A., De Santis, R. & Yasumasu, I. 1994a. High-energy phosphate compounds in spermatozoa of the sea urchin *Arbacia lixula* and *Paracentrotus lividus. Comp. Biochem. Physiol.* 109A: 269-275.

Mita, M., Oguchi. A., Kikuyama. S., De Santis, R. & Nakamura, M. 1994b. Ultrastructural study of endogenous energy substrates in spermatozoa of the sea urchins *Arbacia lixula* and *Paracentrotus lividus. Zool. Sci.* 11: 701 -705.

Mita, M., Oguchi, A., Kikuyama, S., Namiki, H.. Yasumasu, I. & Nakamura, M. 1994c. Comparison of sperm lipid components among four species of sea urchin based on echinoid phylogeny. *Comp. Biochem. Physiol.* 108B: 417-422.

Mita, M., Oguchi, A., Kikuyama, S., Yasumasu, I.. De Santis, R. & Nakamura, M. 1994d. Endogenous substrates for energy metabolism in spermatozoa of the sea urchins *Arbacia lixula* and *Paracentrotus lividus. Biol. Bull.* 186: 285-290.

Mita, M., Yasumasu. I. & Nakamura, M. 1994e.Energy metabolism of spermatozoa of the sand dollar *Clypeaster japonicus*: The endogenous substrate and ultrastructural correlates. *J. Biochem.* 116: 108-113.

Mita, M., Yasumasu, I. & Nakamura, M. 1995. Endogenous substrates for energy metabolism and ultrastructural correlates in spermatozoa of the sea urchin *Diadema setosum. Mol. Reprod. Develop.* 40: 103-109.

Mita, M. & Nakamura, M. 1998. Energy metabolism of sea urchin spermatozoa: An approach based on echinoid phylogeny. *Zool. Sci.* 15: 1-10.

Mohri, H. 1956. Studies on the respiration of sea-urchin spermatozoa. I. The effect of 2,4dinitrophenol and sodium azide. *J. Exp. Biol.* 33: 73-81.

Mohri, H. 1957a. Endogenous substrates of respiration in sea-urchin spermatozoa. *J. Fac. Sci. Tokyo Univ.* IV (8): 51-63.

Mohri, H. 1957b. Fatty acid oxidation in sea-urchin spermatozoa. *Anat. Zool. Jap.* 30: 181- 186.

Mohri, H. 1958. Adenosine triphosphatases of sea urchin spermatozoa. *J. Fac. Sci. Univ. Tokyo IV* (8): 307-315.

Mohri, H. 1959a. Plasmalogen content in sea-urchin gametes. *Sci. Pap. Coll. Gen. Educ. Univ. Tokyo* 9: 263-267.

Mohri, H. 1959b. Enzymic hydrolysis of phospholipids in sea urchin spermatozoa. *Sci. Pap. Coll. Gen. Educ. Univ. Tokyo* 9: 269-278.

Mohri, H. 1961. Column chromatographic separation of phospholipids in sea urchin spermatozoa. *Sci. Pap. Coll. Gen. Educ. Univ. Tokyo* 11: 109-118.

Mohri, H. & Horiuchi, K. 1961. Studies on the respiration of sea-urchin spermatozoa. Ill. Respiratory quotient. *J. Exp. Biol.* 38: 249-257.

Mohri, H. & Yasumasu, I. 1963. Studies on the respiration of sea urchin spermatozoa. V. The effect of pCO_2 *J. Exp. Biol.* 40: 573-586.

Mohri, H. 1964. Phospholipid utilization in sea-urchin spermatozoa. *Pubb. Staz. Zool. Napoli* 34: 53-58.

Mohri, H. & Yasumasu, I. 1966. On the pathway of glucose utilization in sea urchin spermatozoa. *Sci. Pap. Coll. Gen. Educ. Univ. Tokyo* 16: 279-282.

Mohri, H., Ogawa. K. & Miki-Noumura, T. 1979. Localization of dynein in sea urchin sperm flagella. In: Fawcett, D.W. & Bedford, T.M. (eds), *The Spermatozoon*: 119-127. Urban & Schwarzenberg Inc, Baltimore Munich.

Moriwaki, K. 1958. Changes in content of high energy phosphate esters in sea-urchin spermatozoa after dilution. *J. Fac. Sci. Univ. Tokyo* 8: 297-305.

Nishioka, D. & Cross, N. 1978. The role of external sodium in sea urchin fertilization. In: Dirksen, E.R., Prescott, D.M. & Fox, CF. (eds) *Cell Reproduction*. Academic Press, New York, pp. 403-413.

Ohtake, T., Mita, M., Fujiwara, A., Tazawa, E. & Yasumasu, I. 1996. Degeneration of respiratory system in sea urchin spermatozoa during incubation in seawater for long duration. *Zool. Sci.* 13: 857-863.

Peterson, R.N. & Freund, M. 1976. Metabolism of human spermatozoa. In: Hafez, E.S.E. (ed), *Human Semen and Fertility Regulation in Men*: 176-186. The C. V. Mosby Company, St Louis.

Ratto, A., Shapiro. B.M. & Christen, R. 1989. Phosphagen kinase evolution. Expression in echinoderms. *Eur. J. Biochem.* 186: 195-203.

Repaske, D.R. & Garbers, D.L. 1983. A hydrogen ion flux mediates stimulation of respiratory activity by speract in sea urchin spermatozoa. *J. Biol. Chem.* 258: 6025-6029.

Rothschild, Lord 1951. Sea-urchin spermatozoa. *Biol. Rev.* 26: 1-27.

Rothschild, Lord & Cleland, K.W. 1952. The physiology of sea-urchin spermatozoa. The nature and location of the endogenous substrate. *J. Exp. Biol.* 41: 66-71.

Rothschild, Lord 1956. The physiology of sea urchin spermatozoa. Action of pH, dinitrophenol, dinitrophenol + versense, and usnic acid on O_2 uptake. *J. Exp. Biol.* 41: 66-71.

Shapiro, B.M. 1990. Molecular mechanisms of sea urchin sperm activation before fertilization. *J. Reprod. Fertil.* Suppl. 42: 3-8

Shigei, M. 1974. Echinoids. In *The Systematic Zoology 8(2) – Echinoderm.*: 208-232. Nakamura Book Company, Tokyo.

Summers, D.G & Hylander, B.L. 1974. An ultrastructural analysis of early fertilization in the sand dollar, *Echinarachnius parma*. *Cell Tissue Res.* 150: 343-368.

Suzuki, N., Nomura, K., Ohtake, H. & Isaka, S. 1981. Purification and the primary structure of sperm activating peptides from the egg jelly coat of the sea urchin eggs. *Biochem. Biophys. Res. Commun.* 99: 1238-1244.

Suzuki, N. & Garbers, D.L. 1984. Stimulation of sperm respiration rates by speract and resact at alkaline extracellular pH. *Biol. Reprod.* 30: 1167-1174.

Suzuki, N. 1990. Structure and function of sea urchin egg jelly molecules. *Zool. Sci.* 7: 355-370.

Taguchi, S., Yasumasu, I. & Mohri, H. 1963. Changes in the content of adenine nucleotides upon aerobic incubation of sea urchin spermatozoa. *Exp. Cell Res.* 30: 218-223.

Tombes, R.M. & Shapiro, B.M. 1985. Metabolite channeling: A phosphocreatine shuttle to mediate high energy phosphate transport between sperm mitochondrion and tail. *Cell* 41: 325-334.

Tombes, R.M. & Shapiro, B.M. 1987. Enzyme termini of a phosphocreatine shuttle. Purification and characterization of two creatine kinase isozymes from sea urchin sperm. *J. Biol. Chem.* 262: 16011-16019.

Tombes, R.M., Brokaw, C.J. & Shapiro, B.M. 1987. Creatine kinase dependent energy transport in sea urchin spermatozoa: Flagellar wave attention and theoretical analysis ~P diffusion. *Biophys. J.* 52: 75-86.

Yanagisawa, T. 1967. Studies on echinoderm phosphagens. III. Changes in the content of creatine phosphate after stimulation of sperm motility. *Exp. Cell Res.* 46: 348-354.

Yanagisawa, T., Hasegawa, S. & Mohri, H. 1968. The bound nucleotides of the isolated microtubules of sea-urchin sperm flagella and their possible role in flagellar movement. *Exp. Cell Res.* 52: 86-100.

Non-parasitic symbioses between echinoderms and bacteria

CHANTAL DE RIDDER[1] AND TIMOTHY W. FORET[2]
[1]Laboratoire de Biologie marine (C.P. 160/15), Université Libre de Bruxelles, 50 av. F.D. Roosevelt, B-1050 Bruxelles, Belgium
[2]Florida Center for Community Design and Research, University of South Florida,3702 Spectrum Blvd., Suite 180, Tampa, Fl 33612, USA

CONTENTS

1 INTRODUCTION

During the last several decades, symbiosis has been described as a pervasive and evolutionarily significant biological phenomenon that has played a fundamental role in the evolution of life on Earth (Margulis 1976, 1991, Margulis & Fester 1991). The term symbiosis was first defined by the German mycologist Anton De Bary (1879) as "the living together of differently named organisms". Such a broad definition naturally encompasses a wide variety of neutral, beneficial or detrimental interactions between heterospecific individuals and may thus refer to plant pollinisation by insects as well as to plant-herbivore or to prey-predator relationships. By insisting on the duration and contact, several authors sharpened the definition: symbiosis is an interspecific association in which the partners are in intimate contact during an appreciable time relative to their life-spans (e.g., Smith & Douglas 1987). This is the meaning used here. Sapp (1994) reviewed the history of the concept. In symbiosis, the larger partner is usually referred as the host and the smaller one as the symbiont. According to their external or internal position to the host, the symbionts are termed ectosymbiotic or endosymbiotic, respectively. Endosymbionts can be located intracellularly or extracellularly. Extracellular endosymbionts occupy a cavity (e.g., lumen of organs) or intercellular spaces (e.g., intraepithelial spaces). The degree of dependence of the partners (obligate vs facultative symbiosis), their taxonomic specificity (specificity of the symbiosis), the transmission of the symbiont to its host (perpetuation of the symbiosis) and the balance of advantages obtained by the partners (functional significance of the symbiosis) are additional characteristics. In an obligate symbiosis, host and/or symbiont are physiologically (or morphologically) linked to such an extent that they cannot survive outside the association. Alternatively, the symbiosis is facultative (for one or both partners). The degree of specificity of a symbiosis refers to the range of hosts taxa and/or of symbionts taxa involved in a particular symbiosis: the specificity grades from low to high with the decreasing number of taxa involved. The transmission of symbionts from one generation of hosts to the next follows two main modes: the transovarian (or vertical) transmission and the cyclic (or horizontal) transmission. In transovarian transmission, the symbionts are transmitted via the eggs; in cyclic transmission, the symbionts are acquired anew each generation from the environment (indirect transmission) or from conspecific individuals e.g., via contacts between parents and offsprings (direct transmission or reinfection) (Stanier et al. 1986, Douglas 1994, McFall-Ngai 1998). Symbioses encompass various types of interactions between the partners which may derive benefit or harm or be unaffected. Although the interpretation of what is advantageous or disavantageous (cf. limits between benefit and neutrality or between neutrality and pathogenicity) is usually difficult to assess (Smith & Douglas 1987, Morton 1989), the fact that two organisms have evolved a symbiosis implies that at least one partner gains some advantage from the association; the balance of advantages within a symbiosis can be roughly summarized as shown in Table 1. Mutualism refers to a symbiosis involving reciprocal benefit (host and symbiont benefit from the association). Parasitism refers to a symbiosis affecting negatively (more or less severely) the host and positively the symbiont. Another set of symbioses shares the tendency to be harmless for the host and unilaterally beneficial for the symbiont. Giving the nature of the benefit obtained by the symbiont, these symbioses are named: commensalism when the benefit is trophic, phoresis when it refers to provision of transport, epizoism when it refers to provision of surface/support, endoecism when it refers to protection through shelter/burrow sharing and inquilinism when it refers to protection through sheltering within the host body itself. The limits between these categories are not always sharp (see e.g., Stanier et al. 1986, Smith & Douglas 1987, & Morton 1989).

Symbioses are found in terrestrial as well as in aquatic environments. They are much more common and diversified in the marine environment (Morton 1989). Symbioses between unicellular algae and invertebrates, between inver-

Table 1. Balance of advantages in symbioses (*).

Symbiosis	Host	Symbiont
mutualism	+	+
parasitism	−	+
commensalism phoresis epizoism endoecism inquilinism	0	+

(*) positive (+), negative (-) or neutral (0)

tebrates and between invertebrates and vertebrates are clearly the best known. However, symbioses involving procaryotic symbionts and eucaryotic hosts are by far the most abundant but also usually the less well understood (McFall-Ngai 1998, 1999). Bacteria occur in a wide spectrum of natural habitats including not only suitable environments for higher organisms but also extreme environments where few or no higher organisms can live or survive (Brock et al., 1994). The metabolic capabilities and versatility of the bacteria are the keys for such an ecological success. Bacteria are so widespread and metabolically diversified that they can colonize most inert or living substrates. This colonizing aptitude has led procaryotes to develop symbiotic interactions with a variety of organisms. In an evolutionary context of cooperating cells, the establishment/existence of symbioses between eucaryotes and procaryotes is particularly logical. McFall-Ngai (1998) stated two possibilities were present: "animals could develop mechanisms to inhibit interactions with bacterial cells; or, they could take advantage of the already existing trend of interaction and foster colonization by specific subsets of bacteria, availing themselves of bacterial metabolic diversity". In that context, the development of negative interactions (the bacteria becoming parasites/pathogens for the animals) and of real cooperation between bacteria and animals (mutualism) is easily understood. The balance between host defense mechanisms and microbial invasion and, if microbial invasion succeeds, between costs and benefits for both partners has a determining effect in the establishment of a particular symbiosis (Plante et al. 1990, Duncan & Edberg 1995, McFall-Ngai 1998). The difficulties met in the analysis of this delicate balance, together with those met in the isolation and in the metabolic characterization of the bacterial symbionts, explain why the nature of bacterial associations/symbioses is not always clear or easy to assess.

Every living organism continuously encounters bacteria in its immediate surrounding and consequently is susceptible to be invaded by them. This is particularly true in the marine environment as bacteria are widespsread in sea-water and can colonize virtually any submerged substrate, whatever its nature (see e.g., Cooksey & Wigglesworth-Cooksey 1995). For instance, for every planktonic bacterium in an aqueous system it has been estimated that there are between 10^3 and 10^4 bacteria actually attached to surfaces (Watkins & Costerton 1984). In echinoderms, two main anatomical sites where demonstrated or presumed symbiotic non-parasitic bacteria are found are: (1) the cuticle that lines the epidermis and (2) the digestive system which is connected to the outside via one (mouth) or two (mouth and anus) openings. Parasitic symbioses (and bacterial diseases) were reviewed by Jangoux (1990). The bacteria associated with the cuticle are named "subcuticular bacteria" (SCB) (Holland & Nealson 1978), those found within the digestive system are members of the gut microflora (Smith & Douglas 1987, Harris 1993), and will be named here "gut bacteria" (GB). The presence of enigmatic bacteria is also punctually reported in some crinoids within the soft connective tissue of the arms and pinnules (Holland et al.1991), and will be named here "connective tissue bacteria" (CTB). The SCB and the CTB are most probably (endo)symbionts. In contrast, most studies do not answer this point for the GB. A reason for this could be that the GB form complex assemblages in which the discrimination between symbiotic and non symbiotic members is difficult. We shall consider GB as bacteria commonly reported in the gut and suspected or demonstrated to fulfill a particular function within the gut. Discrimination between symbiotic and non symbiotic bacteria will be done when it is reported.

2 SUBCUTICULAR BACTERIA

2.1 *Introduction*

Saffo (1992) stated that many invertebrates are acutely affected by endosymbionts in terms of ecology, physiology, development, and behavior. Subcuticular bacteria (SCB) are endosymbiotic gram-negative bacteria found within the cuticle or within the subcuticular space of all five classes of echinoderms (Holland & Nealson 1978, Féral 1980, McKenzie 1987). SCB are globally distributed symbionts of echinoderms, being reported from a wide variety of geographic locations: Florida (Foret 1999), northeastern Pacific and Atlantic (McKenzie & Kelly 1994, Kelly & McKenzie 1995), northwestern Atlantic (Walker & Lesser 1989), and the southwestern Pacific (Kelly et al. 1995).

The cuticle, or surface coat, of echinoderms is a fibrous and/or granular extracellular matrix associated with the apical branched and unbranched microvilli of epidermal cells (Holland 1984). It envelops the epithelium. Holland and Nealson (1978) and McKenzie (1988) have provided reviews of the cuticle of all echinoderm classes. It generally consists of three layers: 1) an outer filamentous 'fuzzy coat' or glycocalyx, 2) a granular upper cuticle, and 3) a fibrillar, possibly collagenous, lower cuticle. The cuticle consists of glycoprotein (Grigolava & McKenzie 1994), acid and neutral mucopolysaccharides (Holland 1984, Walker & Lesser 1989), and possibly collagen (McKenzie 1988). The cuticle is supported by the microvilli of support cells or non-ciliated cuticle-secreting cells (Walker & Lesser 1989). A network of crypts (the subcuticular space), which may be spacious in those taxa that possess SCB, exists beneath the cuticle. During development, the embryonic hyaline layer produces outpockets which increase the width of the subcuticular lumen, giving the cuticle a lobed appearance (Cerra & Byrne 1998). This space ostensibly accommodates bacterial cells in larvae of symbiotic taxa (Bosch 1992, Cerra et al., 1997, Cerra & Byrne 1998).

The thickness and appearance of the cuticle varies among taxa, suggesting that the cuticles from different classes differ in function (McKenzie 1988). Possible roles of the cuticle include support (Souza Santos & Silva Sasso 1970), protection from abrasion and disease (Coleman 1969, Engster & Brown 1972), surface lubrication and possibly nutrient uptake (McKenzie 1988). Echinoderm surface coats have turnover times of approximately 50 days, although turnover of the outer fuzzy coat may be considerably shorter (Grigolava & McKenzie 1994). Therefore, although the cuticle as a whole is not thought to be anti-fouling, the outer fuzzy coat may serve this function (McKenzie 1988).

2.2 Characterization of subcuticular bacteria

2.2.1 Morphology

SCB are invariably gram-negative rods and spirals. The ultrastructure of SCB has been described for a variety of echinoderms (Féral 1980, Lesser & Blakemore 1990, Kelly et al. 1995, Cerra et al. 1997, Foret 1999). McKenzie and Kelly (1994) further described three distinct bacterial morphologies from ophiuroids using TEM. The first type (T1) are short rods of typically paired bacteria within a common capsule. While these may have complex internal structure and several layers of membranes, they have been only rarely observed (McKenzie et al. 1998). The second type (T2) are longer (~10 μm), thin rods which lack the internal specialization of T1. Often, these are curved or tightly kinked spirals. The third type (T3) is composed of rods with complex, tripartite capsules consisting of electron opaque inner and outer membranes and a well-defined periplasmic space (Walker and Lesser 1989). These may be 1.0 μm in length and have a length: width ratio of 2:1 (Bosch 1992), and may possess vacuoles which are likely poly-β-hydroxybutyrate (PHB) bodies (Kelly & McKenzie 1995, Walker & Lesser 1989). Also, bacteriocytes containing SCB have been observed (see McKenzie et al. 1998).

2.2.2 Location and density of subcuticular bacteria

SCB are found either in the subcuticular space or within the cuticle itself (Holland & Nealson 1978, Féral 1980, McKenzie 1987). SCB are commonly observed within phagosomes of the host epidermal cells (Walker & Lesser 1989, Roberts et al. 1991, Cerra et al. 1997). They are seen less frequently as numerous spirals located within a common vesicle, or bacteriocyte (Bosch 1992, Kelly et al. 1995). The relative distribution of SCB over the surface of the body is assumed to be consistent with the presence of the integument. Tube feet of asteroids, peristomial membranes of regular echinoids, test margins of clypeasteroids, and whole arms of ophiuroids have SCB (Foret 1999). Tube feet are usually selected for study because they typi-

is expensive to perform and is limited to gram-negative symbionts. Direct counting of symbionts from tissue homogenates, although less expensive, may have greater error associated with it. Symbiont abundance is typically expressed in terms of cell numbers per gram wet weight or ash-free dry weight. As these techniques do not consider the variation in cuticle mass that may result from different amounts of tissue and skeletal material between tissues and species, comparisons should be made with caution (Kelly & McKenzie 1992). However, comparisons within taxa should have less error. Other difficulties exist in quantifying SCB using direct counting with epifluorescence microscopy. Using acridine orange as a nucleic acid-binding fluorochrome, emittance shifted from red in log phase growth to green at steady states (Back & Kroll 1991). Current techniques rely on the counting of green-fluorescing cells exclusively (Kelly & McKenzie 1992). As SCB DNA/RNA ratios may be changing, it is possible that the apparent abundance of symbionts using direct counting could change over time. It is important to use caution when interpreting such data found on the basis of fluorescence microscopy observations alone. Squash preparations may not be useful quantitatively as an intact surface area should be used as a reference unit (Largo et al. 1997). Electron microscopic (i.e., stereology and image analysis tools) and molecular biological techniques (i.e., 16S rRNA probes) should be useful in quantifying SCB more precisely and accurately.

In ophiuroids, SCB numbers may vary interspecifically by over an order of magnitude and considerable intraspecific variation has been described (McKenzie & Kelly 1994). Kelly et al. (1995) estimated bacterial loads in regular echinoids ranging from 8.41×10^8 (*Pseudechinus novaezealandiae*) to 4.96×10^9 g^{-1} (*P. huttoni*) ash-free dry weight. Small *P. huttoni* had 2.67×10^9 bacteria. Kelly and McKenzie (1995) found densities of the same order of magnitude for the echinoid *Psammechinus miliaris*, and the asteroid *Anseropoda placenta*. Far fewer SCB were detected from the holothurian *Leptosynapta bergensis*. Foret (1999) found SCB abundance ranging from 1.11×10^9 to 2.21×10^9 cells g^{-1} ash-free dry weight among populations of *Ophiophragmus filograneus* from the east and west coasts of Florida. Loading of SCB from populations of *Luidia clathrata* varied from 1.92×10^8 to 1.77×10^9 cells. McKenzie et al. (2000) found SCB counts to fluctuate over the course of a year from 0.71×10^9 to 1.92×10^9 cells g^{-1} ash-free dry weight for *Ophiothrix fragilis* and from 0.87×10^9 to 2.66×10^9 cells g^{-1} ash-free dry weight for *Amphiura chiajei*. The loading of SCB among some taxa is thus comparable to that in the vestimentiferan trophosome (Cavanaugh et al. 1981). In some species, bacterial counts are considered large enough ($> 10^9$ cells g^{-1} ash-free dry weight (AFDW)) to provide significant amounts of energy to their hosts while the numbers in others are so low that any contribution is unlikely (Kelly & McKenzie 1995). Substantial variation in

118

SCB density may be found within and between populations spatially and temporally (Foret 1999, McKenzie et al. 2000). This suggests that the nature of the interaction between host and symbionts is variable over time and space. Why variation should be great within populations over time and space remains unexplored. However, this variation suggests that examination of species over time and space will likely show that the overall incidence of SCB among echinoderms is higher than is currently represented and may provide useful insight into the factors affecting their occurrence.

2.3 Roles of subcuticular bacteria

2.3.1 Metabolic capabilities
SCB are almost certainly heterotrophic bacteria. Lesser and Blakemore (1990) suggest that SCB belong to the family Vibrionaceae based on guanine: cytosine content, antibiotic sensitivity, other biochemical characters (e.g. lipids), and ultrastructure. Fatty acid profiles of symbiotic and aposymbiotic ophiuroids shows that 16:1ω7 fatty acids comprise a much higher percentage of the host pool in the symbiotic taxa studied, suggesting a potential source of the host's nutrition (McKenzie et al. 2000). Lambda ^{15}N is inversely correlated with symbiont loads, suggesting that SCB possibly contribute to the nitrogen budget of the host. Further evidence of this comes from phylogenetic analysis using PCR which suggests that SCB belong in the α subdivision of the purple non-sulfur Proteobacteria aligning them most closely with the nitrogen-fixing *Rhizobium* (Burnett & McKenzie 1997). No data indicate that SCB are either chemoautotrophic or methylotrophic (Kelly & McKenzie 1995, McKenzie et al. 2000).

2.3.2 Trophic interactions
Although SCB may be abundant, their importance in terms of the nutrition of their hosts is uncertain (Lesser & Walker 1992). Many workers have envisioned a functional role of the symbionts in the nutrition of their hosts largely by circumstantial evidence. Roberts et al. (1991) have implied that SCB could be an important source of dietary supplementation in deep-sea holothurians based on their abundance within the subcuticular space and demonstrated phagocytosis by host cells. Although epidermal transport may be substantial in echinoderms (Bamfield 1982) and other invertebrates (Stephens 1988), marine bacteria generally have greater affinities for dissolved organic matter than eukaryotes (Sepers 1977, Siebers 1979). Therefore, bacteria might be expected to play a role in uptake although empirically documenting the contribution of SCB to the nutrition of their hosts has been challenging. McKenzie et al. (2000) contend that enzyme studies and molecular probes for specific bacterial metabolism genes *in situ* may illuminate the role of SCB to the nutrition of their hosts. The capacity of

SCB to increase production and their contribution to specific tissues remains unknown.

Abundant SCB may augment nutrition during embryonic development (Bosch 1992, Cerra et al. 1997). Lesser and Walker (1992) demonstrated that the SCB-bearing ophiuroid *Amphipholis squamata* had greater than twice the uptake rates of DFAA than the aposymbiotic sympatric species, *Ophiopholis aculeata*, independent of size. They hypothesized that assimilation of dissolved free amino acids (DFAA) into protein is facilitated by SCB, which is subsequently translocated to the brooded embryo. However, the maximum uptake by symbionts accounted for no more than 10% of the total uptake of the adult host. Thus, microbial translocation has not been established as a contributor to the host's nutrition. In contrast to lecithotrophs, planktotrophic larvae may experience variable availability of food in response to ambient conditions that could make SCB beneficial as a stable nutritional source not unlike algal-cnidarian associations (Douglas & Smith 1983).

As Douglas (1994) described, nutritional interactions between symbionts are typically bi-directional and hence more complex than this. For example, as Proteobacteria, SCB may utilize nitrogenous wastes. This may be beneficial given that echinoderms lack discrete, complex excretory systems. The rectal caecum in asteroids functions as a renal organ (Jangoux 1982) although much of the highly diffusible nitrogenous wastes such as ammonia and urea are released directly across the body surface (Stickle 1988). Bosch (1992) noted that SCB numbers in the epidermal regions surrounding the larval stomach are higher than in distal regions of the body, suggesting that a digestive by-product may sustain the bacteria. The uptake of toxic metabolic by-products at the organism-environment interface would be adaptive if it served such a detoxification mechanism. Additionally, specific bacterial metabolism (e.g. glutamine synthase, glutamine synthetase) which incorporates amines into proteins, may serve to benefit the host under specific conditions, similar to that described by Lesser and Walker (1992). In turn, as nitrogen becomes limiting to the hosts, they might 'crop' SCB to supplement nutrition. This process may circumvent the need to feed actively through ingestion. A similar situation could occur in response to a seasonal change in nitrogen content of ingested food. Fong and Mann (1980) have hypothesized this type of role for nitrogen-fixing gut flora of the sea urchin *Strongylocentrotus droebachiensis* in incorporating ammonia and synthesizing essential amino acids that are available to the host as microbial protein.

The potential synergisms involved with this symbiosis remain largely unexplored. Ferguson (1967, 1970) proposed the existence of adaptive transport systems that would overcome the relative isolation of the epidermis by morphological and physiological barriers. It is premature to suggest that SCB serve this function. Yet while epidermal absorption of DOM is sufficient to supply the layer energetically (Ferguson 1980, Stephens et al. 1978), the relative contribution of SCB specifically to these tissues' nutrition may be substantially greater given their densities. During early development, SCB may play a greater relative role in the nutrition of the whole organism. Phagocytosis of SCB in echinoderm larvae supports this possibility (Cameron & Holland 1983, Walker & Lesser 1989, Bosch 1992). While the exact nature of this symbiosis remains unclear, SCB appear to be facultative residents of the subcuticular space and therefore any benefit for one or both of the partners under specific conditions is likely to be opportunistic (Cerra et al. 1997).

2.4 *Perpetuity of the symbiosis*

2.4.1 *Permanence of subcuticular bacteria*
The association between SCB and their hosts does not appear to be permanent. Although SCB have never been positively identified in echinoderm gametes or zygotes, SCB are present early in development (Walker & Lesser 1989, Bosch 1992, Cerra & Byrne 1998). The earliest reported bacterial symbiont observed by Cerra et al. (1997) was from a seven-day-old brachiolaria. They did not find this flora among gastrulae and suggest that introduction of bacteria into the hyaline layer occurs during an early larval stage after hatching.

2.4.2 *Subcuticular bacteria as a sublethal stress indicator*
A practical aspect of the SCB-echinoderm symbiosis is its potential use in bioassays. Biological assays often involve the lethal effects of pollution on organisms, the presence or absence of indicator species, or overall species diversity. Newton and McKenzie (1995, 1998), however, demonstrated a decrease in SCB load in response to oil pollution and suggested the flora may be affected by sublethal stress in the host. Sublethal stress such as static seawater may also reduce the densities of SCB (Newton & McKenzie 1998). Such findings compel further study.

2.5 *Transmission of symbionts*

2.5.1 *Vertical transmission*
SCB appear to represent an unusual group of bacteria capable of circumventing the host's chemical and structural defenses, but it is not known how they are transmitted. As planktotrophic and lecithotrophic larvae possess SCB, infection likely precedes metamorphosis (Cameron & Holland 1983, Cerra et al. 1997, Cerra & Byrne 1998). Two opposing hypotheses currently exist regarding the mechanism of symbiont transfer. The first is that SCB, or their genetic components, are transmitted via gametes to subsequent generations. McKenzie et al. (1998) favor this hypothesis based on the sys-

tematic distribution of SCB among 149 representatives of the phylum and the link between the incidence of SCB and their host's classification. However, endosymbiotic bacteria have not been reported in the many studies of echinoderm gametes, even as casual observations. McKenzie et al. (1998) account for this by noting that bacterial genes have been found in bivalve eggs and larvae (Krueger et al. 1996). Conversely, failure to find SCB in echinoderm eggs led Barker and Kelly (1994) to suggest that either laboratory conditions prevent infection or that SCB are not transmitted vertically through gametes. Indeed, the apparent spatial segregation of SCB and their host's gametes may be expected to be an obstacle to their vertical transmission.

2.5.2 Horizontal transmission

Another hypothesis is that the symbionts are released or escape from the cuticle to reinfect others. Although these bacteria have not been discovered to be free-living, it is not uncommon to find them embedded within the cuticle, suggesting that they can hydrolyze and possibly traverse it. However, de novo infection has not been examined.

Horizontal transmission may be enhanced if it occurred with periodicity during reproductive life stages. SCB occur in both adults and brooding embryos of the ophiuroid Amphipholis squamata (Walker & Lesser 1989, Lesser & Walker 1992). Furthermore, clonal larvae of Luidia possess SCB (Bosch 1992). The squid Euprymna scolopes can regulate the numbers of its symbiont Vibrio fischeri by restricting the amount of oxygen to the light organ. About 90% of these luminescent bacteria are expelled daily, thereby seeding the water column with infective cells for squid hatchlings and other adults (Lee & Ruby 1994). In the western North Atlantic Ocean, Bosch (1992) found 90-100% of the abundant larvae of Luidia harbored SCB by July and August. In January and February, the incidence was lower (89% and 74%, respectively). Foret (1999) found lower bacterial abundance in populations sampled in the winter, potentially suggestive of a seeding mechanism. However, through a yearlong study of ophiuroids, McKenzie et al. (2000) found no obvious seasonal trends in variation of SCB loads.

Considering the spatial separation during gametogenesis between the gonads and the integument in typical broadcast-spawners, the effective contact time between host gamete (vectors) and epithelial symbionts may be fleeting, rendering exchange of genetic material unlikely. In brooding species, on the other hand, the cumulative effect of close contact through generations may be expected to have greater continuity and hence, evolutionary consequences. It is therefore interesting that while Amphipholis squamata and its brood have symbionts, SCB were not found in adult Ophiopholis aculeata sampled during the same period by one group (Lesser & Walker,

1992) only to be later discovered by another (Kelly et al. 1995). These studies imply that SCB numbers are dynamic, which may be evidence of horizontal transmission. Horizontal transfer of symbionts, also known as the "open system", occurs in a variety of marine invertebrates including flatworms, cnidarians, and tridacnid bivalves and may be highly specific between host and symbiont (Trench 1987). Brooding taxa clearly require closer examination. Additional qualitative and quantitative studies of SCB incidence of occurrence and abundance over an entire year should clarify the mechanism of SCB transmission.

2.6 Evolution of the symbiosis

2.6.1 Incidence of subcuticular bacteria

The incidence of SCB in echinoderms is presented in Table 3. This aspect has received much attention in the last decade, particularly for ophiuroids (McKenzie & Kelly 1994). SCB are present in predators, herbivores, detritivores, suspension feeders and deposit feeders. The presence of symbionts appears correlated neither with depth nor with temperature. Therefore, no association has yet been obtained between host ecology and the presence or absence of SCB (Bosch 1992, Kelly & McKenzie 1995, Kelly et al. 1994, Kelly et al. 1995, McKenzie 1994, McKenzie et al. 1998).

Of twenty-two species investigated from Florida (Foret 1999), 55% possessed SCB. Echinoids and ophiuroids possess subcuticular bacteria more frequently (71 and 75%, respectively), while asteroids possess them less (33%). These incidences are similar to that found by Kelly et al. (1995) who surveyed New Zealand echinoderms. Of 63 species examined from the northeast Atlantic, subcuticular bacteria have been described in approximately 40% of the holothurian species examined, and among the other four classes, over 60% of the taxa examined possess SCB (Kelly & McKenzie 1995). However, an important finding with regard to SCB symbiosis in echinoderms is the close taxonomic association between species that possess or lack SCB at the generic and familial levels (McKenzie et al. 1998). Few definitive conclusions are available, as fewer than 5% of the total echinoderm taxa have been examined to date.

2.6.2 The 'Congeneric Species Hypothesis'

An apparent link between host phylogeny and symbiont occurrence and morphology has led to the suggestion that host and symbiont are co-evolving (McKenzie & Kelly 1994). This 'congeneric species hypothesis' has been employed to predict which species possess symbionts, and as importantly, which possess SCB of a specific morphological type. Further analysis of the systematic distribution of the symbiosis has demonstrated that the link between incidence of SCB and phylogeny of their hosts is most evident at

Table 3. Incidence of subcuticular bacteria in echinoderms (adapted from McKenzie et al., 1998).

Taxa	Genus and species	SCB detected (Y :Yes, N :No)	Sources
CRINOIDEA			
Millericrinida			
Isselicrinidae	*Metacrinus rotundus*	Y	Welsch, pers. comm.
	Calamocrinus diomedae	Y	Holland et al. 1990
Cyrtocrinidae	*Gymnocrinus richeri*	Y	Heinzeller & Fechter 1995
Holopodinae	*Holopus rangii*	Y	Grimmer & Holland 1990
Comatulida			
Antedonidae	*Antedon bifida*	Y	McKenzie et al. 1998
	A. petasus	Y	id.
	Leptometra celtica	Y	id.
	L. phalangium	Y	Heinzeller & Welsh 1994
Colobometridae	*Cenometra bella*	N	McKenzie et al. 1998
	Colobometridae perspinosa	N	id.
	Oligometra serripinna	N	id.
	Petasometra clariae	N	id.
	Pontiometra andersoni	N	id.
Comasteridae	*Capillaster multiradiatus*	N	id.
	Comantheria samoanus	N	id.
	C. timorensis	N	id.
	Comanthinus nobilis	N	id.
	Comanthus altermans	N	id.
	C. novazealandiae	N	id.
	C. parvicirrus	N	id.
	Comaster gracilis	N	id.
	C. multibrachiatus	N	id.
	C. multifidis	N	id.
	Comatela nigra	N	id.
	Oxycomanthus benneti	N	id.
Himerometridae	*Himerometra robustipinna*	Y	McKenzie 1992
Mariametridae	*Lampometra palmata*	Y	McKenzie et al. 1998
	Stephanometra oxycantha	Y	McKenzie 1992
Tropiometridae	*Tropiometra afra*	N	McKenzie et al. 1998
Zygometridae	*Zygometra punctata*	Y	McKenzie 1992

Table 3. Continued.

Taxa	Genus and species	SCB detected (Y :Yes, N :No)	Sources

HOLOTHUROIDEA

Aspidochirotida

Synallactidae	*Bathyplotes natans*	Y	Roberts et al. 1991
	Mesothuria intestinalis	Y	McKenzie et al. 1998
	M. lactea	Y	Roberts et al. 1991
	Paroriza pallens	Y	Roberts et al. 1991
	Pseudostichopus spp.	Y	Roberts et al. 1991
Stichopodidae	*Stichopus mollis*	Y	McKenzie et al. 1998
Holothuriidae	*Holothuria forskalii*	N	id.
	Actinopyga agassizii	N	Foret 1999

Elasidopodida

Laetmogonidae	*Benthogone rosea*	Y	Roberts et al. 1991
	Laetmogone violacea	Y	Roberts et al. 1991
Deimatidae	*Deima validum*	Y	Roberts et al. 1991
Psychropotidae	*Benthodytes sordida*	Y	Roberts et al. 1991

Dendrochirotida

Psolidae	*Psolus phantapus*	N	McKenzie et al. 1998
Phyllophoridae	*Neopentadactyla mixta*	N	id.
	Thyone fusus	N	id.
	T. inermis	N	id.
	T. roscovita	Y	id.
Cucumariidae	*Aslia lefevrei*	Y	id.
	Leptopentacta elongata	N	id.
	Ocnus brunneus	N	id.
	O. pygmeus	Y?	Foret 1999 (epizootic?)
	O. lacteus	N	McKenzie et al. 1998
	O. planci	N	id.
	Paracucumaria hyndmani	N	id.
	Pawsonia saxicola	N	id.
	Thyonidium hyalinum	N	id.
	T. drummondii	N	id.

125

Table 3. Continued.

Taxa	Genus and species	SCB detected (Y :Yes, N :No)	Sources
Apodida			
Synaptidae	*Leptosynapta bergensis*	Y	id.
	L. gallienei	Y	Féral 1980
	L. inhaerens	Y	McKenzie et al. 1998
	Labidoplax buskii	Y	id.
	L. digitata	Y	id.
	L. media	Y	id.
Chiridotidae	*Trochodota dunedinensis*	Y	id.
ECHINOIDEA			
Arbaciidae	*Arbacia punctulata*	Y	id.
Cidaroidae	*Goniocidaris umbralaculum*	Y	id.
Parechinidae	*Psammechinus miliaris*	Y	id.
Echinidae	*Echinus acutus (var. norvegicus)*	N	id.
	E. esculentus	N	id.
	Paracentrotus lividus	N	id.
Temnopleuridae	*Pseudechinus novaezealandiae*	Y	id.
	P. albocinctus	Y	id.
	Lytechnius variagatus	Y	Foret 1999
Echinometridae	*Evechinus chloroticus*	N	McKenzie et al. 1998
Clypeasteroida			
Fibulariidae	*Echinocyamus pusillus*		id.
Arachnoidae	*Fellaster zelandiae*	Y	id.
Mellitidae	*Mellita tenuis*	Y	Foret 1999
Spatangoida		Y	
Spatangidae	*Spatangus purpureus*	Y	McKenzie et al. 1998
Loveniidae	*Echinocardium cordatum*	Y	id.
	E. flavescens	Y	id.
	E. pennatifidum	Y	id.
Brissidae	*Brissopsis lyrifera*	Y	id.

126

Table 3. Continued.

Taxa	Genus and species	SCB detected (Y :Yes, N :No)	Sources

ASTEROIDEA

Taxa	Genus and species	SCB detected	Sources
Luidiidae	*Luidia ciliaris*	Y	id.
	L. clathrata	Y	Foret 1999
	L. alternata	Y	Foret 1999
	L. sarsi	Y	McKenzie et al. 1998
	Luidia spp. Larvae	Y	Bosch 1992
Astropectinidae	*Astropecten irregularis*	Y	McKenzie et al. 1998
	A. primigenius	Y	id.
	A. duplicatus	N	Foret 1999
	A. articulatus	N	Foret 1999
	Astropecten spp.	Y	McKenzie et al. 1998
Goniasteridae	*Hippasteria phrygiana*	Y	id.
	Pentagonaster pulchellus	N	id.
Poraniidae	*Porania pulvillus*	Y	id.
Asterinidae	*Asterina gibbosa*	Y	id.
	A. miniata larvae	Y	Cameron and Holland 1983
	A. stellifera	Y	Souza Santos and Sasso 1970
	Anseropoda placenta	Y	McKenzie et al. 1998
	Patiriella calcar larvae	Y	Cerra and Byrne 1997
	P. regularis	Y	McKenzie et al. 1998
Odontasteridae	*Asterodon miliaris*	Y	id.
	Odontaster benhami	Y	id.
Ophidiasteridae	*Ophidiaster kerma*	Y	id.
Solasteridae	*Crossaster paposus*	Y	id.
	Solaster endeca	Y	id.
Echinasteridae	*Echinaster sepositus*	Y	id.
	Echinaster graminicola	N	Foret 1999
	Henricia oculata	N	McKenzie et al. 1998
	Henricia ralphae	N	id.
Asteriidae	*Allostichaster insignis*	N	id.
	A. polyplax	N	id.
	Asterias rubens	N	id.
	Astrostole scabra	N	id.

Table 3. Continued.

Taxa	Genus and species	SCB detected (Y :Yes, N :No)	Sources
	Calvasterias suteri	N	id.
	Coscinasterias calamaria	N	id.
	Leptasterias muelleri	N	id.
	Marthasterias glacialis	N	id.
	Sclerasterias mollis	N	id.
	Stichastrella rosea	N	id.
OPHIUROIDEA			
Asteronychidae	*Asteronyx loveni*	Y	id.
Asteroschematidae	*Astrobrachion constrictum*	Y	id.
Ophiomyxidae	*Ophiomyxa australis*	Y	id.
	O. brevimira	Y	id.
	Ophiogeron supinus	Y	Byrne 1994
Amphilipidae	*Amphilepis ingolfiana*	Y	McKenzie et al. 1998
Amphiuridae	*Amphipholis atra*	Y	id.
	A. gracillima	Y	id.
	A. squamata	Y	Walker and Lesser 1989
	Amphiura abemethyi	Y	McKenzie et al. 1998
	A. amokurae	Y	id.
	A. brachiata	Y	Bell 1974
	A. chiajei	Y	McKenzie et al. 1998
	A. filiformis	Y	id.
Ophiactidae	*Ophiactis balli*	Y	id.
	Ophiopholis aculeate	Y	id.
Ophiotrichidae	*Ophithrix angulata*	Y	id.
	O. fragilis	Y	id.
	O. spiculata	Y	id.
Ophiocomidae	*Ophiocomina nigra*	N	id.
	Ophiocoma bollonsi	Y	id.
	O. wendti	N	id.
	Ophiopteris antipodum	Y	id.
	O. papillosa	Y	id.
Ophionereidae	*Ophionereis fasciata*	N	id.
	O. schayeri	N	id.

Table 3. Continued.

Taxa	Genus and species	SCB detected (Y :Yes, N :No)	Sources
Ophiodermatidae	*Ophioderma longicauda*	Y	Märkel and Röser 1985
	O. panamense	N	McKenzie et al. 1998
	Pectinura maculata	N	id.
Ophiuridae	*Ophiocten gracilis*	Y	id.
	Ophiopleura inermis	Y	id.
	Ophiura affinis	Y	id.
	O. albida	Y	id.
	O. ophiura	Y	id.
	O. sarsi	Y	id.
	Ophiomusium lymani	Y	id.
	Ophioplocus esmarki	Y	id.

the familial and supra-familial levels (McKenzie et al. 1998). As co-evolution may have profound effects on both partners in symbiotic associations, it is of interest to further test this hypothesis on a global scale. Data acquired from a variety of geographical regions should support this assumption if it is indeed valid. The phylogeny of SCB has only begun to be evaluated (Burnett & McKenzie 1997) aligning SCB from the ophiuroid *Ophiactis balli* with the α-Proteobacteria, thus making them an unusual group of marine symbionts closely related to terrestrial symbionts such as *Rhizobium* and the agrobacteria.

2.6.3 *Evolution of the subcuticular bacteria-echinoderm symbiosis*

The evolutionary origin of SCB is of considerable interest. McKenzie et al. (1998) provided data suggesting it is unlikely the symbiosis formed by chance. They also concluded that 1) the loss of SCB is a derived character possibly with profound ultimate consequences for SCB-lacking clades, 2) primitive taxa are more commonly associated with symbionts than "advanced" species, and 3) the absence of SCB in some taxa may represent the ancestral proximate cause that prevented certain lineages from acquiring symbionts. As such, the loss of SCB in lineages such as the Asteriidae may have ultimately resulted in the evolution of more specialized trophic strategies (McKenzie et al. 1998). Unfortunately, few data on the trophic interactions of the symbiosis are available to support this. It may be that SCB in some species are relics of an earlier reliance on a more generalized trophic strategy and that the costs of maintaining the association are too negligible to be selected against. Interestingly, bacterial films are believed responsible for induction of metamorphosis in the sand dollar *Arachnoides placenta* (L.) (Chen & Run 1989), and represent a primary dietary contribution for juvenile irregular echinoids (De Ridder & Lawrence 1982). Alternatively, hosts may have had selection pressure to exclude them. Such would be the case, for example, if specific lineages of bacteria were evolving pathologic traits. Genetic sequencing would help characterize and compare the phylogenies of both partners.

2.7 Conclusion

Although many questions remain concerning the nature of the association, SCB appear to be a unique group of marine microbial symbionts worthy of further study. Phylogenetic similarity between SCB and nitrogen-fixing terrestrial symbionts, the α-Proteobacteria, pose a challenge in explaining the evolutionary origin of the partnership. This, along with their proposed role in protein synthesis, suggest that SCB somehow play a role in nitrogen cycling at the host organism-environment interface. Variability in SCB abundance over time and space and the ability to cure hosts of their symbionts in the lab indicate that the association is dynamic and facultative, and makes SCB an innovative model for studying the interactions and specificity between symbiotic partners. Additionally, the presence of large numbers of bacteria within the integument of some species suggests that the influence of SCB may be substantial in these taxa or under certain conditions. Further investigations using molecular analysis are required to fully understand the relationship between echinoderms and these unusual symbionts.

3 GUT BACTERIA

3.1 Bacterial microflora of the gut: general considerations

The digestive system of many, if not most, animals harbour a particular microflora which appear early during the animals life (i.e., soon after birth or early in the postmetamorphic stages) (Savage 1986, Smith & Douglas 1987). These microfloras are complex communities of various prokaryotic and eukaryotic (Protozoa, Fungi) microorganisms. They encompass transient and permanent members. The latter are gut symbionts (Smith & Douglas 1987). Originally, all these microorganisms enter the digestive tube with food (solid or liquid) or ingested water. Transient microorganisms are those which are killed by the digestive medium (digestive enzymes, low pH, anaerobic or microaerobic conditions) or those which survive and are expelled with the faeces. In contrast, permanent microorganisms proliferate and remain in the gut. This means that they resist the passage of food without being removed with the faeces and that they find good homeostatic growing condition within the gut. Gut microorganisms have various effects on their host. Some are true pathogens (and will not be considered here). Others participate to various degrees to the well-being of their host through their metabolic abilities. Some are essential to digest the food of their phytophagous hosts (Smith & Douglas 1987). Indeed, as bacteria are ecologically interesting "old" partners, they have most usually allowed eucaryotes to widen the spectrum of their metabolic activities, giving them access to new or narrow ecological niches (a way also to challenge biological competition).

The literature on the natural microflora of marine invertebrates is extremely limited. Most studies concern edible molluscs or crustaceans. Echinoderms are next most studied while scant attention has been given to the other invertebrates taxa (Colwell & Liston 1962, Harris 1993). In echinoderms, bibliographical data on the presence of GB are mostly available for (regular) echinoids and holothuroids, with few reports for asteroids and ophiurioids and none for crinoids (Table 4). Reasons for this could be the susceptibility of these two groups to be more dependant on gut bacterial symbioses for food digestion than the other groups (the regular echinoids and holothuroids are respectively herbivorous and detritivorous, i.e., trophic categories which develop symbioses with GB), and their economical interest (regular echinoids and holothuroids are commercially exploited). Historically, the presence of bacteria was first reported in the gut of echinoids by Weese (1926) but investigations on the potential role of the GB started much later with the studies of Lasker and Giese (1954) and of Huang and Giese (1958) who were the first to demonstrate that GB of *Strongylocentrotus purpuratus* degrade algae and some of the algal constituents. The taxonomical/morpho-physiological characterizations of the GB came next and were done during the past three decades starting with Unkles (1977) who isolated *Vibrio*, *Pseudomonas*, *Aeromonas* and *Flavobacterium* from the gut microflora of *Echinus esculentus*. The studies concerned with the GB of holothuroids, asteroids and ophiuroids were done during the past two decades. Most of them included taxonomical/morpho-physiological characterizations of the bacteria. However, although the presence of bacteria has been reported in the gut of various echinoderms, the significance of the associations usually is incompletely understood or even totally unknown. Indeed, there is a lack of data on "who" are the true symbionts, i.e., resident vs transient bacteria and more particularly on "what" are their respective metabolic abilities, their niches within the digestive tube, their abundance (enumeration and turnover rates), their interactions with the other members of the consortium forming the microflora, and with their host. Harris (1993) pointed out in his review of the literature on the gut microbiota of aquatic

131

Table 4. Characters of the gut bacteria in echinoderms.

	Morphotypes/Taxa	Physiological character	Sampling	Sources
ECHINOIDEA				
Diadematidae				
Diadema antillarum	*Vibrio* spp.	N_2-fixing (*), nitrogenase activity	gut homogenate	Guérinot & Patriquin 1981b
Arbaciidae				
Arbacia punctulata	undescribed	digest algae	intestinal contents	Prim & Lawrence 1975
Toxopneustidae				
Lytechinus variegatus	Gram- rods and cocci (7 isolates)	digest algae and plants polysacharides but not cellulose and chitin	intestinal contents	Prim 1973 Prim & Lawrence 1975
Tripneustes ventricosus	Gram - rods, polar flagellum *Vibrio* spp	N_2-fixing (*), nitrogenase activity, facultative anaerobic	gut homogenate	Guérinot & Patriquin 1981 a, b
Echinometridae				
Echinometra lucunter	Gram - *Vibrio* spp.	N_2-fixing (*), nitrogenase activity	gastro-intestinal contents	Guérinot & Patriquin 1981b
Echinidae				
Echinus esculentus	*Vibrio, Pseudomonas, Aeromonas, Flavobacterium*	heterotrophic, mesophilic, aerobic	gut contents	Unkles 1977
Loxechinus albus	*Pseudomonas* (9 strains)	digest agar, manitol, facultative anaerobic	gut contents	Garcia-Tello & Baya 1973

132

Table 4. Continued.

	Morphotypes/Taxa	Physiological character	Sampling	Sources
	Gram - (Pseudomonas-Achromobacter + related bacteria)	sensitivity to Aureomycin and Tetracyclin	gut contents	Garcia-Tello et al. 1978
Parechinus angulosus	filamentous-like bacteria or Fungi?	undescribed	hindgut mucosa	Sweijd et al. 1989
Strongylocentrotidae *Strongylocentrotus droebachiensis*	undescribed	N_2-fixing (*) nitrogenase activity	gut homogenate	Guérinot et al. 1977
	Gram - rods, *Vibrio* spp.	N_2-fixing (*), nitrogenase activity, facultative anaerobic	gut homogenate	Guérinot & Patriquin 1981 a, b
Strongylocentrotus intermedius	undescribed	N_2-fixing (*)	gut contents	Odintsov 1981
	Gram - rods, *Vibrio* spp (96.6% of the strains) *Alteromonas*	alginolytic, fermentative, facultative anaerobic	gut homogenate	Sawabe et al. 1995
Strongylocentrotus nudus	undescribed	N_2-fixing (*)	gut contents	Odintsov 1981
Strongylocentrotus purpuratus	undescribed	digest agar, Iridophycin, and algae (*Iridacea flaccidum*)	gut contents	Lasker & Giese 1954

133

Table 4. Continued.

Morphotypes/Taxa	Physiological character	Sampling	Sources
undescribed	digest laminarin, fucoidin, and entire algae	gut contents	Huang & Giese 1958
undescribed	alginase activity	gut contents	Eppley & Lasker 1959
undescribed	digest algae	intestinal contents (4th and 5th festoons)	Farmanfarmaian & Phillips 1962
Clypeasteroida/Dendrasteridae *Dendraster excentricus*			
undescribed	Iron-reducing, and sulphite-reducing	intestinal caecum (in juveniles)	Chia 1985
Clypeasteroida/Mellitidae *Mellita quinquiesperforata*			
undescribed	unable to digest various algae and seagrasses	gut posterior end contents	Prim & Lawrence 1975
Encope aberrans			
undescribed	unable to digest various algae and seagrasses	gut posterior end contents	Prim & Lawrence 1975
Holasteroida/Pourtalesiidae *Pourtalesia sp*			
undescribed	undescribed	gut contents	Ralijaona & Bianchi 1982
Spatangoida/Asterostomatidae *Heterobrissus niasicus*			
Thiothrix-like bacteria forming layered nodules	sulfur-oxidizing ?	intestinal caecum	De Ridder 1994

Table 4. Continued.

Morphotypes/Taxa	Physiological character	Sampling	Sources
Spatangoida/Schizasteridae			
Schizaster canaliferus			
Thiothrix-like bacteria forming layered nodules	sulfur-oxidizing ?	intestinal caecum	De Ridder 1994
Schizaster lacunosus			
Thiothrix-like bacteria forming layered nodules	sulfur-oxidizing ?	intestinal caecum	De Ridder 1994
Spatangoida/Loveniidae			
Echinocardium cordatum			
Rods, cocci, and filaments forming layered nodules	undescribed	intestinal caecum	De Ridder et al. 1985
Thiothrix spp. (in nodules outer layers)	sulfur-oxidizing	intestinal caecum	Temara et al. 1993, Brigmon & De Ridder 1998
cocci and filaments (+ Fungi, + Protozoa)	presence of fermentative microorganisms	all gut segments and intestinal caecum, filaments lack in stomachal caecum	Thorsen 1999
Echinocardium fenauxi			
Thiothrix-like bacteria forming layered nodules	sulfur-oxidizing ?	intestinal caecum	De Ridder 1994
Echinocardium mediterraneum			
bacterial nodules	sulfur-oxidizing ?	intestinal caecum	Bromley et al. 1995
Echinocardium mortenseni			
Thiothrix-like bacteria forming layered nodules	sulfur-oxidizing ?	intestinal caecum	De Ridder 1994

Table 4. Continued.

Morphotypes/Taxa	Physiological character	Sampling	Sources
Echinocardium pennatifidum			
Thiothrix-like bacteria forming layered nodules	sulfur-oxidizing ?	intestinal caecum	De Ridder 1994
ASTEROIDEA			
Astropectinidae			
Astropecten polyacanthus			
22 strains including: *Vibrio alginolyticus* , *V. damsela*, *Vibrio spp*.,and *Staphylococcus spp.*	production of Tetrodontoxin	intestinal contents	Narita et al. 1987
Solasteridae			
Solaster sp.			
28 strains including:	heterotrophic aerobic	gut contents	Bensoussan et al. 1984 (see also Bensoussan et al. 1979)
Gram- rods Vibrio-like (predominant)	glucose fermentative		
Gram+ rods: Brevibacterium ?	glucose non fermentative		
Gram+ cocci: Micrococcus ?	metabolize benzoate and		
	β hydroxybutyrate		

136

Table 4. Continued.

Morphotypes/Taxa	Physiological character	Sampling	Sources
OPHIUROIDEA			
Amphiuridae			
Ophionema sp.			
Vibrio alginolyticus (14 strains), *V. fluvialis* (2 strains)	no luminescent	gut contents	Dilmore & Hood 1986
HOLOTHUROIDEA			
Aspidochirotida/Holothuriidae			
Holothuria atra			
Proteobacteria (β & γ)	aerobic	gut contents + epithelium (SCB?)	Ward-Rainey et al. 1996
Cytophaga - Flavobacterium-Bacteroides lineage	id.	id.	id.
Actinomycetes	id.	id.	id.
Vibrio spp.	id.	id.	id.
Aspidochirotida/Stichopidae			
Stichopus japonicus			
undescribed	N$_2$-fixing, (*) nitrogenase activity	gut contents	Odintsov 1981
Aspidochirotida/Gephyroturidae			
Pseudostichopus villosus			
Gram- strains predominant	various exoenzymes tested, lipolytic and chitinolytic activities reported	gut contents	Ralijaona & Bianchi 1982

137

Table 4. Continued.

	Morphotypes/Taxa	Physiological character	Sampling	Sources
Pseudostichopus villosus	22 strains including:	heterotrophic aerobic	gut contents	Bensoussan et al. 1984 (see also Bensoussan et al. 1979
	Gram - rods (predominant)	glucose non-fermentative	id.	
	Gram+ cocci: *Micrococcus ?*	metabolize benzoate and β hydroxybutyrate	id.	
	Gram+ cocci: *Staphylococcus ?*	undescribed	id.	
Pseudostichopus sp.	Gram - rods	barophilic	hindgut epithelium (SCB)	Deming & Colwell 1982
Elasipodida/Deimatidae *Deima sp.*	Gram - rods	barophilic	hindgut epithelium (SCB)	Deming & Colwell 1982
Deima validum	Gram - predominant	various exoenzymes tested, lipolytic activities reported	gut contents	Ralijaona & Bianchi 1982
Elasipodida/Psychropodidae *Benthodytes sp.*	*Vibrio marinus* (19 strains) and *V. metschnikovii*	undescribed	gut contents	Dilmore & Hood 1986
Psychropotes sp.	Gram - (6 strains)	barophilic	hindgut epithelium (SCB)	Deming et al. 1981
	Gram- (14 strains)	barotolerant	id.	id.

138

Table 4. Continued.

	Morphotypes/Taxa	Physiological character	Sampling	Sources
Psychropotes sp.	Gram- rods	barophilic	hindgut epithelium (SCB)	Deming & Colwell 1982
	undescribed	metabolize acetate, glutamate	hindgut contents and hindgut epithelium	Tabor et al. 1982
	undescribed	barophilic	hindgut epithelium (SCB)	id.
	31 strains, including 75% Gram+:	heterotrophic, mesophilic	hindgut contents	Doria & Bianchi 1985
	Gram+ cocci : *Micrococcus, Staphylococcus*			
	Gram+ rods: *Corynebacterium, Arthrobacter, Brevibacterium*			
	Gram+ rods (sporulated): *Bacillus*			
Apodida/Synaptidae *Leptosynapta galliennei*	3 morphotypes (coccoid, spindle- and rod-shaped)	undescribed	cuticle of pharynx and esophagus (SCB)	Féral 1980

invertebrates that such a situation seemingly results from the lack of process-oriented studies.

3.2 *Characterization of gut bacteria*

Table 4 lists the reported characters (morphology, taxonomy, physiology) of GB in echinoderms and Table 5, the counts of GB. However, as several methods and units of density have been used, comparisons are difficult. Direct counts (e.g., epifluorescence microscopy) usually provides higher values than plate-viable counts because there is no selective effect of medium, but it does not consider the masking effects of the gut tissue and contents (Harris 1993). All the reported bacteria are extracellular and live in the gut contents (or in subcuticular spaces, see SCB). Their characterization is based on samples that correspond either to suspensions of mixed bacteria directly prepared from the gut contents (or inner wall) or to isolated strains in pure culture. Bacteria isolated from whole animal homogenates (see e.g. Tysset et al. 1961) are not considered.

3.2.1 *Taxonomy*
Several genera of Gram-positive and Gram-negative bacteria have been identified. Most of the Gram-negative genera belong to the Proteobacteria (*Vibrio, Pseudomonas, Aeromonas, Achromobacter, Thiothrix*) and to the bacteria of the *Cytophaga-Flavobacterium-Bacteroides* group. Gram-positive bacteria are *Bacillus, Micrococcus, Staphylococcus, Brevibacterium, Corynebacterium, Arthrobacter. Vibrio* is the most common genus and occurs in echinoids (Unkles 1977, Sawabe et al. 1995, Guérinot & Patriquin 1981a,b), holothuroids (Dilmore & Hood 1986, Ward-Rainey *et al.* 1996), asteroids (Narita et al. 1987) and ophiuroids (Dilmore & Hood 1986). However, most of these identifications are based on the isolation of bacteria in pure culture, i.e., on routine phenotypic methods which are very selective: most bacteria are not recovered by the conventional culture techniques because they do not grow on synthetic media (see e.g., Rozak & Colwell 1987, Brock et al. 1994, Santegoeds et al. 1996). Harris (1993) mentions that often < 10% of the bacteria can be isolated. One could thus expect these taxa to represent only a minor fraction of the existing diversity. This is supported by the study of Ward-Rainey and collaborators (1996) whose phylogenetic analysis and cultivation methods demonstrate that the bacterial flora of *Holothuria atra* (and its immediate surrounding) is widely diversified.

3.2.2 *Host-specificity*
Most GB presumably originate from food or sea-water although one can not reject a direct cyclic transmission (i.e. transmission via conspecific individu-

Table 5. Counts of gut bacteria.

		Authors
ECHINOIDEA		
Echinidae		
Echinus esculentus	$2 \ 10^7$ CFU cm^{-1} section (gut content)	Unkles 1977
Strongylocentrotidae		
Strongylocentrotus droebachiensis	ca. 10^5 (in fall) and ca. 10^{11} (in summer) CFU g^{-1} (gut homogenate) (wet wt) N2-fixing bacteria: ca. 10^5 CFU g^{-1} (gut homogenate) (wet wt)	Guérinot & Patriquin 1981b
Strongylocentrotus intermedius	$3.8 \ 10^5$ - $1.1 \ 10^8$ CFU g^{-1} (gut homogenate) (wet wt)	Sawabe et al. 1995
Holasteroida/Pourtalesiidae		
Pourtalesia sp	$4.6 \ 10^3$ CFU ml^{-1} (gut contents) (wet wt)	Ralijaona & Bianchi 1982
Spatangoida/Loveniidae		
Echinocardium cordatum	10^8 - 10^9 non-filamentous bacteria cm^{-3} (gut content; IC excepted)(*) intestinal caecum (IC): up to $5 \ 10^{10}$ non-filamentous bacteria (*)	Thorsen 1999
	10^4 - 10^7 filamentous bacteria (gut contents; SC and IC excepted)(*) stomachal caecum (SC): 0 filamentous bacteria (*) intestinal caecum (IC): up to 10^9 filamentous bacteria (*)	
ASTEROIDEA		
Porcellanasteridae		
Hyphalaster inermis	up to $9.1 \ 10^6$ CFU ml^{-1} (gut contents)	Ralijaona & Bianchi 1982
Solasteridae		
Solaster sp.	$1.5 \ 10^4$ CFU ml^{-1} (gut contents)	Bianchi et al. 1979
OPHIUROIDEA		
unidentified abyssal species	$1 \ 10^6$ CFU ml^{-1} (gut contents) (wet wt) (heterotrophic, mesophilic bacteria)	Bensoussan et al. 1979
Amphiuridae		
Ophionema sp.	$2.3 \ 10^5$ Vibrio sp g-1 (gut contents) (wet wt)	Dilmore & Hood 1986

Table 5. Continued.

		Authors

HOLOTHUROIDEA
Aspidochirotida/
Holothuriidae

Holothuria atra	bacterial biomass: 0.27 - 0.54 mg C g^{-1} (foregut contents) (dry wt)	Moriarty 1982
	6.2 10^4 CFU g^{-1} (foregut contents)	Ward-Rainey et al.
	1.8 10^6 CFU g^{-1} (hindgut contents)	1996
	2.2 10^4 CFU g^{-1} (foregut wall)	
	5.9 10^4 CFU cm^{-2} (hindgut wall)	
Holothuria tubulosa	bacterial biomass: 50 - 190 μg C g^{-1} (foregut contents) (dry wt)	Amon & Herndl 1991
	3.3 10^9 bacteria g^{-1} (in foregut mucus secretion) (dry wt) (*)	

Aspidochirotida/Stichopidae

Stichopus chloronotus	bacterial biomass: 0.24 - 0.29 mg C g^{-1} (foregut contents) (dry wt)	Moriarty 1982

Aspidochirotida/Gephyroturidae

Pseudostichopus sp	9.38 (± 1.89)-18.5 (± 6.57) 10^8 bacteria g^{-1} (in foregut contents) (dry wt) (*)	Deming & Colwell 1982
Pseudostichopus villosus	up to 6.6 10^3 CFU ml^{-1} (gut contents)	Bensoussan et al. 1979 Bianchi et al. 1979
	up to 6.3 10^8 CFU ml^{-1} (gut contents)	Ralijaona & Bianchi 1982

Elasipodida/Deimatidae

Deima sp	3.57 (± 0.85)-16.7 (± 6.20) 10^8 bacteria g^{-1} (in foregut contents) (dry wt) (*)	Deming & Colwell 1982
Deima validum	up to 7 10^8 CFU ml^{-1} (gut contents) (dry wt)	Ralijaona & Bianchi 1982

Elasipodida/Psychropodidae

Benthodytes sp	7.0 10^5 *Vibrio* sp g^{-1} (gut contents) (wet wt)	Dilmore & Hood 1986
Psychropotes sp	5.7 10^2 CFU ml-1 (gut contents)	Doria & Bianchi 1985

142

Table 5. Continued.

	Authors
ca. 10^5 CFU ml^{-1} (intestinal scrapings suspension) (1 atm) *ca.* 10^6 CFU ml^{-1} (intestinal scrapings suspension) (430 atm)	Deming et al. 1981
2.1 10^9 bacteria g^{-1} (gut contents) (wet wt) (430 atm.) (*) 1.2 10^9 bacteria g^{-1} (gut scrapings) (wet wt) (430 atm.) (*) 200 CFU g^{-1} (gut contents) (wet wt) (430 atm.) 2 10^5 CFU g^{-1} (gut scrapings) (wet wt) (430 atm.)	Tabor et al. 1982

CFU: colony forming units (usually marine agar); (*)= epifluorescence microscopy.

als) or a transovarian transmission (i.e. via the eggs). Diverse and numerous bacteria are regularly ingested by the echinoderms. Once ingested the bacteria are submitted to the selective effect of the gut environment that decreases the bacterial diversity (see e.g., Bensoussan et al. 1984, Ward-Rainey et al. 1996) and produces new assemblages of bacterial taxa (see e.g., Unkles 1977, Ralijaona & Bianchi 1982, Bensoussan et al. 1984, Doria & Bianchi 1985, Sawabe et al. 1995, Plante & Shriver 1998). Consequently, the bacterial community living in the gut differs from that living on/in the food (marine plants, sediments) prior its ingestion. However this difference concerns the community composition and does not imply that the members of the community are symbionts or host-specific.

Species or strains belonging to a relatively limited number of bacterial genera are found in the gut of echinoderms. These genera are common in the marine environment (see e.g. Sawabe et al. 1995). Interestingly, most (e.g. *Pseudomonas*, *Achromobacter*, *Flavobacterium* and *Vibrio*) are also frequent in the gut of other marine invertebrates (Porifera, Cnidaria, Mollusca, Arthropoda, Crustacea and Platyhelminthes) (Colwell & Liston 1962, Dilmore & Hood 1986, Harris 1993), a situation that could indicate that they either better resist the gut conditions (see e.g., Plante & Shriver 1998) or are more frequently associated with the ingested material. Many examples indicate that similar bacteria occur both on the food, i.e., on marine plants or sediment, and in the gut content (Farmanfarmaian & Phillips 1962, Odintsov 1981, Bensoussan et al. 1984, Doria & Bianchi 1985, Ward-Rainey et al. 1996). Some taxa of GB could thus logically characterize a particular trophic category or a particular "behavioural" category of echinoderms (at least in one particular geographic site), but this is poorly documented. For instance, symbiotic sulfur-oxidizing *Thiothrix* bacteria occur in the intestinal caecum of several species of spatangoids. These spatangoids seem to have adopted the same digestive strategy, i.e., a slow digestive transfer relying, at least partly, on the sulfur oxidizing ability of *Thiothrix* to prevent critical accumulation of hydrogen sulfide within the gut (Temara et al. 1993, De Ridder 1994, Brigmon & De Ridder 1998). However, it is not known if the *Thiothrix* species/strain associated with spatangoid are host-specific or not.

Interestingly, endemic bacterial strains occur in the gut of several species of abyssal echinoderms, e.g., in the holothuroids *Psychropotes* sp (Doria & Bianchi 1985), *Pseudostichopus villosus* (Ralijaona & Bianchi 1982, Bensoussan et al. 1984) and *Deima validum* (Ralijaona & Bianchi 1982), and in the asteroid *Solaster* sp (Bensoussan et al. 1984). The development of endemic strains in the gut of abyssal echinoderms could be enhanced by the stable nutrient-rich conditions in the gut compared to the scarcity of nutrients in the abyssal environment (see e.g. Zobell & Morita 1957, Deming et al. 1981, Deming & Colwell 1981, Bensoussan et al. 1984). Moreover, in the deep-sea environment, the digestive tube of detritivorous holothuroids appears to function as a selective environment for barophilic SCB (Deming & Colwell 1981, Deming et al. 1981, Deming & Colwell 1982, Tabor et al. 1982).

3.3 *Roles of gut bacteria*

Three nutritive roles of the GB are commonly reported in echinoderms. First, GB may aid digestion, i.e., degrade some "refractory" food such as structural carbohydrates via hydrolysis and fermentation. Second, they may be used as food (protein source) by detritivorous echinoids and holothuroids. Third, they could contribute to nitrogen conservation by metabolizing the nitrogen wastes of their host.

3.3.1 *Digestive role*
The digestive role of the GB has been investigated extensively in herbivorous echinoids and in detritivorous holothuroids and echinoids but not in asteroids, crinoids and ophiuroids (Table 4). In asteroids, bacteria were found in the intestinal content of *Solaster* sp (Bensoussan et al. 1984) and of *Astropecten polyancanthus* (Narita et al. 1987). They were not found in the pyloric caeca of a variety of species (Araki & Giese 1970, Scheibling 1980). A particular role in the digestive processes of the host is rarely attributed to one particular taxon of bacteria because the variety of metabolic abilities of the isolated bacteria usually overlaps and also because the bacterial diversity

144

exceeds the number of identified taxa. As a rule however, the majority of isolated GB are facultative anaerobes (yet obligate anaerobes or aerobes have been reported too) capable of hydrolysing refractory compounds of plants or detritus.

A. Herbivorous echinoids. Algae and sea-grasses form the primary diet of regular echinoids. Regular echinoids are opportunistic browsers that feed mainly on marine plants and to a less extent on encrusting organisms. Digestion of marine plants and assimilation of their constituents by the echinoids have been widely studied (see e.g. Farmanfarmaian & Phillips 1962, Lasker & Boolootian 1960, Boolootian & Lasker 1964, De Ridder & Lawrence 1982, Frantzis & Grémare 1992, Yano et al. 1993, Greenway 1995). The question raised here is the contribution of the GB to the digestion of plant material in echinoids. Most herbivores (from termites to mammals) lack digestive enzymes to digest plant structural compounds such as cellulose and lignin. They depend on gut microbial metabolism to provide metabolizable end products as, e.g. volatile fatty acids (VFA) (Hungate 1975, Smith & Douglas 1987, Rimmer & Wiebe 1987). In echinoids, evidence indicates that GB have the capacity to digest plant material. But in contrast to the herbivorous insects and mammals, the importance of this bacterial contribution remains unclear. The potential participation of bacteria to the digestion in echinoids was raised when Lasker and Giese (1954) showed that the bacteria of the gut *of Strongylocentrotus purpuratus* digest agar and a red alga (*Iridaea flaccidum*) and when Huang and Giese (1958) pointed out that extracts of the gut wall of the echinoid were unable to degrade algae and some constitutive polysaccharides (laminaran and fucoidan) . The "digestive" capacities of the GB have been more examined in vitro by testing the ability of isolated bacterial strains or suspensions of mixed bacteria from the gut content to digest plants fragments or carbohydrates. The studies indicate:

(1) GB can digest various species of algae and sea-grasses on which the echinoids feed (Lasker & Giese 1954, Prim 1973, Prim & Lawrence 1975, Fong & Mann 1980) (Table 4). Prim and Lawrence (1975) pointed out that some species of intact marine plants are less well hydrolysed than others: e.g., *Caulerpa prolifera* is not digested, *Diplanthera wrightii* and *Thalassia testudinum* poorly digested by the GB of *Lytechinus variegatus*.

(2) GB can hydrolyze several polysaccharides that are the constituents of marine plants: e.g., iridophycin (Lasker & Giese 1954), agar (Garcia-Tello & Baya 1973), alginic acid, agar, starch and carrageenan (Prim 1973), laminaran, carrageenan, starch and agar (Prim & Lawrence 1975), polymannuronate and polyguluronate (Sawabe et al. 1995). These polysaccharides

are either cell wall constituents or storage material (e.g., Dring 1982). Iridophycin is a cell wall constituent of the red alga *Iridaea*. Agar is cell wall structural constituent of red algae and consists mainly of agarose and agaropectin. Alginic acid is a cell wall constituent of brown algae and consists of poly-mannuronate and poly-guluronate. Carrageenan is a cell wall sulphated polysaccharide mucilage found in some red algae and composed of alternating units of modified galactose. Starch is the storage product of green algae and red algae and consists in glucose units (α (1,4) linkages). Laminaran is the storage material of brown algae and consists of glucose units (β (1,3) linkages).

(3) GB can use as a carbon source the monosaccharides constituting the marine plants polysaccharides, e.g. the pentoses: xylose and rhamnose and the hexoses: glucose and galactose (Prim & Lawrence 1975), mannuronate and guluronate (Sawabe et al. 1995), and mannitol (Garcia-Tello & Baya 1973).

GB are usually opportunistic in the utilization of marine plant polysaccharides. In general, the mixed bacterial suspension can use a wide spectrum of carbohydrates (i.e., several polysaccharides can be hydrolysed and several monosaccharides can be used as carbon source) while isolated strains are more specific (Prim & Lawrence 1975, Sawabe et al. 1995). For instance, Prim and Lawrence (1975) observed that each of the strains isolated from *Lytechinus variegatus* had the ability to utilize at least three different polysaccharides while the mixed suspension of bacteria utilized all the polysaccharides provided (laminarin, caragheenin/carageenan, starch and agar).

The discrimination between endogenous (i.e. secreted by the echinoid gut) and exogenous (i.e. bacterial exoenzymes) carbohydrases is a necessary step to clarify the actual contribution of the GB to digestion. However, this aspect is poorly documented as, in most studies, the carbohydrase activities are of unattributed source (Lawrence 1982, Suzuki et al. 1984, Yano et al. 1993). Carbohydrase activities can be deduced from *in vitro* experiments where isolated bacteria are tested for the utilization of particular polysaccharides, suggesting that the GB can be a source of carbohydrases within the gut. There are also indications that both endogenous and exogenous carbohydrases occur in situ in the gut of echinoids. Eppley and Lasker (1959) suggested that the two alginases found in the gut of *Strongylocentrotus purpuratus* have a different origin, the gut wall and the gut contents, and differ in pH optima. Claereboudt and Jangoux (1985) reported endogenous activities for amylase and cellulase in the stomach of *Paracentrotus lividus* and exogenous activities in the intestine. Usually, the available data concern thus the capacity of the GB to utilize marine plants and their main constitu-

146

ents but they do not provide direct evidence of an effective contribution of the bacteria to the digestion of the echinoid. Indeed, although it appears that both the echinoid and its GB can secrete carbohydrases, qualitative and quantitative informations on the sources of enzymes are lacking.

B. Detritivorous holothuroids and echinoids. Although they usually ingest sediment in bulk, the non-suspensivorous holothuroids (e.g. Massin 1982, Moriarty 1982, Moriarty et al. 1985) and the irregular echinoids (e.g. Buchanan 1966, De Ridder & Lawrence 1982, De Ridder et al. 1985b, Hammond 1983, Van Duyl et al. 1992) primarily feed on the sediment associated detritus. Surface sediments, i.e., sediments regularly enriched with organic matter, are more particularly collected by the detritivorous echinoderms.

Because their food is systematically associated with bacteria, detritivores are highly susceptible to harbour bacteria within their gut (e.g., Plante et al. 1990, Harris 1993). Indeed, detritus forms a dynamic microecosystem where particulate organic matter is intimately associated with a variety of microorganisms. These microorganisms are members of complex microbial communities which qualitatively fluctuate with the state of decomposition of the organic matter and with the season (i.e., the nature of the available detritus) (e.g. Brock et al. 1994). Detritus typically supplies the detritivore gut with numerous microorganisms and undergoes lysis prior its ingestion. This means that detritus-associated bacteria may contribute at least twice to the nutrition of the detritivore: (1) as food (part of the ingested microorganisms are digested) and (2) by digesting the ingested detrital organic matter (Khripounoff & Sibuet 1980, Moriarty 1982, Ralijaona & Bianchi 1982, Sibuet 1984, Alberic et al. 1987, Plante & Shriver 1998, Thorsen 1998, 1999). Digestion of bacteria by the detritivore is observed through the rapid decrease of the number of bacteria along the midgut regions (Khripounoff & Sibuet 1980, Moriarty 1982, Ralijaona & Bianchi 1982, Sibuet 1984, Alberic et al. 1987, Plante & Shriver 1998, Thorsen 1999). However, although bacteria are efficiently digested by the detritivores, they constitute only a minor food source (e.g. Plante et al. 1989, 1990). For instance, 2% of the ingested proteins (Yingst 1976, Sibuet et al. 1982) and 3% of the ingested C (Deming & Colwell 1982) are of bacterial origin in abyssal holothuroids. In the littoral species *Holothuria atra*, 10% of the ingested organic C and N and 10% of the assimilated C are of bacterial origin (Moriarty 1982). In contrast, the nutritive contribution of the surviving bacteria could be important: these bacteria participate to the lysis of the ingested detritus and produce metabolites within the echinoderm gut. This digestive contribution (until now never quantified) is supported by the lack of proteases (or the occurrence of weak protease activities) in holothuroids (Sibuet et al. 1982, Massin 1984, Féral 1989) and in irregular echinoids (Kozlovskaya & Vaskovsky 1970, De Ridder 1986), the lack of structural carbohydrases in holothuroids (Féral

1989, Hylleberg Kristensen 1972) and in irregular echinoids (Jeuniaux 1963, Sova et al. 1970, Favarov & Vaskovsky 1971, De Ridder 1986). This is also supported by the the ability of GB to hydrolyse some refractile macromolecules (Ralijaona & Bianchi 1982, Thorsen 1998) or to use a wide variety of organic compounds (Ralijaona & Bianchi 1982, Bensoussan et al. 1979, 1984).

In holothuroids, the foregut (pharynx, esophagus) and to a lesser extent, the hindgut (posterior intestine, rectum, cloaca) contain more bacteria (and organic matter) than the other gut regions and than the surrounding sediment. Foregut sediment (i.e., recently ingested sediment) has a particularly high number of bacteria (e.g., in abyssal holothuroids, 10^2-10^3 higher than in the surrounding, Ralijaona & Bianchi 1982). These significantly higher values are usually interpreted as the result of selective feeding on the organically rich fraction of the sediment (Khripounoff & Sibuet 1980, Moriarty 1982, Sibuet 1984, Alberic et al. 1987). In some abyssal holothuroids this is also interpreted as the result of an intense bacterial proliferation (Bianchi et al. 1979, Ralijaona & Bianchi 1982). However, with the exception of the SCB, GB do not have enough time to proliferate in the foregut as sediment is passing rapidly through this region (Féral 1980, Moriarty 1982). In contrast, bacterial proliferation most obviously occurs in the hindgut where sediment accumulates before to be defecated (Khripounoff & Sibuet 1980, Sibuet et al. 1982, Moriarty 1982, Ralijaona & Bianchi 1982, Deming & Colwell 1982, Sibuet 1984, Alberic et al. 1987). This bacterial proliferation involves both bacteria of the gut contents and SCB (Féral 1980, Deming & Colwell 1981, Deming et al. 1981, Deming & Colwell 1982). According to Deming and Colwell (1981, 1982) and Deming et al. (1981), the hindgut SCB of the abyssal *Psychropotes* sp differs from bacteria of the gut contents by being barophilic rather than barotolerant. This is also supported by the observations of Tabor et al. (1982) who noted the rapid barophilic substrate utilization of these bacteria in *Psychropotes* sp. In abyssal holothuroids, in addition to the intense bacterial proliferation of the foregut and hindgut, there is also a lower though continuous bacterial activity all along the gut length; this has been demonstrated through bacterial counts (Ralijaona & Bianchi 1982) and through measurement of bacterial amino-acids concentrations (glutamic acid and diaminopimelic acid) (Alberic et al. 1987). The bacterial proliferation within the gut of abyssal benthic organisms is usually considered as an adaptation to the oligotrophic surrounding conditions. It is also considered as a significant contribution to nutrient regeneration and account for the recycling of organic material in the deep-sea bottom (Khripounoff & Sibuet 1980, Deming & Colwell 1982, Tabor et al. 1982, Ralijaona & Bianchi 1982, Alberic et al. 1987).

Data on the bacteria occurring in the digestive tube of detritivorous echinoids are limited to spatangoids, clypeasteroids and holasteroids (Table 4). Detailed data on the role of GB are only available for the spatangoid

Echinocardium cordatum (De Ridder et al. 1985a,b, Temara et al. 1991, Temara et al. 1993, De Ridder 1994, Brigmon & De Ridder 1998, Thorsen 1998, 1999). In *E. cordatum*, as in holothuroids, bacteria (and other microorganisms) are particularly abundant in the foregut (pharynx, esophagus, anterior stomachal region) and in the hindgut (rectum). In the foregut, this situation results from the selective feeding behaviour of the echinoid (Buchanan 1966, De Ridder et al. 1985b, Hammond 1983, Van Duyl et al. 1992) but microbial growth is also reported in the anterior stomach (Thorsen 1998). In the hindgut, the increase of the bacterial number is due to bacterial proliferation which is related to the nearby intestinal caecum (De Ridder et al. 1985b, Thorsen 1998, 1999). All spatangoids possess an anterior caecum (the stomachal caecum) and some have a posterior caecum (the intestinal caecum) (e.g. De Ridder & Jangoux 1982, De Ridder 1994). *Echinocardium cordatum* has both caeca. Thorsen (1998) observed that these caeca are the sites of an intense microbial activity as indicated by their high oxygen fluxes and consumption rates, low C/N ratio and high amount of particulate C and N. These results were confirmed by microbial counts (Thorsen 1999). Each caecum contains a distinct microbial community characterized by different metabolic abilities. Fungi and Protozoa proliferate in the stomachal caecum and are fermenters that seemingly allow the echinoid to metabolize refractile carbohydrates (Thorsen 1998, 1999). High production rates of short-chained fatty acids (acetate, proprionate) indicate that fermentation is particularly intense in this caecum. Similar fermenters also occur in the intestinal caecum but their activity is less. Indeed, the intestinal caecum is the site of proliferation of a variety of bacteria (filaments, rods and cocci), Fungi and Protozoa (De Ridder et al. 1985a, Thorsen 1999), but sulfur-oxidizing chemolithotrophic filamentous bacteria of the genus *Thiothrix* predominate (Temara et al. 1993, De Ridder 1994, Brigmon & De Ridder 1998). These filamentous bacteria build nodules by forming mats around large detrital particles (plant fragment, shell, organic aggregates) (De Ridder et al. 1985a, De Ridder 1994). Microscopical observations indicate that each nodule is a complex microbial community (De Ridder et al. 1985a) where mutualistic interactions most presumably occur between the microorganisms. These multilayered nodules reach ca. 5mm in diameter and may be numerous. They seemingly are kept in the caecum for a long period of time (weeks) before to be expelled with the feces (De Ridder *et al.*1985a, Temara & De Ridder 1992). The sulfo-oxidizing *Thiothrix* spp. of the intestinal caecum utilize the sulfide that regularly forms both in the nodule and in the intestinal content as a result of degradation of organic matter (Temara et al. 1993, Brigmon & De Ridder 1998, Thorsen 1998). These processes leading to anoxic conditions and the production of sulfide are common in the (marine) sediment (e.g. Fenchel & Finley 1995). Moreover, sulfide is particularly toxic and its removal throught the activities of symbiotic sulfur–oxidizing bacteria can be considered as a host defense mechanism (Vismann 1991). Such defense mechanisms are reported in symbiotic invertebrates living in hydrothermal vents sites (e.g., Prieur et al. 1990) or in anoxic sediments (e.g. Dubilier et al. 1995, Gillan & De Ridder 1997). In *Echinocardium cordatum*, removal of sulfide from the intestinal content and from the nodules by the bacteria could allow the sediment to be retained within the intestine for a longer time and increase hydrolysis of the organic matter (De Ridder 1994). Several species of spatangoids have developed a symbiosis with *Thiothrix*-like bacteria (Table 3). This symbiosis appears unique to the order Spatangoida in the class Echinoidea and occurs as synapomorphies in three phylogenetically dispersed families (Loveniidae, Schizasteridae and Asterostomatidae). Its occurrence in phylogenetically dispersed families could be an adaptative response to similar functional pressures, e.g., a particular digestive strategy (De Ridder 1994). According to Bromley et al. (1995), the symbiosis *Thiothrix*-spatangoids would allow the animal to burrow deeply in the sediment (below the RPD layer) and to feed on anoxic sediment, because the bacteria can supply their host with nutrients through chemosynthesis. The way the spatangoids acquire their symbionts (bacteria, Protozoa or Fungi) remains unknown. A cyclic indirect transmission is probable for *Thiothrix* bacteria (*Thiothrix* species commonly occur in the sediment at the oxic-anoxic boundary, e.g. Fenchel & Finley 1995). The intestinal caecum occurs only in the few spatangoid species harbouring symbiotic *Thiothrix* or *Thiothrix*-like bacteria. This situation suggests that the development of this organ is induced by the presence of the bacteria. Such an effect is known in the squid-*Vibrio* symbioses where significant morphological changes are induced by the bacteria during the development of the light organ in juvenile hosts (Montgomery & McFall-Ngai 1995, McFall-Ngai 1999)

A complex (multi-pouched) intestinal caecum occurs as a synapomorhy in post-metamorphic juveniles clypeasteroids of the suborder Scutellina (Mooi & Chen 1996). As it is filled up with iron oxide (magnetite) coated grains and progressively lost during growth, this caecum was suggested to be a weight belt that stabilize the juveniles in a shifting environment (Chia 1973, 1985, Chen & Chen 1994). Chia (1985) hypothesized that the heavy sand particles are selected at the entrance of the mouth by the buccal podia and in the buccal cavity during their transfer, and that the selection is based either on the density of the grains or on the associated microflora. Iron-reducing bacteria occur on the grains and seemingly induce a progressive dissolution of the grains owing to acid production by the bacteria (Chia 1985). Although this has not been demonstrated, this caecum appears to be a good site to culture symbiotic bacteria (and other microorganisms) which could participate to the nutrition of juvenile clypeasteroids.

3.3.2 *Conservation of nitrogen*

N_2-fixation by symbiotic GB occur in a variety of plant-eating animals from mammals to insects whose diet is usually poor in protein (Smith & Douglas 1987). These bacteria metabolize waste nitrogenous compounds produced by their host (e.g. urea, uric acid) or atmospheric nitrogen and contribute to the protein requirements of their host by synthetizing amino-acids and other nitrogenous compounds that can be absorbed by their host. When the bacteria occur and multiply in the foregut, the nitrogen assimilated by the bacteria also can be recovered by the host through digestion (e.g. rumen symbionts). The method used to determine whether bacteria are N_2-fixing or not is the measurement (through gas chromatography) of acetylene reduction (Stewart et al. 1967, Guérinot et al. 1977, Odintsov 1981). This method is based on the bacterial reduction of N_2 to ammonium and convertion of ammonium into an organic form. This reduction process is catalyzed by a complex nitrogenase which is not specific for N_2 and which also reduces other nitrogen compounds as acetylene (C_2H_2) (Brock et al. 1994). Bacterial nitrogenase activity (acetylene reduction) occurs in the digestive tube of six species belonging to four different families of regular echinoids (Table 4) (Guérinot et al. 1977, Guérinot & Patriquin 1981a,b, Odintsov 1981) and one species of holothuroid (Odintsov 1981). N_2-fixing GB presumably occur in most herbivorous echinoids (Guérinot & Patriquin 1981b).

N_2-fixation by GB supplies the echinoids with nitrogen. Microbially fixed 15_N is transferred to the tissue of *Strongylocentrotus droebachiensis* (Guérinot & Patriquin 1981b). However, the mechanism of transfer remains unknown. It could rely on the phagocytosis of bacteria or on the assimilation of metabolites released by the bacteria in the gut. The actual contribution of N_2 fixing bacteria to the nitrogen needs of the echinoids is unknown. Guérinot *et al.* (1977) have calculated that *S. droebachiensis* in kelp beds could receive about 8-15% of its daily nitrogen requirements from N_2 fixation. Guérinot and Patriquin (1981b) have also estimated that the nitrogen needs for growth are only 29% of the amount of nitrogen fixed yearly by an echinoid. They suggested that the yearly rate of N_2 fixing could satisfy the yearly nitrogen needs of the echinoids. In contrast, Odintsov (1981) suggested the contribution of daily nitrogen requirements from N_2 fixation calculated by Guérinot and Patriquin (1981b) are overestimated. This author reported a much lower contribution for *Strongylocentrotus nudus* and *Strongylocentrotus intermedius*. He also reported that the holothuroid *Stichopus japonicus* has a much lower (10 to 30 times lower) fixing rate than the echinoids. Odintsov (1981) concluded that nitrogen fixation in the gut is not significant for the nitrogen balance of echinoderms (at least during winter). The contribution of bacterially fixed nitrogen to the herbivorous echinoids (or other echinoderms) nitrogen balance remains controversal and no clear answer can be provided as only little informations are available on echinoid in nitrogen requirements for growth and maintenance.

Bacterial nitrogenase activity is influenced by the nitrogen contents of the marine plants eaten by the echinoids. The nitrogen contents of food and the nitrogenase activity of the GB are inversely related (Guérinot & Patriquin 1981b). These authors suggest that the bacterial nitrogenase activity is completely or partly inhibited by nitrogen metabolites released by the food. These links between the bacterial nitrogenase activity and the food nitrogen content would explain why seasonal fluctuations of N_2 fixation rates may occur in one species of echinoid that feeds on the same marine plants throughout the year. For instance, the C:N ratio of *Laminaria* spp fluctuates from 16:1 up to 50:1 during the year. The lower values being observed in summer and fall. This periodic decrease in dietary nitrogen is correlated with a periodic increase of N_2 fixing rate (Guérinot & Patriquin 1981b). The influence of the dietary nitrogen contents on the nitrogenase activity is also well illustrated by tropical echinoids. Guérinot and Patriquin (1981b) found that *Diadema antillarum* and *Echinometra lucunter* that feed mostly on corals and encrusting algae have lower N_2 fixing rates than *Tripneustes ventricosus* that feed on sea-grasses. This is related to the respective diets of the echinoids: *D. antillarum* and *E. lucunter* feed mostly on coral polyps which have higher N contents than macrophytes such as seagrasses and kelp eaten by *T. ventricosus*.

The N_2 fixing bacteria isolated both from temperate (*S. droebachiensis*) and tropical (*D. antillarum, E. lucunter, T. ventricosus*) echinoids were identified as facultatively anaerobic sucrose-positive *Vibrio* spp. These observations were the first reports of N_2 fixation in the Vibrionaceae (Guérinot & Patriquin 1981a,b). These authors expect N_2 fixation by gut *Vibrio* spp to occur in many other herbivorous invertebrates. *Vibrio* spp are predominant members of the bacterial microflora in the gut of other echinoids as *Echinus esculentus* (Unkles 1977) and of other herbivorous invertebrates (Colwell & Liston 1962, Dilmore & Hood 1986).

The N_2-fixing bacteria in the gut of echinoids and holothuroids appear to be transient rather than permanent (i.e., symbiotic). According to Odintsov (1981), the N_2-fixing bacteria in the gut of *S. purpuratus, S. nudus* and of *Stichopus japonicus* are transient. The N_2- fixation in the gut simply continues the activity of N-fixing bacteria occuring in the environment and that were ingested with the food. In contrast, Guérinot and Patriquin (1981b) consider that some of these bacteria could be permanent as unfed *S. droebachiensis* contain N_2-fixing bacteria in the gut. These bacteria are not numerous and are inactive because they lack a carbon source. Guérinot and Patriquin (1981b) also reported than antibiotic-treated echinoids become reinfected only after a long period (18 days) of exposure to seawater and kelp. They suggest that the relationship between N_2-fixing bacteria and echinoids is (1) not obligatory on either part and (2) of mutual benefit, with the echinoid providing the bacteria with a carbon source and a suitable habitat (near neutral pH, presence of N_2, anaerobic milieu) and the bacteria providing the echinoid with fixed nitrogen.

3.4 *Functional aspects of the gut and occurrence of bacteria*

Hungate (1975) predicted interactions of microorganisms-animals on the basis of the nature of the food consumed. One can expect cooperation between gut microflora and animal when high carbohydrate fibrous food is consumed, i.e., when the animal cannot efficiently digest its food (the microflora breaking down the ingested refractile material and providing the animal with assimilable metabolites), and competition when the food consists of proteins or easily digested carbohydrates, i.e., when the animal and its bacteria both efficiently digest the food or utilize the same metabolite. For instance, some plant constituents as mannitol are assimilated by herbivorous echinoids (Boolootian & Lasker 1964), but this monosaccharide is also used as a carbon source by the bacteria (Garcia-Tello & Baya 1973). Cooperation and competition often occur simultaneously, the balance between the two interactions determining whether or not the host benefits from the association (e.g., Plante et al. 1990). This last aspect is often difficult to analyze in echinoderms-GB associations because the nutritional interactions are usually not quantified. In this context, the modeling approach of Plante and co-workers (1990) on the interactions of bacteria–detritivores is particularly interesting as it provides keys to analyze a particular association.

To understand how echinoderms and GB may cooperate, it is necessary to consider the architecture and some functional aspects of the digestive tube, e.g., its mechanic. This means that, in addition to the gut anatomy, the "packaging" of the ingested food, the pH, pO_2 and Eh conditions along the digestive tube, the aspect of the gut contents during its digestive transit, and the gut transit time have to be examined. Such a functional approach provides information on the most appropriate sites in the digestive tube for the development of symbiotic, or at least cooperating, bacteria. This approach has been tentatively initiated by Hylleberg Kristensen (1972), considered by Prim and Lawrence (1975) and analyzed in the light of the optimal digestion theory and the cost-benefit concept by Plante and co-workers (1990).

As seen above, GB may contribute to the nutrition of echinoderms. However, whether or not these bacteria are transients or residents is usually unknown. The contribution of the GB (transient or permanent) to the nutrition of their host naturally relies on their metabolic abilities. The nutritional contribution of the transient bacteria also relies on the time spent in the gut. In contrast, because of their permanence, the residents can be particularly well integrated to their host's biology and become a kind of "necessary tool"

153

for its well being. Resident bacteria resist food movement through different ways: (1) a high division rate to compensate for the loss of cells, (2) adherance to the gut wall (between folds, setae or microvilli), (3) sheltering within a particular organ (caecum or outpocket) of the gut (Hylleberg Kristensen 1972, Smith & Douglas 1987).

According to Plante et al. (1990), enclosure within a gut enhances microbial digestion via exoenzymes, and thus the release of accessible metabolites to the co-occurring microorganisms (and to the host animal). A fortiori, these effects are reinforced if the bacteria are located within a narrow space (e.g. in a caecum, in a pellet or in subcuticular spaces). If nutrients/substrates are available to the bacteria, enclosure would logically induce their proliferation (with the development of successive consortia and mutualistic interactions between the microbes), an increase of the metabolites concentration and the establishment of a gradient that facilitates the diffusion of the bacterial metabolites to the host. One could expect transitory structures – such as pellets – to enhance the proliferation of transient bacteria while permanent structures – such as caeca or subcuticular spaces – will most probably harbour symbionts. In that context, building pellets appears as a subtle way to exploit the transient bacteria. They indeed are enclosed with their substrate (plants or detritus) and thus lytic activities are virtually ensured. Moreover, as the pellets are moved along the gut, or accumulated in some regions, they will be a kind of permanent source of metabolites during gut transfer. However, although food enclosure facilitates exogenous enzymatic activities, it raises the problem of access of the endogenous enzymes. Endogenous enzymes should be mixed with the food prior it is packed into pellets. Some endogenous enzymes could act on bacterial metabolites diffusing from the pellets. In holothuroids and echinoids, the whole gut is lined with an absorbing epithelium, but enzymatic secretion mostly occur in the anterior regions (anterior intestine and stomach, respectively). Secretory regions are particularly well localized in spatangoids as their digestive tube is compartmentalized in highly specialized regions (De Ridder & Jangoux 1993). But there is little detailed information on the nature and the importance of these secretions. All these considerations underline the importance of the gut anatomy (e.g. occurrence of sheltered areas for bacteria), of the food packing (e.g. occurrence of pellets), and of the digestive mechanics (e.g. retention time of food, feeding rhythms). These aspects are considered hereafter for echinoids and holothuroids. The general anatomical features of the digestive tube and mechanics are described in De Ridder and Jangoux (1982) for the echinoids and in Féral and Massin (1982) for the holothuroids.

Anoxic conditions seemingly prevail in the digestive tube of holothuroids and echinoids (e.g. Claereboudt & Jangoux 1985, Plante & Jumars 1992, Thorsen 1998). In detritivores, radial profiles indicate that some oxygenation occurs in the periphery of the gut lumen close to the gut wall as a result of oxygen diffusion from the coelomic cavity (Plante & Jumars 1992, Thorsen 1998). In spatangoids, the wall of the intestine (i.e., the second loop of the gut) and of the intestinal caecum face or even contact (intestinal caecum) the ampullae of the respiratory tube feet; this arrangement naturally facilitates oxygen diffusion towards the gut lumen (Buchanan et al. 1980, De Ridder 1994, Thorsen 1998). With a few exceptions of very acid values (usually occurring then in foregut regions), the pH levels are usually 6-8 (e.g., Lawrence 1982, Plante & Jumars 1992, Claereboudt & Jangoux 1985, Thorsen 1998). Low pH values can increase mucus fluidity and thus facilitate diffusion across a mucus layer, rendering mucus-packed food more accessible for endogenous enzymes or for absorption. In detritivorous echinoderms, the axial profiles of pO_2 and pH indicate rather constant values that mimic those of the surrounding sediment while Eh values may undergo significant changes along the gut length (Plante & Jumars 1992). These authors observed that the Eh axial profiles of *Molpadia intermedia* and *Brisaster latifrons* decrease between the foregut and midgut and increase in the hindgut. Eh decrease is attributed to microbial activity whereas its increase in the hindgut seemingly is due to anal intake of aerated water. Eh changes are clearly influenced by the nature of food and by the residence time of the sediment, and they can be determinant in the occurrence of different populations of microorganisms along the gut length (Plante & Jumars 1992).

3.4.1 *The digestive tube of echinoids*
Differences in gut anatomy, food packing and digestive mechanic markedly occur between herbivorous (regular) echinoids and detritivorous (irregular) echinoids. The digestive tube of echinoids forms two superimposed horizontal loops that run in opposite directions; the first loop (stomach) is connected to the mouth and the second (intestine) to the anus via two vertical segments, one corresponding to the buccal cavity (lacking in spatangoids), the pharynx and esophagus, and the other to the rectum. A siphon usually bypasses the first loop. The digestive anatomy of regular echinoids is homogenous throughout the group, the single obvious change being the development of a siphon which is lacking in primitive forms (e.g., De Ridder & Jangoux 1982). In contrast, the digestive anatomy of irregular echinoids shows variations, especially in spatangoids where highly differentiated siphons and conspicuous caeca occur.

In herbivorous echinoids, the food (living or dead plant fragments) is packed into rounded pellets inside the buccal cavity; each pellet is coated with a thick mucus layer while crossing the pharynx and is transferred in this state to the anus (Buchanan 1969). Such packing ensures an intimate contact between transient bacteria and their substrate during the whole digestive transfer, and can enhances bacterial growth and the diffusion of metabolites towards the gut lumen. However the localization of the gut bacteria in relation to the pellet is poorly documented. Most transient bacteria presumably occur inside the pellets but bacteria also occur on the pellet's surface on/in its covering mucus sheath (Farmanfarmaian & Phillips 1962). The access to food for endogenous enzymes (mostly secreted in the stomach where the food is already packed into pellets) remains puzzling: pH could influence locally the permeability of the pellets to endogenous enzymes; some endogenous enzymes could act on the bacterial metabolites diffusing from the pellets. However, qualitative and quantitative data on enzymes (endogenous versus exogenous) are lacking to answer this point. Interestingly, during a feeding cycle, pellets regularly accumulate in the intestine which is spacious and absorbing (De Ridder & Jangoux 1982). This could allow the echinoid to benefit from a more advanced digestion of the plant material. This idea is supported by the fact starving animals store pellets in their intestine (e.g. Lasker & Giese 1954, Fuji 1967).

A siphon in herbivorous echinoids might be related to mechanical constraints as it may maintain a constant gut volume (e.g. Buchanan 1969). It could also influence the maintenance of GB in the intestine (where the oldest pellets accumulate), as it allows some circulation of water between the pellets and may supply dissolved nutrients and some oxygen.

In herbivorous echinoids, formation of pellets might be considered basically as an adaptation for gardening transient bacteria. These echinoids have a tubular gut and no caecum. Such a caecum indeed would not be necessary, as the pellets themselves perform the gardening of bacteria. In such a functional context, resident (symbiotic) bacteria have less opportunity to occur. Comparative studies of gut bacteria from fed, starved and refed echinoids using controlled diets would provide some answer on the origin of the bacteria, and thus whether or not some may persist in the gut lumen.

The digestive tube of spatangoids contains bulky sediments that are never compacted into pellets. Mucus coats the newly ingested sediment molting them into a column through the action of a particularly well-developed musculature. Mucus secretion occurs intensively in the anterior part of the digestive tube (esophagus, and anterior stomach) (De Ridder & Jangoux 1993). However, the sediment is never ensheathed in mucus in the subsequent regions. Sediment usually accumulates and occupies the entire gut lumen at the distal extremity of each loop as a result of mechanical constraints (i.e. before ascending regions of the gut traject). Yet the filling stage of the stomach and the intestine fluctuates (De Ridder 1982, De Ridder & Jangoux 1985). Feeding and defecation are discontinuous and seemingly alternate. Before expulsion, sediment accumulates in the whole intestine and rectum, filling their lumens (De Ridder 1982).

As spatangoids never form permanent pellets, one could hypothesize they are less efficient in their exploitation of the transient bacteria compared to holothuroids or to herbivorous echinoids. In contrast, they cultivate symbiotic microorganisms within their caeca (a stomachal caecum and, in some species, an additional intestinal caecum). The stomachal caecum is a fermentation chamber owing to the activities of Fungi and Protozoa (Thorsen 1998, 1999). The intestinal caecum contains a variety of microorganisms, among which sulfur-oxidizing bacteria (Temara et al. 1993) and fermenters Fungi and Protozoa (Thorsen 1999) play proeminent functions. The occurrence of a foregut and a hindgut caecum containing active microbial fermenters led Thorsen (1998) to compare the digestive tube of *Echinocardium cordatum* to that of ruminants (foregut fermenters), where microbial digestion of the food particles occur anteriorly in some stomachal chambers (e.g., the rumen) with some residual fermentation in an hindgut caecum. As no sediment penetrates these caecums, how do adequate substrates reach the caecal microorganisms? The stomachal caecum opens into the anterior stomach where the sediment is packed into a mucus-ensheathed column. Packed sediment allows sea-water to enter the mouth and to circulate in the esophagus and anterior stomach between the sediment column and the digestive wall. The water originates from the bottom superficial layer and thus moves dissolved organic matter and fine detrital particles. Excess water which is not pumped into the siphon can reach the stomachal area and supplies the caecal microorganisms with nutrients. The anterior stomach and the stomachal caecum are the main sites of the gut where digestive enzymes are secreted (De Ridder 1986, De Ridder & Jangoux 1993). These enzymes presumably act on the particulate organic matter supplied by the circulating water. Diffusion of metabolites from the sediment column is presumably not important in this region because the column is continuously moved forwards and replaced by newly ingested sediment (De Ridder 1982, De Ridder & Jangoux 1985). Thorsen (1998) suggests that the digestive tube of *Echinocardium cordatum* functions as a CSTR-PFR reactor (see Penry & Jumars 1987,1990, Jumars & Penry 1989), the tubular gut segment acting as PFR (plug-flow reactor) and the caecums as CSTR-reactors (continuous-strirred-tank-reactor). However a plug reactor model would fit only the most anterior and posterior tubular gut segments as the sediment in the mid segments may fill partly the gut lumen and thus undergo some axial mixing. Such a mixing seemingly occurs periodically. This could be a way to avoid sulfide accumulation in the mid gut regions and could be complementary to the detoxifying effect of the symbiotic sulfur-oxidizing bacteria of the intestinal caecum. The removal of sulfide by the symbiotic bacteria could have two main nutritional consequences for the spatangoid host. First, at the level of the caecum, it allows maintenance of an active microflora known to produce nutrients, e.g., essential polyunsaturated fatty acids (linolenic acid) (Temara et al. 1991) and short-chained fatty acids (Thorsen 1998). Second, at the level of the intestine, it facilitates a longer retention time of the sediments and consequently a more advanced hydrolysis of the organic matter (De Ridder 1994, Brigmon & De Ridder 1998, Thorsen 1999).

3.4.2 *The digestive tube of holothuroids*

The digestive tube of holothuroids is tubular and, with the exception of some synaptids and elasipods, very long and looped. It consists mostly of an absorbing intestine. There is no caecum except in the elasipods which possess a rectal pocket connected to the cloaca. The anterior part of the digestive tube (buccal cavity, esophagus, stomach -when present- and anterior intestine) is a mixing zone where sediment grains are loose. Sediment is progressively compacted and coated with mucus at the beginning of the posterior intestine to form a series of saucage-like pellets that are attached to each other through the thick continuous mucus sheath. These pellets thus occur in most of the gut, i.e., the posterior intestine and rectum (and cloaca in species lacking respiratory trees) and are moved by peristalsis. In elasipods, the rectal pocket also accumulates pellets (Féral & Massin 1982).

Penry (1989) used the gut reactor models to describe the patterns of particle movement in the gut and observed that the holothuroid gut functions as a (modified) plug-flow model in which the sediment particles are not mixed into the plug (pellets) (either radially or axially) (see also Penry & Jumars 1986, 1987). She suggests that the (endogenous) enzymes must be mixed with the sediment before it is compacted in the intestine. The nutrients derived from digestion would then diffuse towards the gut lumen and reach the absorbing enterocytes. According to Penry (1989), the compacted sediment (plug) improve diffusion (mixing would disrupt a diffusion gradient). In such a functional context, most of the ingested bacteria (transients attached to the detritus or free) would occur inside the plug. If the bacteria that survive the microenvironment within the plug pursue their metabolic activities, they will contribute along the whole gut length to digestion (Plante et al. 1990). The occurrence of distinct microbial consortia along the gut length of abyssal holothuroids (e.g., Ralijaona & Bianchi 1982) would be related to the different microenvironmental conditions that develop within the pellets with time (e.g., prevailing anoxic conditions in the oldest pellets).

Whether or not bacteria occur outside the pellets in the digestive fluids and contribute to the nutrition of the holothuroid is questionable because they meet less static conditions than those living in the pellets and face the endogenous enzymes. For instance, in the foregut regions (mixing zone), part of the ingested bacteria die throught the activity of endogenous enzymes (and benefit the holothuroid). In aspidochirote holothuroids, the digestive transfer lasts 2 to 7 hours (*in situ* measurements). As defecation is discontinuous while feeding and pellet formation is continuous, the pellets accumulate regularly in the second half of the posterior intestine (e.g. Yamanuchi 1939, Massin 1978, Féral & Massin 1982). This explains why microbial proliferation systematically occurs in the hindgut. The only resident bacteria of holothuroids could be the SCB of the foregut and hindgut regions. They must function as enclosed bacteria and would be particularly efficient in creating metabolite gradients which can benefit their host. Yet, their role remains enigmatic. There is no digestive function known for the rectal pocket of elasipods. However, this caecum could be a site of intense bacterial activity as pellets accumulate there and as it possesses a cuticle (and thus presumably SCB). Hansen (in Féral & Massin 1982) suggested that the caecum acts as a balast regulating the buyancy of these "swimming holothuroids".

3. 5 *Conclusions*

There is clearly a need for studies of the microflora diversity and role in the gut of echinoderms. The actual contribution of the different members of the bacterial microflora to the echinoderm's nutrition remains indeed unclear. Qualitative and quantitative informations on the metabolic activities are fragmentary and consequently insufficient to answer the real nature of the GB-echinoderms interactions. A multidisciplinary approach would be ideal, integrating studies of the microbial/bacterial diversity (taxonomic and metabolic) through cultivation techniques and molecular analyses, and studies of the feeding biology of the echinoderm host. Because they consider the functional aspects (abilities and needs, limits and constraints) of "the associated organisms", the cost and benefit-based analyses and the digestion-absorption models (chemical reactors) developped by Plante, Jumars, Penry and co-workers open new perspectives to understand how GB and host can interact. Such analyses and models provide consistent frames for future researchs.

4 THE BACTERIA OF THE CONNECTIVE TISSUE OF CRINOIDS

Spherical bodies (*ca.* 100 μm in diameter) filled with Gram-negative rod-shaped bacteria (0.8 μm in diameter, 3 μm in length) occur in the deep connective tissues of males and females of *Calamocrinus diomedae* (Holland et al. 1991). The bodies are restricted to the arms and pinnules and are scattered in the connective tissues and in the stereom spaces. Each body is limited by an external lamina (*ca.* 20 nm thick). There is little information on the benefits both partners obtain from this association. The symbiotic bacteria were suggested to participate to the nutrition of their host: (1) through chemosynthesis and/or (2) through bioluminescence which allow to attract some planktonic preys. Until now this symbiosis is only reported in one deep-sea species (*C. diomedae*). More studies are needed to understand the relationships between the symbionts and to conduct a survey of the occurrence of the symbiosis in the crinoids class.

ACKNOWLEDGMENTS

We would like to thank John Lawrence and Michel Jangoux for helpful discussions and critical reading of the manuscript, and E. Bricourt for her help in typing. Work supported by an FRFC grant to C. De Ridder (convention 2. 4515.99).

REFERENCES

Albéric, P. & A. Khripounoff 1984. Relations entre les compositions en acides aminés des particules en voie de sédimentation, du contenu intestinal des holothuries abyssales et du sédiment environnant. *Mar. Chem.* 14: 379-394.

Albéric, P., J.-P. Féral & M. Sibuet 1987. Les acides aminés libres, reflet de l'activité bactériennes – dans les contenus digestifs des holothuries : différence entre zones abyssale et littorale. *C.R. Acad. Sci. Paris* 305(6): 203-206.

Amon, R.W. & G.J. Herndl. 1991. Deposit feeding and sediment: I. Interrelationship between *Holothuria tubulosa* (Holothuroida, Echinodermata) and the sediment microbial community. *Mar. Ecol.* 12: 163-174.

Araki, G.S. & A.C. Giese 1970. Carbohydrases in sea stars. *Physiol. Zool.* 43: 296-305.

Back, J.P. & R.G. Kroll 1991. The differential fluorescence of bacteria stained with acridine orange and the effects of heat. *J. Appl. Bacteriol.* 71:51-58.

Bamfield, D. 1982. Epithelial absorption. In: *Echinoderm Nutrition.* M. Jangoux & J.M. Lawrence (eds), Balkema, Rotterdam: 317-330.

Barker, M.F. & M.S. Kelly 1994. The occurrence and transmission of subcuticular bacteria in echinoderms. In: B. David, A. Guille, J.P. Féral & M. Roux (eds), *Echinoderms Through Time*: 13. Balkema, Rotterdam.

Bell, A.C. 1974. Histology and ultrastructure of *Acrocnida brachiata*. Ph.D thesis. Queen's University, Belfast.

Bensoussan, M.G., P.M. Scoditti & A.J.M. Bianchi 1979. Etude comparative des potentialités cataboliques de microflores entériques d'échinodermes et de bactéries du sédiment superficiel prélevées en milieu abyssal. *C.R. Acad. Sc. Paris* 289: 437-440.

Bensoussan, M.G., P.M. Scoditti & A.J.M. Bianchi 1984. Bacterial flora from echinoderm guts and associated sediment in the abyssal Verna Fault. *Mar. Biol.* 79: 1-10.

Bianchi, A.J.M., P. Scoditti & M.G. Bensoussan 1979. Distribution des populations bactériennes hétérotrophes et les tractus digestifs d'animaux benthiques recueillis dans la faille Vema et les plaines abyssales du Demerara et de Gambie. *Vie Marine* 1: 7-12.

Boolootian, R.A. & R. Lasker 1964. Digestion of brown algae and the distribution of nutrients in the purple sea urchin *Stronglylocentrotus purpuratus. Comp. Biochem. Physiol.* 11: 273-289.

Bosch, I. 1992. Symbiosis between bacteria and oceanic clonal sea star larvae in the western North Atlantic Ocean. *Mar. Biol.* 114: 495-502.

Brigmon, R.L. & C. De Ridder, 1998. Symbiotic relationship of *Thiothrix* spp. with an echinoderm. *Appl. Environm. Microbiol.* 64: 3491-3495.

Brock, T.D., M.T. Madigan, J.M. Martinko & J. Parker 1994. *Biology of microorganisms.* Prentice-Hall International: 909 p.

Bromley, R.G., Jensen M. & V. Asgaard 1995. Spatangoid echinoids: deep-tier trace fossils and chemosymbiosis. *Neues Jahr. Geol. und Pal. Abh.* 195(1-3): 25-35

Buchanan, J.B. 1966. The biology of *Echinocardium cordatum* (Echinodermata: Spatangoidea) from different habitats. *J. Mar. Biol. Ass. U.K.* 46: 97-114.

Buchanan, J.B. 1969. Feeding and the control of volume within the tests of regular sea-urchins. *J. Zool. Lond.* 159: 51-64.

Buchanan, J.B., Brown B.E., Coombs T.L. Pirie B.J.S. & J.A. Allen 1980. The accumulation of ferric iron in the guts of some spatangoid echinoderms. *J. Mar. Biol.* 60: 631-640.

Burnett, W.J. & J.D. McKenzie 1997. Subcuticular bacteria from the brittlestar *Ophiactis balli* (Echinodermata: Ophiuroidea) represent a new lineage of extracellular marine symbionts in the α-subdivision of the class Proteobacteria. *Appl. Environ. Microbiol.* 63: 1721-1724.

Cameron, R.A, & N.D. Holland 1983. Electron microscopy of extracellular material during the development of the seastar, *Patiria miniata* (Echinodermata: Asteroidea). *Cell Tissue Res.* 234: 193-200.

Cavanaugh, C.M., S.C. Gardiner, M.L.S. Jones, H.W. Jannasch & J.B. Waterbury 1981. Prokaryotic cells in the hydrothermal vent tube worm *Riftia pachyptila*; possible chemoautotrophic symbionts. *Science* 213: 340-342.

Cerra, A. and M. Byrne 1998. Cell surface features and the extracellular matrix around the larvae of *Patiriella* species with planktonic, benthic and intragonadal development: Implications for larval nutrition. In: R. Mooi & M. Telford (eds), *Echinoderms: San Francisco*: 221-226. Balkema, Rotterdam.

Cerra, A., M. Byrne, O. Hoegh-Guldberg 1997. Development of the hyaline layer around the planktonic embryos and larvae of the asteroid *Patiriella calcar* and the presence of associated bacteria. *Invert. Reprod. Dev.* 31: 337-343.

Chen, C.-P. & B.-Y. Chen 1994. Diverticulum sand in a miniature sand dollar *Sinaechinocyamus mai* (Echinodermata: Echinoidea). *Mar. Biol.* 119: 605-609.

Chen, C.-P. & J. Run 1989. Larval growth and bacteria-induced metamorphosis of *Arachnoides placenta* (L.) (Echinodermata:Echinoidea). In: J.S. Ryland & P.A. Tyler (eds), *Reproduction, Genetics, and Distributions of Marine Organisms*: 55-59. Olsen & Olsen, Fredensburg, Denmark.

Chia, F.S. 1973. Sand dollars: a weight belt for the juvenile. *Science* 181: 73-74.

Chia, F.S. 1985. Selection, storage and elimination of heavy sand particles by the juveniles sand dollar, *Dendraster excentricus* (Eschscholtz). In: Keegan B.F. & B.D.S. O'Connor (eds), *Echinodermata*: 215-221. Balkema, Rotterdam.

Claereboudt, M. and M. Jangoux 1985. Conditions de digestion et activités de quelques polysaccharidases dans le tube digestif de l'oursin *Paracentrotus lividus* (Lamarck) (Echinodermata). *Biochim. Syst. Ecol.* 13:51-54.

Coleman, R. 1969. Ultrastructure of the tube foot sucker of a regular echinoid *Diadema antillarum* Philippi. *Z. Zellforsch* 96: 162-172.

Colwell, R.R. & J. Liston 1962. The natural bacterial flora of certain marine invertebrates. *J. Insect. Pathol.* 4: 23-33.

161

Cooksey, K.E. & B. Wigglesworth-Cooksey 1995. Adhesion of bacteria and diatoms to surfaces in the sea: review. *Aquat. Microb. Ecol.* 9: 87-96.

De Bary, H.A. 1879. Die Erscheinung der Symbiose. In: Trübner R.J., *Vortag, gehalten auf der Versammlung Deutscher Naturforscher und Aertze zu Cassel.* Strassburg: 1-30.

Deming, J.W. & R.R. Colwell 1981. Barophilic bacteria associated with deep-sea animals. *Bioscience* 31: 507-511.

Deming, J.W. & R.R. Colwell 1982. Barophilic bacteria associated with digestive tracts of abyssal holothurians. *Appl. Environm. Microbiol.* 44: 1221-1230.

Deming, J.W., Tabor P.S. & R.R. Colwell 1981. Barophilic growth of bactera from intestinal tracts of deep-sea invertebrates. *Microb. Ecol.* 7: 85-94.

De Ridder, C., 1982. Feeding and some aspects of the gut structure in the spatangoid echinoid, *Echinocardium cordatum* (Pennant). In: J.M. Lawrence (ed.), *Echinoderms: Proceedings of the International Conference, Tampa Bay:* 5-9. J.M. Lawrence (ed.), Balkema, Rotterdam.

De Ridder, C. 1986. *La nutrition chez les échinodermes psammivores. Etude particulière du spatangide fouisseur, Echinocardium cordatum (Pennant) (Echinodermata, Echinoidea).* PhD thesis, Université Libre de Bruxelles: 275 pp.

De Ridder, C. 1994. Symbioses between spatangoids (Echinoidea) and *Thiothrix*-like bacteria (Beggiatoales). In: David B., Guille A., Féral J.-P. & M. Roux (eds), *Echinoderms through time:* 619-625. Balkema, Rotterdam.

De Ridder, C. & M. Jangoux 1982. Structure and functions of digestive organs: Echinoidea (Echinodermata). In: M. Jangoux & J.M. Lawrence (eds), *Echinoderm Nutrition:* 213-214. Balkema, Rotterdam.

De Ridder, C. & M. Jangoux, 1985. Origine des sédiments ingérés et durée du transit digestif chez l'oursin spatangide, *Echinocardium cordatum* (Pennant) (Echinodermata). *Annls Inst. océanogr. Paris* 61(1): 51-58.

De Ridder, C. & M. Jangoux 1993. The digestive tract of the spatangoid echinoid *Echinocardium cordatum* (Echinodermata): morphofunctional study. *Acta Zoologica (Stockholm)*, 74(4): 337-351.

De Ridder, C. & J.M. Lawrence 1982. Food and feeding mechanisms in echinoids (Echinodermata). In: M. Jangoux & J.M. Lawrence (eds), *Echinoderm Nutrition:* 57-115. Balkema, Rotterdam.

De Ridder, C., Jangoux M. & L. De Vos 1985(a). Description and significance of a peculiar intradigestive symbiosis between bacteria and a deposit-feeding echinoid. *J. Exp. Mar. Biol. Ecol.* 91: 65-76.

De Ridder C., Jangoux M. & E. Van Impe, 1985(b). Food selection and absorption efficiency in spatangoid echinoid, *Echinocardium cordatum* (Echinodermata). B. Keegan & B. O' Connor (eds), *Echinodermata:* 245-251. Balkema, Rotterdam.

Dilmore, L.A. & M.A. Hood 1986. Vibrios of some deep-water invertebrates. *FEMS Microb. Letters* 35: 221-224.

Doria, V. & E. Bianchi 1985. Bactériologie des sédiments superficiels et des contenus de tractus digestifs d'invertébrés. In: L. Laubier & C. Monniot (eds), *Les Peuplements profonds du Golfe de Gascogne:* 183-192. IFREMER, Brest.

Douglas, A.E. 1994. *Symbiotic interactions.* Oxford Univ. Press: 148 pp.

Douglas, A.E. & D.C. Smith 1983. The costs of symbiosis to the host in green hydra. *Endocytobiology* 2: 631-647.

Dring, M.J. 1982. *The biology of marine Plants.* Contemporary Biology, Arnold E., London: 199 pp.

Dubilier, N., Giere O. & M.K. Grieshaber 1995. Morphological and ecophysiological adaptations of the marine oligochaete *Tubificoides benedii* to sulfidic sediments. *Am. Zool.* 35: 163-173.

Duncan, H.E. & S.C. Edberg 1995. Host-microbe interaction in the gastro-intestinal tract. *Crit. Rev. Microbiol.* 21: 85-100.

Engster, M.S. and S.C. Brown 1972. Histology and ultrastructure of the tube foot epithelium in the phanerozoan starfish, *Astropecten. Tissue Cell* 4: 503-518.

Eppley, R.W. & R. Lasker 1959. Alginan in the sea urchin *Strongylocentrotus purpuratus. Science* 129: 214-215.

Farmanfarmaian, A. & J.H. Phillips 1962. Digestion, storage, and translocation of nutrients in the purple sea urchin (*Strongylocentrotus purpuratus*). *Biol. Bull.* 123: 105-120.

Favarov, V.V. & V.E. Vaskovsky 1971. Alginases of marine invertebrates. *Comp. Biochem. Physiol.* 38B: 689-696.

Fenchel, T. & B.J. Finley 1995. *Ecology and evolution in anoxic worlds.* Oxford Series in Ecology and Evolution, Oxford Univ. Press: 276 pp.

Féral, J.-P. 1980. Cuticule et bactéries associées des épidermes digestif et tégumentaire de *Leptosynapta galliennei* (Herapath) (Holothurioidea: Apoda). Premières données. In: Jangoux M. (ed.), *Echinoderms: Present and Past*: 285-290. Balkema, Rotterdam.

Féral, J.-P. 1989. Activity of the principal digestive enzymes in the detritivorous apodous holothuroid *Leptosynapta galliennei* and two other shallow water holothuroids. *Mar. Biol.* 101: 367-379.

Féral, J.-P. & C. Massin 1982. Digestive systems: Holothuroidea. In: M. Jangoux & J.M. Lawrence (eds), *Echinoderm Nutrition*: 191-212. M. Jangoux & J.M. Lawrence (eds), Balkema, Rotterdam.

Ferguson, J.C. 1967. An autoradiographic study of the utilization of free exogenous amino acids by starfishes. *Biol. Bull.* 133:317-329.

Ferguson, J.C. 1970. An autoradiographic study of the translocation and utilization of amino acids by starfish. *Biol. Bull.* 138:14-25.

Ferguson, J.C. 1980. Fluxes of dissolved amino acids between seawater and *Echinaster. Comp. Biochem. Physiol.* 65A:291-295.

Fong, W. & K.H. Mann 1980. Role of gut flora in the transfer of amino acids through a marine food chain. *Can. J. Fish. aquat. Sci.* 37: 88-96.

Foret, T.W. 1999. Subcuticular bacteria in Florida echinoderms: incidence of occurrence and variation in symbiont densities over time and space. Master Thesis. University of South Florida.

Frantzis, A. & A. Grémare 1992. Ingestion, absorption, and growth rates of *Paracentrotus lividus* (Echinodermata: Echinoidea) fed different macrophytes. *Mar. Ecol. Prog. Ser.* 95:169-183.

Fuji, A. 1967. Ecological studies on the growth and food consumption of japanese common littoral sea urchin, *Strongylocentrotus intermedius* (A. Agassiz). *Mem. Fac. Fish. Hokkaido Univ.* 15: 83-160.

Garcia-Tello, P. & A.M. Baya 1973. Acerca de la posible funcion de bacterias agaroliticas del erizo blanco. *Mus. Nac. Hist. Nat.* 15: 3-8.

Garcia-Tello, P., Campos V. & A.M. Baya 1978. Sensibilidad a algunos antibioticos de bacterias aisladas de *Loxechinus albus* (Mol.) y *Merluccius gayi. Mus. nac. Hist. nat. Chile* 22(260): 4-6.

Gillan, D. & C. De Ridder, 1997. Morphology of an iron oxide-encrusted biofilm forming on the shell of a burrowing bivalve (Mollusca). *Aquat. microb. Ecol.* 12: 1-10.

Greenway, M. 1995. Trophic relationships of macrofauna within a Jamaican seagrass meadow and the role of the echinoid *Lytechinus variegatus* (Lamarck). *Bull. Mar. Sc.* 56: 719-736.

Grigolava, I. & J.D. McKenzie 1994. Biochemical studies of echinoderm surface coats using radiolabelling techniques. In: B. David, A. Guille, J.P. Féral, & M. Roux (eds.), *Echinoderms through Time*: 685-690. Balkema, Rotterdam.

Grimmer, J.C. & N. Holland 1990. The structure of a sessile, stalkless crinoid (*Holopus rangii*). *Acta Zool.* 71(2): 61-67.

Guérinot, M.L. & D.G. Patriquin 1981a. N_2-fixing vibrios isolated from the gastrointestinal tract of sea urchins. *Can. J. Microbiol.* 27: 311-317.

Guérinot, M.L. & D.G. Patriquin 1981b. The association of N_2-fixing bacteria with sea-urchins. *Mar. Biol.* 62: 197-207.

Guérinot, M.L., Fong W. & D.G. Patriquin 1977. Nitrogen fixation (acetylene reduction) associated with sea urchins (*Strongylocentrotus droebachiensis*) feeding on seaweeds and eelgrass. *J. Fish. Res. Board Can.* 34: 416-420.

Hammond, L.S. 1983. Nutrition of deposit-feeding holothuroids and echinoids (Echinodermata) from a shallow reef lagoon, Discovery Bay, Jamaica. *Mar. Ecol. Prog. Ser.* 10: 297-305.

Harris, J.M. 1993. The presence, nature and role of gut microflora in aquatic invertebrates : a synthesis. *Microb. Ecol.* 25: 195-231.

Heinzeller, T. & H. Fechter 1995. Microscopical anatomy of the Cyrtocrinid *Cyathidium meteorensis* (sive *foresti*) (Echinodermata, Crinoidea). *Acta Zool. (Stockholm)* 76: 9-148.

Heinzeller, T. & U. Welsh 1994. Crinoidea. In: Harrison, F.W. & F.-S. Chia (eds), *Microscopic Anatomy of Invertebrates. Vol. 14. Echinodermata*: 9-148. Wiley-Liss, New-York.

Holland, N.D. 1984. Echinodermata: Epidermal cells. In: J. Bereiter-Hahn, A.G. Matoltsy, & K. Silvia Richards (eds), *Biology of the Integument 1: Invertebrates*: Springer Verlag, New York: 769 pp.

Holland, N.D. & K.H. Nealson 1978. The fine structure of the echinoderm cuticle and the subcuticular bacteria of echinoderms. *Acta Zool.* 59: 169-185.

Holland, N.D., Grimmer J.C. & K. Wiegmann. 1991. The structure of the sea lily *Calamocrinus diomedae*, with special reference to the articulations, skeletal microstructure, symbiotic bacteria, axial organs, and stalk tissues (Crinoida, Millericrinida). *Zoomorphology* 110: 115-132.

Huang, H. & A.C. Giese 1958. Test for digestion of algal polysaccharides by some marine herbivores. *Science* 127: 475.

Hungate, R.E. 1975. The rumen microbial ecosystem. *Ann. Rev. Microbiol.* 29: 39-66.

Hylleberg Kristensen, J. 1972. Carbohydrattes of some marine invertebrates with notes on their food and on the natural occurence of the carbohydrates studied. *Mar. Biol.* 14: 130-142.

Jangoux, M. 1982. Excretion. In: M. Jangoux & J.M. Lawrence (eds), *Echinoderm nutrition*: 437-445. Balkema, Rotterdam.

Jangoux, M. 1990. Diseases of Echinodermata. In: O. Kinne (ed.), *Diseases of marine Animals* 3: 439-567. Biologische Anstalt Helgoland, Hambourg.

Jeuniaux, C. 1963. *Chitine et chitinolyse*. Masson & Cie, Paris, 181 pp.

Jumars, P.A. & D.L. Penry 1989. Digestion theory applied to deposit feeding. In: Lopez G., G. Taghon, & J. Levinton (eds), *Ecology of marine deposit feeders*: 114-128. Springer Verlag, Heidelberg.

Kelly, M.S. & J.D. McKenzie 1992. The quantification of subcuticular bacteria in echinoderms. In: L. Scalera-Liaci & C. Canicatti (eds.), *Echinoderm Research*: 225-228. Balkema, Rotterdam.

Kelly, M.S., & J. D. McKenzie 1995. Survey of the occurrence and morphology of subcuticular bacteria in shelf echinoderms from the northeast Atlantic Ocean. *Mar. Biol.* 123: 741-756.

Kelly, M.S., J.D. McKenzie & M.F. Barker 1994. Sub-cuticular bacteria: Their incidence in echinoderms. In: B. David, A. Guille, J.P. Féral & M. Roux (eds), *Echinoderms through Time:* Balkema, Rotterdam.

Kelly, M.S., M.F. Barker, J.D. McKenzie & J. Powell 1995. The incidence and morphology of subcuticular bacteria in echinoderm fauna of New Zealand. *Biol. Bull.* 189: 91-105.

Khripounoff, A. & M. Sibuet 1980. La nutrition d'échinodermes abyssaux I. Alimentation des holothuries. *Mar. Biol.* 60: 17-26.

Kozlovskaya, E.P. & V.E. Vaskovsky 1970. A comparative study of proteinases of marine invertebrates. *Comp. Biochem. Physiol.* 34: 137-142.

Krueger, D.M., Gustafson, R.G. & Cavanaugh, C.M. 1996. Vertical transmission of chemoautotrophic symbionts in the bivalve *Solemya velum* (Bivalvia: Protobranchia). *Biol. Bull.* 190:195-202.

Largo, D.B., K. Fukami, M. Adachi & T. Nishijima 1997. Direct enumeration of bacteria from macroalgae by epifluorescence as applied to the fleshy red algae *Kappaphycus alvarezii* and *Gracilaria* spp. (Rhodophyta). *J. Phycol.* 33: 554-557.

Lasker, R. & A.C. Giese 1954. Nutrition of the sea urchin *Strongylocentrotus purpuratus. Biol. Bull.* 106: 328-340.

Lasker, R. & R.A. Boolootian 1960. Digestion of the alga, *Macrocystis pyrifera,* by the sea urchin, *Strongylocentrotus purpuratus. Nature* 188: 1130.

Lawrence, J.M. 1982. Digestion. In: M. Jangoux & J.M. Lawrence (eds), *Echinoderm Nutrition*: 283-316. Balkema, Rotterdam.

Lee, K. and E.G. Ruby 1994. Effect of the squid host on the abundance and distribution of symbiotic *Vibrio fischeri* in nature. *Appl. Environ. Microbiol.* 60: 1565-1571.

Lesser, M.P. & R.P. Blakemore 1990. Description of a novel symbiotic bacterium from the brittlestar, *Amphipholis squamata. Appl. Environ. Microbiol.* 56: 2436-2440.

Lesser, M.P. & C.W. Walker 1992. Comparative study of the uptake of dissolved amino acids in sympatric brittlestars with and without endosymbiotic bacteria. *Comp. Biochem. Physiol.* 101B: 217-223.

Margulis, L. 1976. A review: genetic and evolutionary consequences of symbiosis. *Exp. Parasitol.* 39: 277-349.

Margulis, L. 1991. Symbiogenesis and symbionticism. In: *Symbiosis as a source of evolutionary innovation.* MIT Press, Cambridge, Massachusetts: 454 pp.

Margulis, L. & R. Fester 1991. *Symbiosis as a source of evolutionary innovation.* MIT Press, Cambridge, Massachusetts: 454 pp.

Markel, K. & U. Röser 1985. Comparative morphology of echinoderm calcified tissues: histology and ultrastructure of ophiuroid scales (Echinodermata, Ophiuroida). *Zoomorphology* 105: 197-207.

Massin, C. 1978. *Etude de la nutrition chez les holothuries aspidochirotes (Echinodermata).* PhD thesis, Université Libre de Bruxelles: 204 pp.

Massin, C. 1982. Food and feeding mechanisms: Holothuroidea. In: M. Jangoux & J.M. Lawrence (eds), *Echinoderm Nutrition*: 43-55. Balkema, Rotterdam.

Massin, C. 1984. Structures digestives d'holothuries Elasipoda (Echinodermata): *Benthogone rosea* Koehler et *Oneirophanta mutabilis* Theel. *Arch. Biol.* (Bruxelles) 95: 153-185.

McFall–Ngai, M.J. 1998. The development of cooperative associations between animals and bacteria: establishing détente among domains. *Amer. Zool.* 38: 593-608.

McFall–Ngai, M.J. 1999. Consequence of evolving with bacterial symbionts: insights from the squid-*Vibrio* Associations. *Annu. Rev. Ecol. Syst.* 30: 235-256.

McKenzie, J.D. 1987. The ultrastructure of the tentacles of eleven species of dendrochirote holothurians studies with special reference to the surface coats and papillae. *Cell Tissue Res.* 248:187-199.

McKenzie, J.D. 1988. Echinoderm surface coats: Their ultrastructure, function, and significance. In: R.D. Burke, P.V. Mladenov, P. Lambert & R.L. Parsley (eds), *Echinoderm Biology*: 697-706. Balkema, Rotterdam.

McKenzie, J.D. 1992. Comparative morphology of crinoid tube feet. In: Scalera-Liaci L. & C. Canicatti (eds), *Echinoderm Research*: 73-79. Balkema, Rotterdam.

McKenzie, J.D. 1994. Using the very small to comment on the very large. In: B. David, A. Guille, J.P. Féral & M. Roux (eds), *Echinoderms through Time*: 73-85. Balkema, Rotterdam.

McKenzie, J.D. & M.S. Kelly 1994. Comparative study of subcuticular bacteria in brittlestars (Echinodermata: Ophiuroidea). *Mar. Biol.* 120: 65-80.

McKenzie, J.D., K.D. Black, M.S. Kelly, L.C. Newton, L.L. Handley, C.S. Scrimgeour, J.A. Raven & R.J. Henderson 2000. Comparisons of fatty acid and stable isotope ratios in symbiotic and non-symbiotic brittlestars from Oban Bay, Scotland. *J. Mar. Biol. Ass. U.K.* 80:311-320.

McKenzie, J.D., W.J. Burnett, & M.S. Kelly 1998. Systematic distribution of subcuticular bacteria in echinoderms. In: R. Mooi & M. Telford (eds), *Echinoderms: San Francisco*: 53-59. Balkema, Rotterdam.

Montgomery, M.K. & M.J. McFall-Ngai 1995. The inductive role of bacterial symbionts in the morphogenesis of a squid light organ. *Amer. Zool.* 35: 372-380.

Mooi, R. & C.-H. Chen 1996. Weight belts, diverticula, and the phylogeny of the sand dollars. *Bull. mar. Sci.* 58: 186-195.

Moriarty, D.J.W. 1982. Feeding of *Holothuria atra* and *Stichopus chloronotus* on bacteria, organic carbon and organic nitrogen in sediments of the Great Barrier Reef. *Aust. J. mar. freshwater Res.* 33: 255-263.

Moriarty, D.J.W., Pollard P.C., Hunt W.G., Moriarty C.M. & T.J. Wassenberg 1985. Productivity of bacteria and microalgae and the effect of grazing by holothurians in sediments on a coral reef flat. *Mar. Biol.* 85: 293-300.

Morton, B. 1989. *Partnerships in the sea : Hong Kong's marine symbioses.* Hong Kong Univ. Press: 124 pp.

Narita, H., S. Matsubara, N. Miwa, S. Akahane, M. Murakami, T. Goto, M. Nara, T. Noguchi, T. Saito, Y. Shida & K. Hashimoto 1987. *Vibrio alginolyticus*, a TTX-producing bacterium isolated from the starfish *Astropecten polyacanthus*. *Nippon Suisan Gakkaishi* 53: 617-621.

Newton, L.C. & J.D. McKenzie 1995. Echinoderms and oil pollution: A potential stress assay using bacterial symbionts. *Mar. Poll. Bull.* 31: 453-456.

Newton, L.C. & J.D. McKenzie 1998. Development and evaluation of echinoderm pollution assays. In: R. Mooi and M. Telford (eds), *Echinoderms: San Francisco*: 405-410.

Odintsov, V.S. 1981. Nitrogen fixation (acetylene reduction) in the digestive tract of some echinoderms from Vostok Bay in the Sea of Japan. *Mar. Biol. Lett.* 2: 259-263.

166

Penry, D.L. 1989. Tests of kinematic models for deposit-feeders' guts: patterns of sediment processing by *Parastichopus californicus* (Stimpson) (Holothuroidea) and *Amphicteis scaphobranchiata* Moore (Polychaeta). *J. Exp. Mar. Biol. Ecol.* 128: 127-146.

Penry, D.L. & P.A. Jumars 1986. Chemical reactor analysis and optimal digestion theory. *BioScience* 36: 310-315.

Penry, D.L. & P.A. Jumars 1987. Modeling guts as chemical reactors. *Am. Nat.* 129: 69-96.

Penry, D. L. & P. A. Jumars 1990. Gut architecture, digestive constraints and feeding ecology of deposit-feeding and carnivorous polychaetes. *Oecologia* 82: 1-11.

Plante, C.J. & P.A. Jumars 1992. The microbial environment of marine deposit-feeder guts characterized via microelectrodes. *Microb. Ecol.* 23: 257-277.

Plante, C.J. & A.G. Schriver 1998. Patterns of differential digestion of bacteria in deposit-feeders: a test of ressource partitioning. *Mar. Ecol. Prog. Ser.* 163: 253-258.

Plante, C.J., P.A. Jumars & J.A. Baross 1989. Rapid bacterial growth in the hindgut of a marine deposit feeder. *Microb. Ecol.* 18: 29-44.

Plante, C.J., P.A. Jumars & J.A. Baross 1990. Digestive associations between marine detritivores and bacteria. *Annu. Rev. Ecol. Syst.* 21: 93-127.

Prieur, D., Chamroux S., Durand P., Erauso G., Fera P., Jeanthon C., Le Borgne L., Mével G. & P. Vincent 1990. Metabolic diversity in epibiotic microflora asscociated with the Pompeii worms *Alvinella pompejana* and *A. caudata* (Polychaetae: Annelida) from deep-sea hydrothermal vents. *Mar. Biol.* 106: 361-367.

Prim, P. 1973. Utilization of marine polysacharide by bacteria of the gut of the sea-urchin, *Lytechinus variegatus. Quart. J. Fl. Acad. Sci.* 36(1): 11.

Prim, P. & J.M. Lawrence 1975. Utilization of marine plants and their constituents by bacteria isolated from the gut of echinoids (Echinodermata). *Mar. Biol.* 33: 167-173.

Ralijaona, C. & A. Bianchi 1982. Comparaison de la structure et des potentialités métaboliques des communautés bactériennes du contenu du tractus digestif d'holothuries abyssales et du sédiment environnant. *Bull. Cent. Etud. Rech. Sci. Biarritz* 14: 199-214.

Rimmer, D.W. & W.J. Wiebe 1987. Fermentative microbial digestion in herbivorous fishes. *J. Fish. Biol.* 31: 229-236.

Roberts, D., D.S.M. Billet, G. McCartney & G.E. Hayes 1991. Procaryotes on the tentacles of deep-sea holothurians: a novel form of dietary supplementation. *Limnol. Oceanogr.* 36: 1447-1452.

Rozak, D.B. & R.R. Colwell 1987. Survival strategies of bacteria in natural environment. *Microbiol. Rev.* 51: 365-379.

Saffo, M.B. 1992. Invertebrates in endosymbiotic associations. *Amer. Zool.* 32: 557-565.

Santegoeds, C.M., Nold S.C. & D.M. Wards 1996. Denaturating gradient gel electrophoresis used to monitor the enrichment culture of aerobic chemoorganotrophic bacteria from a hot spring cyanobacterial mat. *Appl. Environm. Microbiol.* 62: 3922-3928.

Sapp, J. 1994. *Evolution by association. A history of symbiosis.* Oxford Univ. Press: 255 pp.

Savage, D. 1986. Gastrointestinal microflora in mammalian nutrition. *Ann. Rev. Nutr.* 6: 155-178.

Sawabe, T., Y. Oda, Y. Shiomi & Y. Ezura 1995. Alginate degradation by bacteria isolated from the gut of sea urchins and abalones. *Microb. Ecol.* 30: 193-202.

Scheibling, R.E. 1980. Carbohydrases of pyloric caeca of *Oreaster reticulatus* (L.) (Echinodermata: Asteroidea). *Comp. Biochem. Physiol.* 67B: 297-300.

Sepers, A.D. 1977. The utilization of dissolved organic compounds in aquatic environments. *Hydrobiologia* 52: 39-54.

Sibuet, M. 1984. Les invertébrés détritivores dans l'écosystème abyssal. Sélection de la nourriture et régime alimentaire chez les holothuries. *Océanis* 10 (6): 623-639.

Sibuet, M., Khripounoff A., Deming J., Colwell R. & A. Dinet 1982. Modification of the gut contents in the digestive tract of abyssal holothurians. In: J.M. Lawrence (ed.), *Echinoderms: Proceedings of the international conference, Tampa Bay*: 421-428. Balkema, Rotterdam.

Siebers, D. 1979. Transintegumentary uptake of dissolved amino acids in the sea star *Asterias rubens*. A reassessment of its nutritional role with special reference to the significance of heterotrophic bacteria. *Mar. Ecol. Prog. Ser.* 1: 169-177.

Smith, D.C. & A.E. Douglas 1987. *The biology of Symbiosis*. Contemporary Biology, Arnold E. Publ., London: 302 pp.

Souza Santos, H. & W. Silva Sasso 1970. Ultrastructure and histochemical studies on the epithelium revestment layer in the tube feet of the starfish *Asterina stellifera*. *J. Morphol.* 130: 287-296.

Sova, V.V., Elyakova L.A. & V.E. Vaskovsky 1970. The distribution of laminarinases in marine invertebrates. *Comp. Biochem. Physiol.* 32: 459-464.

Stanier, R.Y., J.L. Ingraham, M.L. Wheelis & P.R. Painter 1986. *General Microbiology*. MacMillan Educ. Ltd: 689 pp.

Stephens, G.C. 1988. Epidermal amino acid transport in marine invertebrates. *Biochim. Biophys. Acta* 947: 113-138.

Stephens, G.C., M.J. Volk, S.H. Wright & P.S. Backlund 1978. Transepidermal accumulation of naturally occurring amino acids in the sand dollar *Dendraster excentricus*. *Biol. Bull.* 154: 335-347.

Stewart, W.D.P., Fitzgerald G.P. & R.H. Buris 1967. In situ studies on nitrogen fixation using the acetylene reduction technique. *Proc. Nat. Acad. Sci.* 58: 2071-2078.

Stickle, W.B. 1988. Pattern of nitrogen excretion in the phylum Echinodermata. *Comp. Biochem. Physiol.* 91A: 317-321.

Suzuki, M., R. Kikuchi & T. Ohnishi 1984. The polysaccharide degradation activity in the digestive tract of sea urchin *Strongylocentrotus nudus*. *Bull. Jpn Soc. Sci. Fish.* 50 : 1255-1260.

Sweijd, N.A., D. Pillay, C.D. McQuaid, V.H. Bandu & A.A.W. Baecker 1989. Filamentous structures associated with the gut mucosa of the sea urchin *Parechinus angulosus*. *Electron Microsc. Soc. Southern Africa Proc.* 19: 99-100.

Tabor, P.S., Deming J.W., Ohawada K.A. & R.R. Colwell 1982. Activity and growth of microbial populations in pressurized deep-sea sediment and animal gut samples. *Appl. Envir. Microbiol.* 44(2): 413-422.

Temara, A. & C. De Ridder, 1992. Nodule formation by the intradigestive symbionts of the spatangoid echinoid *Echinocardium cordatum* (Echinodermata). Scalera-Liaci L. & C. Canicatti (eds), *Proc. 3rd Eur. Echinoderm Conf.*: 237-239. Balkema, Rotterdam.

Temara, A., De Ridder C. & M. Kaisin, 1991. Presence of an essential polyunsaturated fatty acid in intradigestive bacterial symbionts of a deposit-feeder echinoid (Echinodermata). *Comp. Biochem. Physiol.* 100 B: 503-505.

Temara, A., De Ridder C., Kuenen J.G. & L.A. Robertson, 1993. Sulfide-oxidizing bacteria in the burrowing echinoid, *Echinocardium cordatum* (Echinodermata). *Mar. Biol.* 115: 179-185.

168

Thorsen, M.S.1998. Microbial activity, oxygen status and fermentation in the gut of the irregular sea urchin *Echinocardium cordatum* (Spatangoida: Echinodermata). *Mar. Biol.* 132: 423-433.

Thorsen, M.S. 1999. Abundance and biomass of the gut-living microorganisms (bacteria, protozoa and fungi) in the irregular sea urchin *Echinocardium cordatum* (Spatangoida: Echinodermata). *Mar. Biol.* 133: 353-360.

Trench, R.K. 1987. Dinoflagellates in non-parasitic symbioses. In: F.J.R. Taylor (ed), *The Biology of Dinoflagellates*: 531-570. Blackwell, Oxford.

Tysset, C., Mailloux M., Brisou J. & M. Jullian 1961. Sur la microflore normale de l'oursin violet des côtes algéroises (*Paracentrotus lividus* Lamk.). *Arch. Inst. Pasteur Algérie* 29: 271-286.

Unkles, S.E. 1977. Bacterial flora of the sea urchin *Echinus esculentus. Appl. Envir. Microbiol.* 34: 347-350.

Van Duyl, F.C., A.J. Kop, A. Kok & A.J.J. 1992. The impact of organic matter and macrozoobenthos on bacterial and oxygen variables in marine sediment boxcosms. *Neth. J. Sea Res.* 29: 343-355.

Vismann, B. 1991. Sulfide tolerance: physiological mechanisms and ecological implications. *Ophelia* 34: 1-27.

Walker, C.W. & M.P. Lesser 1989. Nutrition and development of brooded embryos in the brittlestar *Amphipholis squamata*: Do endosymbiotic bacteria play a role? *Mar. Biol.* 103: 519-530.

Ward-Rainey, N., Rainey F.A. & E. Stackebrandt 1996. A study of the bacterial flora associated with *Holothuria atra. J. Exp. Mar. Biol. Ecol.* 203: 11-26.

Watkins, L. & J.W. Costerton 1984. Growth and biocide resistance of bacterial biofilms in industrial systems. *Chemical Times and Trends*: 35-40.

Weese, A.O. 1926. The food and digestive processes of *Strongylocentrotus droebachiensis. Publ. Puget Sound Lab.* 5: 165-178.

Yamanouchi, T. 1939. Ecological and physiological studies on the holothurians in the coral reef of Palao Islands. *Stud. Palao trop. Biol. Sta.* 4: 603-636.

Yano, Y., Y. Machiguchi & Y. Sakai 1993. Digestive ability of *Strongylocentrotus intermedius. Bull. Jpn Soc. Sci. Fish.* 59: 733.

Yingst, J.Y. 1976. The utilzation of organic matter in shallow marine sediments by an epibenthic deposit-feeding holothurian. *J. Exp. Mar. Biol. Ecol.* 23: 55-69.

Zobell, C.E. & R.Y. Morita 1957. Barophilic bacteria in some deep sea sediments. *J. Bacteriol.* 73: 563-568.

Echinoderm taphonomy

WILLIAM I. AUSICH

Department of Geological Sciences, 155 South Oval Mall, The Ohio State University, Columbus, OH 43210, USA

CONTENTS

1 INTRODUCTION

Taphonomy (Greek: "burial law") was originally defined as the "study of the transition (in all details) of animal remains from the biosphere into the lithosphere" (Efremov 1940, p. 85), and in modern study as "the study of processes of preservation and how they affect information in the fossil record" (Behrensmeyer & Kidwell, 1985). This is the interval between the death of an organism and its exposure as a fossil at the Earth's surface, where weathering processes occur.

All modern paleontological study must include taphonomic considerations, either explicitly or implicitly, because the quality of available data may limit the inferences that can be made about ancient organisms and ancient ecosystems. Taphonomic studies include the nature of preservation, which differs for major groups because of differing proportions of hard parts to soft parts, composition of hard parts, and robustness of hard parts. Taphonomic processes, such as decomposition, disarticulation, transportation, and diagenesis are also important (Fig. 1). Normal decay and decomposition lead to a progressive loss of biological information, from an organism's genotype to its morphology (Fig. 2). Thus, taphonomic inquiry also examines the data biases that result from differential preservation and are now a part of the fossil record.

Fortunately, this normal progression of information loss is arrested in many cases. The most common, of course, is that the skeleton of many organisms is durable so that it can withstand decay and common sedimentary and diagenetic processes (Fig. 3). Thus, organisms with hard parts, like the endoskeleton of echinoderms, are relatively easy to preserve and form the vast majority of fossils in the geologic record. Special circumstances are necessary for complete preservation of an organism with many parts, especially important for echinoderms, and special geochemical conditions can even preserve soft part mophology or aspects of the original biochemistry, such as proteins. Even elements of an organism's genotype have survived fossilization, in exceptional circumstances (Austin et al. 1997, Poinar 1999), although to date no fossil DNA has been extracted from echinoderms.

Taphonomy is divided into two parts: biostratinomy, between death and final burial; and diagenesis, between final burial and exposure at Earth's surface (Lawrence 1968, Müller 1979) (Fig. 1). Biostratinomy begins with the death struggle of an organism and ends with final burial (Müller 1979). Necrosis, death processes, is an essential part of understanding echinoderm taphonomy, because to a large extent the mode of death determines echinoderm preservation. This is the case because echinoderms are constructed of multiplated endoskeletons. Individual plates are composed of high-magnesium calcite (10 to 20 mole % $MgCO_3$) (Clark & Wheeler 1992, Weber 1969, MacQueen et al. 1974, Donovan 1991), and they are bound together by ligaments, muscles, cement, or some combination of these. Ligamentary and muscular tissues decay rapidly after death, so complete echinoderm preservation typically requires that an echinoderm be buried alive.

During the past two decades, three reviews of echinoderm taphonomy have been completed, Lewis (1980), Donovan (1991), and Brett et al. (1997). Each of these important summaries have a different focus. Lewis (1980) principally discussed the controls on disarticulation and the processes by which disarticulation is halted to bring about good preservation. Donovan (1991) concentrated on the important processes for echinoderm preservation, and Brett et al. (1997) primarily developed a model of typical echinoderm preservational types and a model for echinoderm taphofacies.

An additional discussion of echinoderm taphonomy is warranted both to synthesize the diverse literature on echinoderm taphonomy as well as to integrate a variety of taphonomic approaches and interpretations that have recently developed. This contribution will 1) consider the history of echinoderm taphonomic studies; 2) summarize echinoderm taphonomic processes; and 3) discuss contributions of echinoderm taphonomy in answering questions in echinoderm anatomy, echinoderm paleoecology, and sedimentation and stratigraphy.

2 EARLY HISTORY OF ECHINODERM TAPHONOMY

Taphonomic research has flourished only relatively recently despite the fact, as pointed out by Cadée (1991), the first taphonomic study was completed by

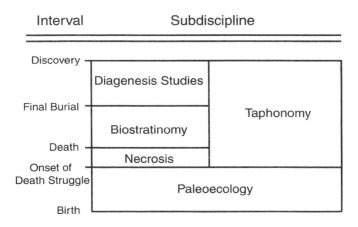

Figure 1. Paleoecology and taphonomy with taphonomy subdisciplines (modified from Lawrence 1968).

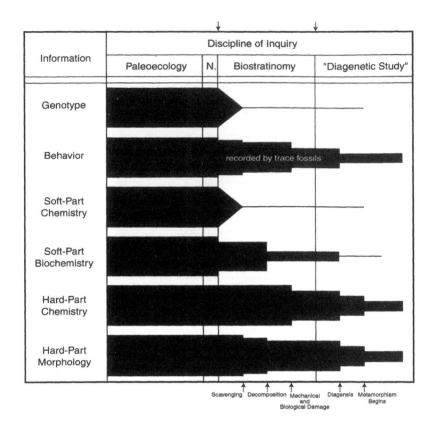

Figure 2. Information loss through the various stages of taphonomy. With exceptional preservation, even the genotype may be preserved, however with enough metamorphism all traces of the fossil record are eliminated. N, necrosis.

Leonardo da Vinci (see MacCurdy 1938, Mather & Mason 1939). Taphonomic concerns and thinking have a long, rich history prior to the definition of this subdiscipline by Efremov (1940) (Olson 1980, Behrensmeyer & Kidwell 1985, Wilson 1988, Cadée 1991, Martin 1999).

The earliest publications that incorporated taphonomy dealt with preservation patterns to help infer the mode of life of unusual crinoids. Typical echinoderm taphonomic research began in the German tradition of *Aktuopaläontologie* and biostratinomy, as exemplified by Seilacher (1960), Schwartzacher (1961, 1963), Schäfer (1962, 1972), Chave (1964), and Linck (1965). However, this "observational and experimental taphonomy" was not the beginning of echinoderm taphonomic thinking. A practical or "inferential taphonomy" has been understood for echinoderms for probably as long

174

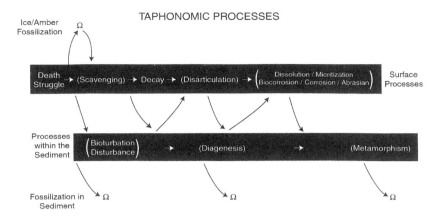

Figure 3. Taphonomic processes illustrated principally at and below the sediment surface with potential transfer positions noted. ( ), indicates process may or may not occur; S, indicates fossilization.

as collectors have seriously sought fossil echinoderm specimens. Generally, collectors do not want isolated echinoid spines, asteroid ossicles, or crinoid columnals. They want echinoderm specimens as complete as possible. Thus, collectors have intuitively understood the basic disarticulation sequence of fossil echinoderms and the most likely stratigraphic and sedimentologic circumstances in which complete fossil echinoderm specimens can be found. This very early echinoderm taphonomic thinking is documented by T. Austin, Sr. and T. Austin, Jr. where, in an unpublished manuscript (1855, see Ausich et al., in press), they recognized both the decomposition of Lower Mississippian crinoids and their contribution to limestone formation.

The bibliography presented here is a listing of cited papers that concern some aspect of echinoderm taphonomy. It is clear that the beginnings of echinoderm taphonomic study began during the early 1960s. Work steadily increased during the 1960s and 1970s (Fig. 4), with a burgeoning of research during the 1980s and 1990s.

3 BIOSTRATINOMY AND DIAGENSIS

3.1 *Necrosis and biostratinomy*

The prevailing means by which benthic invertebrates die is not known, but these should include partial or total consumption during predation (Gale

175

1986, Nebelsick 1999a), being swept by currents from a living site to an environment where life can no longer be sustained, disease (Lessios et al. 1984, Hughes et al. 1985, Aronson 1987, Lessios 1988), being killed by a sudden flush of water that cannot sustain life (i.e., anoxia, high temperature, hyposaline water), senescence, and rapid burial by sediment (see Müller 1979, Chestnut & Ettensohn 1988, Lawrence 1996 for further discussion). These necrotic processes introduce echinoderms, partial or complete, to a variety of taphonomic processes. Even ingested echinoderm hard parts pass through a predator's digestive system and are returned to the benthic environment (Wright & Wright 1940, Rasmussen 1950, Zangrel & Richardson 1963, Malzahn 1968, Janicke 1970, Schäfer 1972, Gale 1986, Watkins 1991,

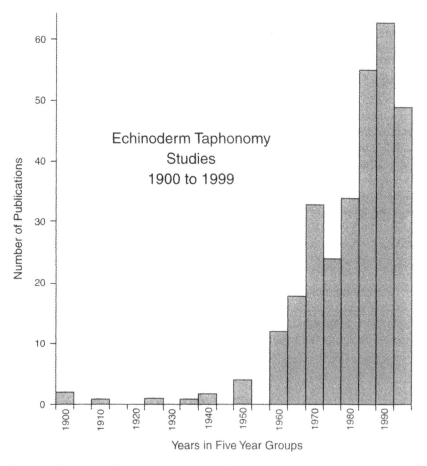

Figure 4. Histogram illustrating history of echinoderm taphonomic study cited in this review. 1985 is for studies published from 1985-1989, 1990 is for 1990-1994, etc.

Breton 1992). Echinoderms are more dense than sea water, so they remain on the sea floor immediately after death. Some have speculated that echinoderms, especially stalked echinoderms, were positively buoyant so that they floated after death (Müller 1953, Marcher 1962, Schwartzacher 1963, Seilacher 1968, Ruhrmann 1971a,b, Seilacher 1973, Gluchowski 1987). However, Blyth Cain (1968) and Meyer and Meyer (1986) reported otherwise for comatulid crinoids. If a living stalked crinoid crown is severed from its stem, it falls to the substratum (Messing et al. 1988). No evidence among living echinoderms supports post-mortem buoyancy, except among echinoids. After an animal dies and decomposition gases are generated, the buoyancy of an echinoderm is altered, so that, for example, the gas-filled body of *Asterias rubens* is distended and can be moved easily along the substratum (Schäfer 1962, 1972). Similarly, bodies of holothuroids fill with gases and roll along the substratum (Müller 1979). However, decaying echinoids (*Diadema antillarum*) are the only echinoderms that have been observed floating after death (Reyment 1986; personal observation). One way, presumably, is that these dead tests achieved positive buoyancy for a short time and sank to the bottom when further decomposition prevented trapping of gases. However, Reyment (1986) interpreted the specimen that he found floating as having been desiccated and "mummified" prior to floating. In this state, a regular echinoid could float for a long time (Reyment 1986).

Dead echinoderms at the sediment-water interface are subjected to a variety of processes that promote disarticulation of endoskeletons and scattering of individuals, partial individuals, or isolated ossicles. Breton (1992) listed the following factors that contribute to the normal progression of disarticulation: 1) predation; 2) scavenging; 3) bioturbation; and 4) currents. The timing of tissue decay in this sequence of biological and mechanical manipulation is also critical. Tissue decay is commonly immediate and has significant impact on preservation. Echinoderm plates are held together by muscular tissue, ligamentary tissue, interlocking stereom, cementation, or some combination of these. Various studies in experimental taphonomy have documented the decay sequence of echinoderms, and it is instructive to review specific examples of decomposition for each class. Examples given below should be applied with caution to other taxa and to fossils. The decay sequence for a given species under similar conditions is relatively uniform. However, within a class, genera with different constructions may behave differently. Even the same taxon exposed to different environmental conditions may behave differently. Most important, the rate of disarticulation is affected by the microbial community present, factors affecting the microbial community, temperature, oxygen content, turbulence, and the presence of scavengers (Plotnick 1986, Allison 1988, 1990, Kidwell & Baumiller 1989, 1990).

All things being equal, echinoderms should disarticulate according to the predominant type(s) of connection binding adjacent ossicles. Ligamentary tissue is a refractory tissue and should decay less rapidly than muscular tissue. As decay of a single tissue type is an area-dependent process, thinner ligamentary tissue should decay more rapidly than thicker ligamentary tissues. Cemented articulations do not decay, and for at least thin-plated specimens, fresh ligament is stronger than calcite (Smith 1984, Kidwell & Baumiller 1990). However, cemented articulations of thicker plated specimens and cemented articulations with interlocking stereom should be robust and resist breakage. This theoretical sequence of tissue decay leads to a predictable sequence of progressive plate disarticulation where contrasting articulations occur on a single animal. This has been demonstrated for echinoids (Schäfer 1972, Kidwell & Baumiller 1990) and crinoids (Baumiller & Ausich 1992, Ausich & Baumiller 1993).

A critical factor in understanding echinoderm disarticulation is that the dominant or only soft-tissue that binds adjacent plates is ligamentary. Kidwell and Baumiller (1990) demonstrated a threshold effect in the decay of echinoid ligamentary tissue. Ligamentary tissue and only slightly decayed ligamentary tissue are quite strong. Thus, disarticulation patterns in live echinoids are basically the same as those of dead echinoids that have not decayed sufficiently to reach the threshold where ligamentary tissue is weakened. During this window, ligamentary articulations may even be stronger than calcite (Smith 1984, Kidwell & Baumiller 1990). Thus, prior to reaching this threshold, echinoids may break across plates. After attaining the threshold, breakage typically occurs at plate sutures. The principal factor dictating the period of time necessary to reach the threshold for ligaments is temperature (Kidwell & Baumiller 1990). This conclusion for echinoids may be reasonably extended to the ligamentary tissues of other echinoderms.

How might an animal be expected to decay and disarticulate if left undisturbed on the sea floor, and what factors influence the disarticulation progression? A problem with the decay observations and experiments on modern echinoderms discussed below is that relatively few species have been studied. Thus, for example, one or two "model" living asteroids are used to represent the decay and disarticulation of all asteroids. Undoubtedly, one or two "model" asteroids fail to account for the range of constructional morphologies among all living asteroids, and all living and fossil asteroids would increase the diversity of morphologies even more. At least in some instances, fossil morphologies have been considered for crinoids (Meyer et al. 1989) and asteroids (Breton 1992). Furthermore, decay and disarticulation have only been calibrated in a limited number of cases against environmental variables, such as temperature, oxygen content, presence of scavengers, and burial. More cases of calibration are needed, and other variables, such as microbial floras and Eh/pH conditions, need to be considered. Despite the limitations of "model" organisms, this must be the starting point, and it is instructive to consider each living class.

3.2 *Asteroid decay and disarticulation*

Decomposition among North Sea asteroids differs radically based on constructional morphology (Schäfer 1972); he contrasted *Astropecten irregularis* and *Asterias rubens*. *A. irregularis* is larger and has heavier dorsal plating. During decomposition, the body is only bloated slightly, the arms do not become flexible, and the dorsal body wall remained intact until decomposition reached an advanced stage (Schäfer 1972). This is in sharp contrast to *A. rubens* with a more lightly constructed dorsal body wall. In *A. rubens* the following decomposition and disarticulation sequence was documented. (Schäfer 1972).

1) Colors began to fade and the body bloated as the accumulation of decomposition gases began soon after death. The body was flexible only during this stage or preceding death.
2) After five days the carcass was bloated; arms were straight, body was rigid, skin was stretched. The bloated carcass was easily moved, but it remained negatively buoyant and on the substratum.
3) After six days, the dorsal body wall began to separate from the remainder of the animal beginning at the distal arm tips. This continued for the next eleven days resulting in large areas of the dorsal body wall, soft tissue and ossicles, sloughing and subject to transportation by currents.
4) After approximately twelve days, decomposition of the digestive organs, water vascular system and ventral musculature was complete, but the ventral skeleton remains articulated (Schäfer 1972).

Eventually, the oral skeleton also disarticulated. A carcass of *Asterias rubens* preserved at stage 4 can be easily distinguished from one that was buried rapidly because the ossicles of the aboral body wall are absent. If an *A. rubens* is buried rapidly and preserved, the dorsal skeleton would be preserved superimposed on the ventral skeleton, but a general rule cannot be developed from this result because of contrasting constructional styles among asteroids.

Glynn (1984) reported nearly complete decay and disarticulation of the asteroid *Acanthaster planci* in four days and in eight days at Panama and Guam, respectively. He attributed the faster rates in Panama to be the result of a greater diversity and abundance of scavengers and predators (Glynn 1984). Moran (1992) reported a similar pattern of decay in *Acanthaster planci* from the Great Barrier Reef:

1) After three days the body wall and skeleton collapsed and various organs and tissues extended through the body wall.

2) After four days, bacterial decay and decomposition were significant with a film over the body.

3) After nine days disarticulation was nearly complete.

Scavengers played a significant role in the decomposition and disarticulation of *Acanthaster* (Moran 1992). LeClair (1993) reported similar findings for *Solaster* and *Pisaster*.

Allison (1990) examined decomposition and disarticulation among asteroids from the Puget Sound area, including *Pisaster ochraceus, Luidia foliolata,* and *Henricia leviuscula*. The decay experiments were completed at a water temperature of 6°C. Decay proceeded much slower than that reported by Schäfer (1972). Although preserved intact, *P. ochraceus* and *L. foliolata* fell apart completely when touched after 48 or 98 and 48 days, respectively. In contrast, *H. leviuscula* survived the entire 98-day experiment.

Living asteroids used in taphonomic experiments do not explain the taphonomy of all asteroids in the fossil record, because additional morphological architectures occur among fossil asteroids (Breton 1992). In particular, Breton (1992) considered the taphonomy of goniasterid asteroids largely from Cretaceous chalk facies. Goniasterids have robust marginal plates around the entire periphery of the test. Smaller, poorer articulated plates, actinolaterals, abactinals, ambulacrals, and adambulacals, cover the majority of the test. Breton (1992) concluded that after death the actinolaterals and abactinals collapsed first, perhaps accelerated by scavenger activity on this weak part of the test. The marginals were generally much more resistant, but eventually disarticulated next, beginning at the arm tips.

3.3 *Ophiuroid decay and disarticulation*

Ophiuroids are much more susceptible to rapid decay and disarticulation than asteroids because of the construction of the arms. Disarticulation of North Sea ophiuroids begins only 15 hours after death (Schäfer 1972). Ophiuroid carcasses have not been observed floating during decomposition (Schäfer 1972).

Meyer (1971) examined controlled, *in situ* ophiuroid disarticulation in tidal pools and subtidal settings in Panama. Results with *Ophiocoma echinata* and *Ophioderma cinereum* differed. Disarticulation was very rapid in both cases, with isopod scavenging contributing to ophiuroid decay and disarticulation. *O. echinata* was nearly completely disarticulated into discrete ossicles within three to five days, whereas short arm segments could remain articulated in *O. cinereum* after six days. Lewis (1986, 1987) completed laboratory disarticulation experiments with *Hemipholis elongata* and *Microphiopholis atra* from the Gulf of Mexico. Three stages of disarticulation and decay were recognized.

1) Gradual loss of body color, body fluids, and decomposition gases. Some stiffness occurred in arms.
2) Arms became flexible. No disarticulation occurred, although the dorsal disk body wall may slough off.
3) Disarticulation began. (Lewis 1987).

M. astra lost its disk integument within one week and largely disarticulated within two weeks; whereas during this time *H. elongata* had small spines disarticulate but was otherwise intact (Lewis, 1986).

3.4 *Echinoid decay and disarticulation*

Among living echinoids, regular and irregular echinoids have quite different preservational potential because of both constructional morphology and mode of life (Kier 1977, Seilacher 1979, Smith 1984, Kidwell & Baumiller 1990). Regular echinoids are epifaunal and typically have long spines, although spines are short in some regular echinoids like *Lytechinus*. In the North Sea, drooping spines on the regular echinoid *Echinus* were considered indicative of a sick, dying echinoid, and after death all of the spines drooped (Schäfer 1972). The impact of taphonomy on echinoid preservation is quite important, as demonstrated by Greenstein (1990) who concluded that the distribution of living echinoids was not mirrored in the sediment.

The following disarticulation sequence was described in *Echinus* by Schäfer (1972):
1) After seven days, the spine muscles were decomposed and spines and pedicellaria began to detach, with pedicellaria first (Smith 1984). Spines and pedicellaria fell from the aboral side of the test first.
2) After 12 days, the periproctal membrane decomposed, and the apical system typically collapsed into the test interior. The peristomal membrane also became disarticulated and, thus, the teeth of the lantern became exposed.
3) After 17 days, the lantern disarticulated but stayed within the corona.
4) After weeks, the echinoid test finally began to disarticulate, separating along plate sutures typically into ambulacral and interambulacral fields.

Irregular echinoids have short thin spines with relatively little soft tissue. Schäfer (1972) stated that these spines disarticulated very soon after decomposition began, and Smith (1984) indicated that irregular echinoid spines disarticulated within hours of death. Irregular echinoids also tend to have more robust tests, due to interlocking stereom across sutures and, perhaps, some cementation (Régis 1977, Smith 1980, 1984, Greenstein 1991).

Kidwell and Baumiller (1989, 1990) evaluated the disarticulation of regular echinoids under controlled, experimental conditions. Spine disarticulation in *Arbacia punctulata* was strongly dependent on water temperature but not by aeration (Kidwell & Baumiller 1989). The sequence of progressive disarticulation proposed by Schäfer (1972) was confirmed in the disarticulation-transportation experiments of Kidwell and Baumiller (1990), with the progressive disarticulation sequence dependent on decay of ligamentary tissues. However, if the ligamentary decay threshold is not met when an echinoid is subjected to transportation or impact, it does not follow expectations of Schäfer (1972). With ligaments intact, the corona breaks across plates. Both *Strongylocentrotus purpuratus* and *Strongylocentrotus droebachiensis* broke equatorially if tumbled prior to the ligamentary decay threshold (Kidwell & Baumiller 1990). Again, temperature is the important factor determining length of time necessary to decay ligamentary tissue.

Greenstein (1991) performed decay and burial experiments on the Caribbean echinoids *Diadema antillarium*, *Eucidaris tribuloides*, and *Echinometra lucunter*. Decay experiments were conducted at 22°C and yielded decay progression in *D. antillarium* and *E. tribuloides* that was similar to that of other regular echinoids (Schäfer 1972; and Kidwell & Baumiller 1990). However, the less robust *D. antillarium* began to lose spines rapidly, within one day, and the corona was typically disarticulated into fragments within seven days (Greenstein 1991). *E. tribuloides* decay and disarticulation were more correlated to the ligamentary decay threshold of Kidwell and Baumiller (1990). Little spine loss or disarticulation occurred before seven days, but after seven days spine loss and corona fragmentation occurred rapidly (Greenstein 1991). In contrast, the irregular echinoid *E. lucunter* has a robustly cemented corona. Therefore, despite the fact that spines were significantly disarticulated after three days, the corona remained intact throughout the experiment (Greenstein 1991). Decay and tumbling experiments further demonstrated these distinctions (Greenstein 1991). Burial delayed but did not prevent decay. Tumbling experiments demonstrated that appreciable disarticulation and fragmentation occurred within one hour, but further appreciable disintegration did not occur with additional tumbling (Greenstein 1991). The relative degree of disarticulation among the echinoids studied was correlated to the amount of stereom interlocking across plate sutures (Greenstein 1991).

In decay experiments on Red Sea clypeasteroids, *Clypeaster humilis* and *Echinodisus auritus*, Nebelsick and Kampfer (1994) reported that spine disarticulation began during the first day and was completed by the fourth day; and jaws began to disarticulate during the first five days but did not separate from the corona. In the Red Sea, dead clypeasteroids commonly remained articulated, but once fragmentation along plate sutures began, it proceeds rapidly. Corona disarticulation was probably induced by bioerosion or scavengers (Nebelsick & Kampfer 1994).

Sadler and Lewis (1996) reported decay rates for the heart urchin *Meoma ventricosa* during *in situ* decay experiments at the sediment-water interface:

1) Significant spine loss began during fourth to fifth day.
2) All spines disarticulated within two weeks and at approximately two weeks the corona began to disarticulate.
3) After one month 40% of coronas were partially disarticulated but 50% remained intact for up to twelve months.

Burial experiments delayed disarticulation; but as with those left to decay on the sediment-water interface, 50% of buried specimens remained intact for up to twelve months (Sadler & Lewis 1996).

Many Paleozoic echinoids had imbricated plating on the test, and they undoubtedly had even poorer preservation potential, perhaps similar to the diadematids (Kidwell & Baumiller 1990, Table 3), although this aspect of alternative echinoid constructions requires more study.

3.5 *Holothuroid decay and disarticulation*

Despite the fact that holothuroids lack an articulated skeleton, their decomposition is similar to asteroids. North Sea holothuroid decomposition was described by Schäfer (1972) as follows:
1) Gases generated during decomposition bloated the body, which could not float but moved freely along the substratum by bottom turbulence.
2) After several weeks, the body wall became completely decomposed, and embedded calcareous ossicles were freed.

Allison (1990) reported that decomposition and disarticulation rates among four holothuroids from Puget Sound had an order of magnitude variation. In all cases the body swelled due to decomposition gases similar to the results of Schäfer (1972). The integrity of the cylindrical body was maintained during swelling, but at some point the body disintegrated during sampling into fragment sheets of gelatinous body wall or complete disarticulation occurred. *Leptosynapta clarki* completely disarticulated after 11 days, whereas the body of *Cucumaria miniata* was still more-or-less intact with the body partially cylindrical (Allison 1990, Table 2) after 98 days.

3.6 *Crinoid decay and disarticulation*

Blyth Cain (1968) reported disarticulation within two days for the comatulid *Antedon bifida*, regardless of oxygenation conditions. Both Meyer (1971) and Liddell (1975) considered comatulid crinoids in low latitude settings in various shallow-water, controlled habitats. Meyer (1971) studied *Comactinia echinoptera* in Panama, and reported that
1) After two to three days, the cirri and majority of arms had disarticulated.
2) After six days, only the isolated calyx and a few arm fragments remained articulated.

Liddell (1975) reported similar results for *Nemaster rubiginosa* in experiments with specimens at the sediment-water interface.

1) After one to two days, the ventral disk was lost; coloration faded; distal arms, many pinnules, and most cirri were disarticulated; articulated arm fragments of 1 to 3 cm occurred; aboral flexure of articulated arms was common.

2) After six days, only the isolated calyx and a few arm fragments remained articulated.

Liddell (1975) also reported that specimens decayed in an agitated setting completely disarticulated in the time frame noted above. Scavengers accelerated decomposition and disarticulation. In contrast, specimens recovered after six days of burial were fully articulated and retained their living coloration (Liddell 1975).

Crinoid columnals and pluricolumnals may be introduced as sedimentary particles either by autotomy or after death. Column autotomy on distal synostosis nodal facets of living stalked crinoids is reported by Baumiller (1994) and Baumiller et al. (1991). Baumiller (1994) used various physical and chemical stresses to examine the detailed sequence of disarticulation among comatulids and stalked crinoids. The sequence of disarticulation was distal-most syzygies in brachials, pinnules, medial arm syzygies, proximal arm syzygies, stalk synostoses, cirri, proximal columnals, radials, and basals (Baumiller 1994, Hagdorn & Baumiller 1998). Baumiller and Ausich (1992), Baumiller and Hagdorn (1995), and Baumiller et al. (1995) examined column disarticulation, and Ausich and Baumiller (1993) examined disarticulation of the column in relation to the crown. Among living isocrinids the sequence of disarticulation recognized for stalks in laboratory experiments (Baumiller & Ausich 1992) was:

1) After five days, the column disarticulated into noditaxes, nearly equal length column segments composed of internodals (relatively constant number) and the distal nodal.

2) After 19 days the noditaxes remained intact.

3) After 22 days, the column segments had disarticulated into isolated columnals.

Temperature was shown to be important by Meyer and Oji (1992). Crinoid disarticulation was significantly slowed at low temperatures.

As with asteroids, fossil crinoids, especially Paleozoic crinoids, had a much broader range of constructional morphology than living crinoids, so a strict actualistic approach must be used with caution. For example most Paleozoic crinoids lacked muscular tissue in the arms. Crinoids with only ligamentary tissue in the arms had a different disarticulation sequence than isocrinids (Ausich & Baumiller 1993).

1) Arms, pinnules, and cirri began to disarticulate approximately when the column began to disarticulate into nearly equal-length segments.

2) Lightly constructed aboral cups/calyxes would disarticulate.
3) The column segments disarticulated into isolated columnals.
4) More robustly constructed aboral cups/calyxes would finally begin to disarticulate.

Disarticulation of the column into equal-sized segments prior to complete disarticulation into isolated columnals, as experimentally documented in living isocrinids, is also recognized among fossil isocrinids (Baumiller et al. 1995), fossil holocrinids (Baumiller & Hagdorn 1995, Hagdorn & Bamiller 1998) and among some Paleozoic crinoids (Baumiller & Ausich 1992, Ausich & Baumiller 1998).

Crinoids also produce sedimentary particles through periodic autotomy of the distal stem. This is demonstrated among living isocrinids because the distal stalk columnal is most commonly large and the distal facet is a synostosis (Conan et al. 1981, Fujita et al. 1987). This behavior is also suggested for Triassic holocrinids (Hagdorn & Baumiller, 1998) and for fossil isocrinids.

3.7 *Echinoderms as sedimentary particles*

Complete specimens, individual ossicles, and fragments of any size can be transported along the sediment-water interface. Because of the strength of ligamentary tissue, complete specimens can be transported short distances (Okulitch & Tovell 1941) or considerable distances, if it occurs while the animal is still alive or shortly after death (Seilacher 1968, Kesling 1969, Kidwell & Baumiller 1990, Meyer et al. 1989). Echinoids with spines absorb impact well while being transported (Strathmann 1981), thus protecting the corona from breakage.

Evidence of whole animal transportation is provided by abraded whole asteroids (Blake 1967, Gale 1986); by aligned organisms such as ophiuroids (Müller 1963), eocrinoids (Sprinkle 1973), and echinoids (Nebelsick & Kroh 1999); by aligned organisms and drag traces (Seilacher 1960, Goldring & Langenstrassen 1979); and by lenticular concentration of echinoids (Néraudeau & Breton 1993). Crinoids provide numerous examples of current orientation of complete organisms (Lane 1973, Müller 1979 and many other studies). Roman and Fabre (1986) interpreted a Lower Cretaceous echinoid bed as a beach deposit, to which whole echinoid individuals were transported and accumulated. Lawrence (1996) discussed several examples in which whole organism transportation occurred during echinoderm mass mortality events.

Aslin (1968) distinguished rapidly buried echinoids from those that had lain on the sediment-water interface, because the former had spines, lantern, and apical system and tests hollow or filled with cement. In contrast, echinoids that had spent some time after death on the bottom lacked spines, lantern, and apical system, and the tests were filled with sediment other than the enclosing sediment in which they were preserved. Breton et al. (1994) recognized nearly whole and partial asteroids as being transported because of poor preservational details.

Preserved posture of complete echinoderms may be as much a factor of when an animal was transported as the flexibility of an animal during life or the turbulence during transportation. Schäfer (1972) reported that asteroid arms could only be bent immediately preceding or immediately after death. Once decay began, the asteroid body became rigid and arms remained extended.

Crinoid pluricolumnals were common sedimentary grains in many habitats, and long pluricolumnals commonly have preferred orientations. Schwartzacher (1963) reported that crinoid pluricolumnals began to move when current velocities reached 2 to 10 cm/sec. Schwartzacher (1961, 1963) also demonstrated that crinoid pluricolumnals have a wide array of orientations when in traction bedload, but the predominant orientation is one that is a few degrees from normal to the current direction. This orientation is the result because long pluricolumnals roll and slide along the substratum but bottom roughness and other factors cause deviations from rolling normal to the current direction. Pluricolumnal orientations plotted on a symmetrical rose diagram yield a bimodal distribution, with random scatter, and the bisector of the obtuse angle indicates the recorded current sense (Schwartzacher 1961, 1963, Ausich & Lane 1980). Nagle (1967) also measured crinoid pluricolumnal orientations. His results had a unimodal distribution perpendicular to currents rather than the bimodal distribution reported by Schwartzacher (1961, 1963) and Ausich and Lane (1980). However, it is not possible to know whether the results of Nagle (1967) are different or the result of small sample size.

After soft tissue decay and disarticulation of the endoskeleton, isolated echinoderm ossicles become part of the traction bedload at the sediment-water interface. Because echinoderm skeletons have a three-dimensional skeletal mesh called stereomic microstructure, the specific gravity of ossicles is much less than is apparent. Savarese et al. (1997) measured the porosity and specific gravity of both living and fossil crinoid columnals. They determined that porosity values ranged from 52 to 72% in living crinoids, and fossils ranged from 39 to 69%. The specific gravity of calcite is 2.72, but because of porous stereom these columnals with soft tissue absent yielded specific gravities of 1.47 to 1.83 among modern examples and 1.54 to 2.10 among fossils. Blyth Cain (1968) reported slightly lower specific gravities for *Antedon bifida*. Ignoring the influences of drag, which affected both settling rates and entrainment characteristics, a crinoid columnal behaved hydrodynamically like a quartz sphere with 0.1 times the diameter (Savarese et al. 1997). Thus, disarticulated echinoderm ossicles should be highly susceptible to transportation, especially prior to diagenesis when the stereomic porosity can be occluded with cement. Transportation of crinoid pluricolumnals is demonstrated because they are recognized as making tool marks on sediment surfaces (Benton & Gray 1981). However, evidence for extensive dispersal is mixed, and a distinction between allochthonous and parautochthonous postmortem transportation must be made. Allochthonous transportation is transportation away from the facies or habitat where an individual lived, whereas a parautochthonous occurrence is transportation from the living site but within the facies or habitat where it lived. In contrast, autochthonous preservation is *in situ* preservation.

Allochthonous distribution of echinoderm ossicles was clearly documented by Schäfer (1972). Ruhrmann (1971a, b) and Brett (1985) documented that skeletal elements from the same organism were preserved in different facies. Brett (1985) showed that pelmatozoan living sites were on Silurian bioherms indicated by the *in situ* preservation of holdfasts, but columnal and calyx remains were typically preserved in bioherm flank beds. Kobluk and Lysenko (1984), Kobluk and Mielczarek (1984), and Greenstein (1989) documented that slight postmortem movement of diademid echinoids occurred down the reef slope on Bonaire in the Netherlands Antilles.

Even echinoderm mass mortality without burial will not necessarily leave a mark in the sedimentary record. The best known recent example is from the mass mortality of *Diadema antillarum* that occurred throughout the Caribbean during 1983 (Lessios et al. 1984, Hughes et al. 1985, Lessios 1988). Despite the rapid, catastrophic death of thousands of individuals, Greenstein and Meyer (1988) and Greenstein (1989) did not find a significant increase in *Diadema* remains in Caribbean sediments. Similarly, mass mortalities of the asteroid *Acanthaster planci* on the Great Barrier Reef left no signature in the sediment (Frankel 1978, Moran 1992, Moran et al. 1986, Greenstein et al. 1995). Greenstein et al. (1995) found that the abundance of recoverable ossicles decreased dramatically four weeks after the mortality event. Another example of a mass mortality was reported by Häntzschel (1936), where complete *Echinocardium cordatum* transported into the wash zone on the islands of Norderney (Schäfer 1972). These echinoids were piled as deep as 30 cm along a beach width of 8 m and extended for 5 km along the shore. However, within a few weeks the tests were fragmented, and no record of this incredible accumulation remained.

Despite the relative ease with which disarticulated echinoderm ossicles can be transported (Savarese et al. 1997), they are commonly not widely dispersed, rather they are reported as parautochthonous preservation. Among living crinoids, this has been demonstrated both for shallow-water and for deep-water settings. Meyer et al. (1984b), Meyer and Meyer (1986), Lewis (1997), and Meyer (1997) reported that poorly sorted, disarticulated crinoi-

dal material occurred in the sediment beneath living shallow-water comatulid crinoids. Messing (1997), Fujita et al. (1987), Ameziane-Cominardi
and Roux (1987), and Messing and Rankin (1995) reported the same for
living deep-sea, stalked crinoids. Glynn (1984) concluded that ossicles of
Acanthaster planci were not widely distributed after death, as did Nebelsick (1999b) for *Clypeaster* in the Red Sea. Among fossils, Broadhurst and
Simpson (1973) reported only minor transportation for Lower Carboniferous reef flank beds, and Watkins (1991) reported very little sorting of crinoidal skeletal elements in Silurian crinoidal, reef-related sediments. Contrary
to other studies Blyth Cain (1968) concluded that isolated crinoid ossicles
would have to undergo considerable transportation before significant size
sorting would occur. Messing et al. (1990) and Llewellyn and Messing
(1993) reported local transportation (parautochthonous) among deep-sea
crinoids and echinoids. With parautochthonous preservation, disarticulated
remains that can be identified to species have great utility for paleoecological studies (Sprinkle 1973a, Lane & Webster 1980, Kammer 1984, Holterhoff 1997a, 1997b).

In the *Aktuopaläontologie* tradition, Nebelsick (1992a, 1992b, 1994,
1995a,b,c) completed a series of studies on the distribution of live echinoids
and their remains in the Red Sea that parallel and extend work in the Caribbean Sea by Greenstein (1991). In the Red Sea, complete dead echinoid coronas present at the bottom where they lived were rare for regular echinoids,
common for clypeasteroids, and absent for spatangoids (Nebelsick 1992a,
1992b). However, fragments of these echinoids had a fairly high fidelity to
that of the living animals (Nebelsick 1992a,b), similar to Greenstein (1991)
and Néraudeau et al. (1998). Comparison of preservation among five clypeasteroids in the Red Sea indicated three primary factors leading to better preservation potential: 1) corona size; plate thickness; and presence of internal,
paired, radially oriented buttresses and pillars around the circumference; 2)
predation, both occurrence and type; and 3) life habitat (Nebelsick 1995c).

The geologic record is replete with examples of reefs or carbonate buildups whose lithologic composition contrasts with the enclosing sediments
(Brett 1985, Ausich & Meyer 1990 and many others), and many Paleozoic
and Mesozoic examples supported large pelmatozoan populations. The fact
that echinoderm ossicles are commonly confined to the buildup facies and
not distributed to enclosing sediments provides further evidence of restricted
postmortem transportation of isolated ossicles, even if redistributed some-
what among buildup facies (Manten 1971, Franzén 1977, Brett, 1985). The
ultimate transportation of individual echinoderm ossicles depends on many
factors, including size, turbulence of the living site, and rate of deposition
at the living site.

One of the more amazing discoveries in echinoderm biology in recent
years was by Oji and Amemiya (1998), who discovered that distal crinoid

stalks (both naturally autotomized and surgically detached) and isolated pluricolumnals could live free from the crown for at least 13 months. This discovery on the living crinoid *Metacrinus rotundus* suggests that modern sediments with pluricolumnal fragments and numerous crinoidal limestones from the rock record may be or may have been composed of living bioclastic debris. This may account for the numerous pluricolumnals common in ancient sediments, and it may delay or counteract the ease of echinoderm ossicle transportation noted by Savarese et al. (1997). Furthermore, crinoid columns that lose their crown may not always become sedimentary particles immediately. Messing and Llewellyn (1992) reported *in situ*, crownless, erect crinoids for more than six months.

At the sea floor, echinoderm ossicles are subject to a wide variety of physical, chemical and biological taphonomic processes (Dodd et al. 1985, Kidwell & Bosence 1991). Chemical processes include dissolution and corrosion, reported for echinoids (Peterson 1976, Flessa & Brown 1983, Nebelsick 1994), shallow-water crinoids (Lewis & Peebles 1988, Lewis et al. 1990), deep-water crinoids (Llewellyn & Messing 1994), and ophiuroids (Jensen & Thomsen 1987). Dissolution experiments of echinoderm high-Mg calcite generally indicated intermediate rates of dissolution if compared to other biogenic forms of calcium carbonate (Flessa & Brown 1983).

The physical behavior of echinoderm ossicles in bedload has had contrasting interpretations. Seilacher (1973) believed that the highly porous and low density ossicles would be less likely to be abraded physically. However, Chave (1964) demonstrated with tumbling experiments that echinoderm calcite is much less durable than that of robust molluscs, which would counter balance Seilacher (1973), and Blake (1967), Gale (1986), and Breton (1992) reported abraded asteroid ossicles. In any event, Meyer and Meyer (1986) and Lewis et al. (1990) reported abrasion on ossicles from shallow-water habitats, Llewellyn and Messing (1993) reported abrasion in deep-water habitats, and Nebelsick (1994, 1999b) reported abrasion on cylpeasteroids. Llewellyn and Messing (1993) reported a correlation between abrasion and distance from the living population (Messing, 1997).

Biological processes are varied. One obvious biological process is that crinoid pluricolumnals and echinoid and crinoid tests are sedimentary particles that become sites of larval attachment. Numerous organisms, from algae to foraminifera and from sponges to other echinoderms, attach to echinoderm surfaces (Ernst et al. 1973, Liddell & Brett 1982, Meyer & Ausich 1983, Gale 1986, Powers & Ausich 1990, Breton 1992, Kudrewicz 1992, Nebelsick 1994, 1996a, 1999b, Nebelsick et al. 1997). This includes both macroscopic encrusters, as well as algal, bacterial, and fungal microborings that cause bioerosion, including micritization (Roux 1975, Ameziane-Cominardi & Roux 1987, Lewis et al. 1990, Breton 1992, Lewis & Echols 1994, Nebelsick 1994, Echols & Lewis 1995, Bourseau et al. 1997). Simon and Poulicek (1990) and Simon et al. (1990) demonstrated that bioerosion occurred both in aerobic and aerobic conditions. Liddell and Brett (1982) used the occurrence of epizoans to conclude that some well-sutured camerates and blastoids must have been intact on the sea floor for a few years in order to accumulate the epizoan assemblages present. Also, bite marks occur on echinoderm grains (Lewis et al. 1990), but healed bite marks record unsuccessful predation rather than a taphonomic process (Nebelsick 1994). Breton (1992) reported biting traces on asteroid ossicles that could be either due to predation, scavenging, or manipulation of isolated ossicles.

Echinoderms can also be transported vertically through a floating ice shelf system by a process called cryogenic taphonomy (Ausich & Hart 1990). This was identified on the Antarctic shelf where the bottom fauna became incorporated in ice, floated, and became frozen into the McMurdo Ice Shelf. Eventually, the "fossils" thawed in surface meltwater, and disarticulated ossicles were transported back to the benthic sediment (Ausich & Hart 1990). Crinoids that passed through this cryogenic cycle were identifiable due to fractured surface stereom on ossicles and, perhaps, by fractured ossicles, which are caused by the freeze and thaw prevalent of this setting. Although Ausich and Hart (1990) only considered crinoids, the same fractured texture should be a taphonomic signature left on other echinoderm ossicle surfaces that pass through this very special transportation.

3.8 *Diagenesis processes*

Early during diagenesis, the high-magnesium calcite of echinoderms, which is metastable, is converted to low-magnesium calcite. As discussed in Donovan (1991), this is due to replacement of magnesium by calcium cations with no change in the calcite crystal structure (Land 1967, Bathurst 1971, Smith 1984). An entire echinoderm ossicle is a single crystal with optical continuity and commonly becomes the seed for syntaxial, low-magnesium calcite cement (Evamy & Shearman 1965, 1969, Bathurst 1971). Syntaxial cement either occludes intra-ossicle porosity, increases ossicle size, or both; and this can potentially occur at the sediment-water interface but, probably more commonly, after burial. Once buried, intra-ossicle porosity can also be filled, commonly with micrite, siliciclastic muds, or glauconite (Roux 1975). Given the appropriate diagenetic microenvironment, pyrite can also form, either as pyrite framboids within or on ossicles (Gaspard & Roux 1974, Jensen & Thomsen 1987, Feldman 1989, Breton 1992, Bartels et al. 1998, Bourseau et al. 1998) or even replacing soft tissue (see below). Améziane-Cominardi and Roux (1987) reported that growth of pyrite framboids contributed to the fracturing of crinoid ossicles weakened by biocorrosion and micritization. Also, Feldman (1989) reported dolomitization of echinoderm ossicles following pyritization during early diagenesis. Phosphatization of echinoderms may also occur under appropriate diagenetic conditions (Kudrewicz 1992, Breton 1992), as will replacement of echinoderm calcite by limonite and silica (Breton 1992, Simms 1994).

Compaction and breakage of echinoderms can also occur during diagenesis. Although this is probably common, it is little studied. Two exceptions are Ramsey (1967) and Durham (1978), who studied irregular echinoids. Corrosion can also occur after burial as the result of pressure solution. Breton (1992) reported dissolution on echinoderm ossicles when in contact with siliciclastic sediment grains. In some instances, this dissolution produced pits that mimic plate sculpturing.

4 SHORT-CIRCUITING TAPHONOMIC PROCESSES

4.1 Complete skeletal preservation

Given the rapidity of echinoderm decay and disarticulation, echinoderms should be preserved only rarely. In a study of a subtidal community, Schopf (1978) predicted that asteroids and ophiuroids should only be preserved as isolated ossicles and, following Kier (1977), that infaunal echinoids should be preserved more commonly than epifaunal echinoids. Yet, complete and nearly complete echinoderms are remarkably common in the fossil record. Thus, for preservation of complete echinoderm fossils, the expected progression of decomposition and disarticulation must have been terminated. The most common means by which this occurred was through rapid, relatively deep burial. Depending on the circumstances, burial that is too shallow may not result in complete preservation, because a vagile individual may be able to burrow from beneath the blanket of sediment, bioturbation may disturb or scatter the skeletal elements (Bell 1976, Gale 1986, Maples & Archer 1989), or subsequent bottom turbulence may exhume the specimen. Even only partial burial improves preservation on the down side of the organism (Kauffman 1981, Brett & Baird 1986, Simms 1986, Hall 1991). However, if exhumation occurs after soft-tissue decomposition within the sediment, disarticulation will occur rapidly (Schäfer 1972).

Schäfer (1972) reported that the asteroid *Asterias rubens* cannot escape from burial of 60 cm or more. In the North Sea, both this species and *Astropecten irregularis* apparently have only limited ability to move vertically in sediment and cannot overcome either slow or rapid burial. Death occurs due to suffocation (Schäfer 1972). Rapid burial yields a complete animal, but the animal will respond to slow burial by crawling horizontally. Resistance of the accumulating sediment imposes drag on the animal so that the individual moves with one or two leading arms or with all arms trailing, which has been mistaken for current orientation (Schäfer 1972, Fig. 56). Animals that

191

experience a gradual deterioration of environmental conditions preceding death may autotomize their arms, with the largest individuals reacting first (Schäfer 1972).

Some North Sea ophiuroids can move upward through accumulating sediment; if deposition is not too rapid. However, others can be buried by as little as five cm of rapidly deposited sediment (Schäfer 1972). Species with long, flexible arms (e.g., *Amphiura filiformis*) have a trauma posture (and preserved death posture) with tightly twisted, coiled arms. In contrast, species with short, stiff arms like *Ophiura texturata* have a trauma posture with outstretched arms (Schäfer 1972), and this becomes the preserved posture.

Echinoids that live on hardgrounds or among cobbles are more typically crushed than buried by sediment, which accelerates disarticulation. Those that live on soft substrata are susceptible to burial (Schäfer 1972). *Psammechinus miliaris* lives on sand substrata in the North Sea and must be buried by more than 20 cm for entombment (Schäfer 1972). With less than five cm of sediment deposited on a *P. miliaris*, it will maintain its horizontal attitude and move up to the new sediment-water interface. If buried by 5 to 20 cm of sediment, a *P. miliaris* animal will rotate upward and burrow to the surface (Schäfer 1972).

Irregular echinoids live infaunally and are subject to both burial and exhumation. Schäfer (1972) reported that *Echinocardium cordatum* were entombed after being buried rapidly by 30 cm of fine sand. In rare instances, crinoids have been buried in an upright, erect posture (Müller 1953, Donovan & Pickerall 1995), undoubtedly indicative of rapid burial.

Some of the same characteristics that are used to infer catastrophic burial were once considered evidence for quiet-water settings. For example, consider Laudon and Beane (1937, p. 238): "Because of the excellence of preservation and complete absence of orientation in their long directions, it is assumed that they lived in quiet water. The thin bedded evenly laminated dolomitic limestone suggests that they were not rapidly buried." Even as late as 1968, the rapidity of "catastrophic burial" was not fully appreciated in all cases. For example, "ultimately an influx of calcareous mud smothered the living community. The mud filled in the channelways between the limestone knobs and then encroached upon the upper surfaces of the knobs. The low-lying edrioasteroids, corals, bryozoans and worm tubs were the first to be covered and finally the cystoids succumbed. Continued catastrophic deposition resulted in a thin blanket-like layer of mud above the knobs which prevented disarticulation and/or destruction of the organisms by scavengers." (Koch & Strimple 1968).

Rapid burial is now recognized as being essential for excellent echinoderm preservation (Spencer & Wright 1966, Blake 1967, Sprinkle & Gutschick 1967, Aslin 1968, Blyth Cain 1968, Kier 1968, Bantz 1969, Kessling 1969, Lane 1971, Rosenkrantz 1971, Hess 1972a, Schäfer 1972, Lane 1973,

Brower 1974, Blake 1975, Brower & Veinus 1978, Durham 1978, Blake 1980, Meyer et al. 1981, Smith & Paul 1982, Sprinkle 1982a, Hess 1983, Welch 1984, Hess 1985, Brett & Baird 1986, Gale 1986, Brett et al. 1991, Durham 1993, Whiteley et al. 1993, Lane & Ausich 1995, Bourseau et al. 1998, and others). In fact, the opposite is also generally recognized, complete preservation of echinoderms or other taphonomically liable organisms is typically an indication of rapid sedimentation (Brett & Baird 1993), although some exceptions occur (Blake 1967). Rapid burial is recognized as important in most environmental settings, but it is essential on hardground surfaces (Brett & Liddell 1989, Guensburg 1984). More recently, storm deposits, tempestites, have been recognized as the most common process for this rapid burial. The role of tempestites in echinoderm preservation has been discussed by numerous workers. Perhaps the first to attribute good echinoderm preservation to storms was Hawkins and Hampton (1927). More recent advocates of tempestite preservation include Rosenkrantz (1971), Kier (1972), Goldring and Stevenson (1972), Bloos (1973), Watkins and Hurst (1977), Lewis (1980), Brett and Eckert (1982), Franzén (1982), Kolata and Jollie (1982), Sprinkle and Longman (1982) Schumacher and Ausich (1983), Guensburg (1984), C.A. Meyer (1984), Aigner (1985), Hess and Holenweg (1985), Schumacher and Ausich (1985), Schumacher (1986), Chestnut and Ettensohn (1988), Courville et al. (1988), C.A. Meyer (1988a, 1988b), Miller et al. (1988), Parsons et al. (1988), Lehman and Pope (1989), Meyer et al. (1989), Brett and Seilacher (1991), Hess (1991a, 1991b), Schubert et al. (1992), Hess (1994), O'Brien et al. (1994), Brett (1995), Hess and Blake (1995), Daley (1996), Stilwell, et al. (1996), Brett and Taylor (1997), Hagdorn and Baumiller (1998), and others. More generally, rapid burial, where the bottom is overwhelmed with sediment, is called an obrusion deposit (Seilacher 1970, Brett 1990); and tempestites are among the most common type. In addition to simple burial, Sass and Condrate (1985) suggested that a large influx of freshwater with a tempestite in a nearshore setting may have anaesthetized ophiuroids prior to burial, thus leading to exceptional preservation.

Processes other than tempestites may also result in rapid burial. Echinoderms originally interpreted as part of a tempestite deposit (Kolata & Jollie 1982) have been reinterpreted as possibly being smothered by the remains of a plankton bloom (Jacobson et al. 1988, Brower & Kile 1994). Sprinkle (1973) suggested that turbidites were an important process for eocrinoid preservation. Hess and Blake (1995) interpreted a storm as responsible for burial of infaunal asteroids and echinoids, but final burial did not occur until these infauna were exhumed, dragged across the sea floor, and reburied.

Echinoderms preserved in ancient channels are commonly interpreted to be autochthonous, Whereas these echinoderms could have been buried during a storm, it is equally possible that channel deposition unrelated to storms was responsible for rapid burial and preservation. Examples of channel-dwelling echinoderms include a Mississippian blastoid (Sprinkle & Gut-schick 1967), Ordovician crinoids (Haugh 1979, Eckert 1987, LeMenn & Spjeldnaes 1996), Mississippian crinoids (Ausich et al. 1979), and post-Paleozoic asteroids (Hess 1972).

Erdtmann and Prezindowski (1994) and LaDuca and Brett (1997) concluded that anoxia was responsible for preservation of the unusual Silurian crinoids from the Mississinewa Shale of Indiana (Lane & Ausich 1995). Diett and Mundlos (1972) suggested that an ophiuroid swam into oxygen-poor waters that enabled exceptional preservation of ophiuroids in the LaVoulte beds of southern France. Kesling (1971) suggested that Devonian asteroids were entrapped in low salinity waters, as was suggested by Bourseau et al. (1991). Bourseau et al. (1991) examined the preservation of ophiuroids from Late Jurassic lithographic limestones of Cerin, France. They concluded that ophiuroids were swept into a lagoon by storms. In the lagoon, they were killed by either a salinity or a temperature contrast. Individuals may have been preserved under a microbial film (Bourseau et al. 1991). Reduced salinity was also suggested as a factor in the death and preservation of asteroids (Roman & Strougo 1987, Breton 1997).

Even with preservation of a complete animal, the resultant fossil may not be ideal. This is well documented by the preservation of the Devonian asteroid, *Michiganaster*. In unraveling the reasons for the preservational condition of this multi-armed asteroid, Kesling (1971) identified the following taphonomic processes and considerations: 1, the disk was not robustly constructed, so the disk collapsed after death and was distorted further after burial and compaction; 2, distortion due to muscular contractions associated with rigor mortis; 3, loss of some plates due to decay of the loosely constructed disk; 4, uneven replacement of plates; and 5, dissolution of plates during later diagenesis or weathering. Blake and Zinsmeister (1979) reported excellently preserved asteroids with pedicellaria lacking. The test of edrioasteroids, both the oral surface and peduncle, is composed primarily of imbricate plates that are highly susceptible to collapse (Bell 1976, Meyer 1990), and complete edrioasteroids were commonly disturbed by bioturbation if the entombing shales were thin (Bell 1976).

Numerous examples of well-preserved echinoderms that were preserved through rapid burial can be cited, in addition to those above. A few additional examples include Breton (1997) for asteroids, Kesling and LeVasseur (1971) for ophiuroids, Meyer and Weaver (1980) for crinoids, Rozenkranz (1971) for echinoids, and Tetreault (1995) for echinoids and ophiuroids.

4.2 *Soft-part and molecular fossil preservation in echinoderms*

Soft-part preservation in echinoderms is very rare. In addition to rapid burial without subsequent physical or biological disturbance, special early diagenetic conditions must be present. Examples include pyritization, as in the Hunsrück Slate (Devonian) of Germany, where asteroids, ophiuroids, and perhaps crinoids display soft-part preservation (Bergström 1990, Bartels & Blind 1995, Briggs et al. 1996). Holothurian preservation is perhaps best known from the Mazon Creek area of Illinois (Middle Pennsylvanian) (Baird et al. 1986, Sroka 1988), but soft-part holothurian preservation is also known from other localities (Pawson 1980, Smith & Gallemí 1991). Preservation of tube feet occurred as molds in an Ordovician asteroid (Gale, 1987), and plated tube feet preservation occurred among ophiocistioids (Jell, 1983). Calcium phosphate replacement, which may yield exceptional preservational detail, is only reported for echinoderms from the Jurassic of Italy, where ophiuroids are preserved (Pinna 1985).

Debate on the phylogenetic affinity of *Echmatocrinus brachiatus* (Middle Cambrian, Burgess Shale) will decide whether an early example of tube feet preservation occurs among echinoderms. Sprinkle (1973) and Sprinkle and Collins (1999) believed this Cambrian fossil is a crinoid, whereas Conway Morris (1993) and Ausich and Babcock (1998, in press) regard it as a cnidarian.

In rare instances, molecular fossils have been reported from Mesozoic crinoids from Europe (Blumer 1951, 1960, 1962a, 1962b, 1965, Thomas & Blumer 1964, Hess 1972b, Hess et al. 1999). This results from special diagenetic circumstances that preserve the organic molecules called fringilites within the fossilized calcite. Preliminary results also indicate that molecular fossils may also be preserved in Paleozoic crinoids (Ausich & Chin, unpubl.).

5 MODES OF ECHINODERM PRESERVATION

Brower (1974, p. 269) identified three common preservational states for the Late Ordovician crinoids that he studied: 1) Disarticulated and isolated columnals and plates; 2) complete calyxes with arms and column absent; and 3) complete crowns with or without column attached. In general, these preservational states were correlated to rapidity of burial, with state 1 slow and state 3 rapid (Brower 1974), which is consistent with disarticulation experiments cited above.

Also with the taphonomic considerations discussed above, Smith (1984) recognized five typical echinoid preservational styles that have inferred taphonomic histories: 1) Whole corona complete with spines, pedicellaria, and lantern; 2) whole corona with or without apical plates but with peristomal plates, periproctal plates, spines, and lantern absent; 3) coronal segments disarticulated along plate sutures and isolated plates; 4) discrete accumulation of disarticulated coronal plates, spines, and lantern plates; 5) coronal fragments broken across plates, with or without spines. Preservational mode (1) occurred due to rapid, final burial of the animal when it was alive, thus preserving every detail. Preservational mode (2) resulted if the animal had been dead on the sea floor long enough for the first stages of preservation to occur. Perhaps an algal coating or other factor delayed coronal disarticulation, and the degree of interlocking among plates was also a factor. Burial occurred prior to corona disarticulation. Preservational mode (3) represents mode (2) with continued or more rapid disarticulation at the sea floor. Preservational mode (4) records even further decomposition and disarticulation than in (2) and (3), however it would have to have occurred in a low turbulence setting so that disarticulated plates were not transported away. Finally, preservational mode (5) indicates that a live animal, or one immediately after death and prior to ligament decomposition, was crushed and broken, probably by either predation or turbulence (Smith 1984).

For stalked echinoderms, especially crinoids, differential resistance to disarticulation has been recognized among various groups. Guensburg (1984) recognized three categories of preservational robustness in his study of Middle Ordovician echinoderms. The most resistant to disarticulation were disparids and thick-plated camerates, less resistant were hybocrinids and paracrinoids, and thin-plated camerates were most likely to be disarticulated. Among late Paleozoic crinoids, differential resistance to disarticulation was described for Lower Mississippian crinoids by Hudson et al. (1966), Meyer et al. (1989), and Ausich and Sevastopulo (1994) and for Pennsylvanian crinoids by Lewis (1986) and Holterhoff (1996). With some exceptions, Paleozoic crinoid clades display a range of resistance to disarticulation as outlined in Table 1. These distinctions were relatively consistent throughout the Paleozoic, and most articulate (post-Paleozoic) crinoids probably behaved analogous to cladids. At least for disparids, their resistance to disarticulation is present even at the microcrinoid size (Sevastopulo & Lane 1988), which has allowed for a detailed understanding of disparid ontogeny. Finally, among asteroids, constructional differences correlated to preservation have been postulated for an Eocene asteroid fauna (Blake & Zinsmeister 1988). All of these preservational assumptions require further testing.

From comparison of cyclocystoids in various stages of preservation, Smith and Paul (1982) inferred the sequence of progressive disarticulation for a cylcocystoid lying dead exposed at the sediment-water interface. From first to last to disarticulate , the sequence of lost plates was predicted to be dorsal annular plates, cover plates collapse onto radials and lost, peripheral skirt, radial and interradial plating of the disc, and, last, the marginal ring.

More generally, Brett et al. (1997) developed a three-part taphonomic grade scale for fossil echinoderms based on comparison to living echino-

196

derms (holothurians not included). Comparative states of preservation can be correlated, theoretically, to disarticulation resistance among living echinoderms. Type 1 echinoderms have very poorly sutured plates that are thin, imbricated, with muscular articulations, with relatively insignificant ligamentary articulations, or with skeletal elements enclosed in mesodermal and epidermal tissue. These echinoderms occur typically as complete specimens in a Konservat-Lagerstätte or as isolated ossicles (Brett et al. 1997). Type 2 echinoderms have an endoskeleton with contrasting preservational potential; part of the body will disarticulate with ease (similar to Type 1), and the remainder has well-sutured plates (similar to Type 3). Brett et al. (1997) discussed the range of fossil occurrences that typify this category. Type 3 echinoderms have securely sutured plates over the majority of the body so that it is resistant to disarticulation. This strength results from interlocking stereom across sutures or ankylosed sutures. Among living echinoderms, asteroids ophiuroids, and comatulid crinoids are Type 1; most regular echinoids are Type 2, and irregular echinoids are Type 3 (Brett et al. 1997). Figure 5 is an expanded diagram from Brett et al. (1997) that estimates the distribution of fossil echinoderms among these three broad preservational categories.

6 TAPHONOMIC BIASES IN THE GEOLOGIC RECORD

A disparity between fossil regular and irregular echinoid preservation (Fisher 1952, Wigley & Stinton, 1973) has been recognized for some time, and echinoids have been considered in theoretical simulations of fossil preservation (Lasker 1976). Fisher (1952) pointed out the disparity in construction between regular and irregular echinoids, thus explaining why irregulars were more commonly preserved. However, Kier (1977) was the first to begin serious study of the impact of taphonomic differences on the preserved fossil record of echinoderms. Kier (1977) outlined the basic constructional and environmental distinctions responsible for this preservational bias, and others have subsequently tested, elaborated, and further demonstrated his main points. In particular, Kier (1977) attributed the higher preservation potential of irregular echinoids to the following: 1) Experimental disarticulation trials; 2) infaunal habit of irregular echinoids; 3) constructional morphology; 4) deposit feeding trophic habit of infaunal irregular echinoids that leaves a dead corona sediment filled; 5) regular echinoids live in more turbulent habitats where tests can be broken; and 6) regular echinoids live in habitats where erosion rather than deposition occurs. Kier (1977) demonstrated the geological bias for preferential preservation of irregular echinoids by a number of comparative measures between regular and irregular echinoids during the present and through time.

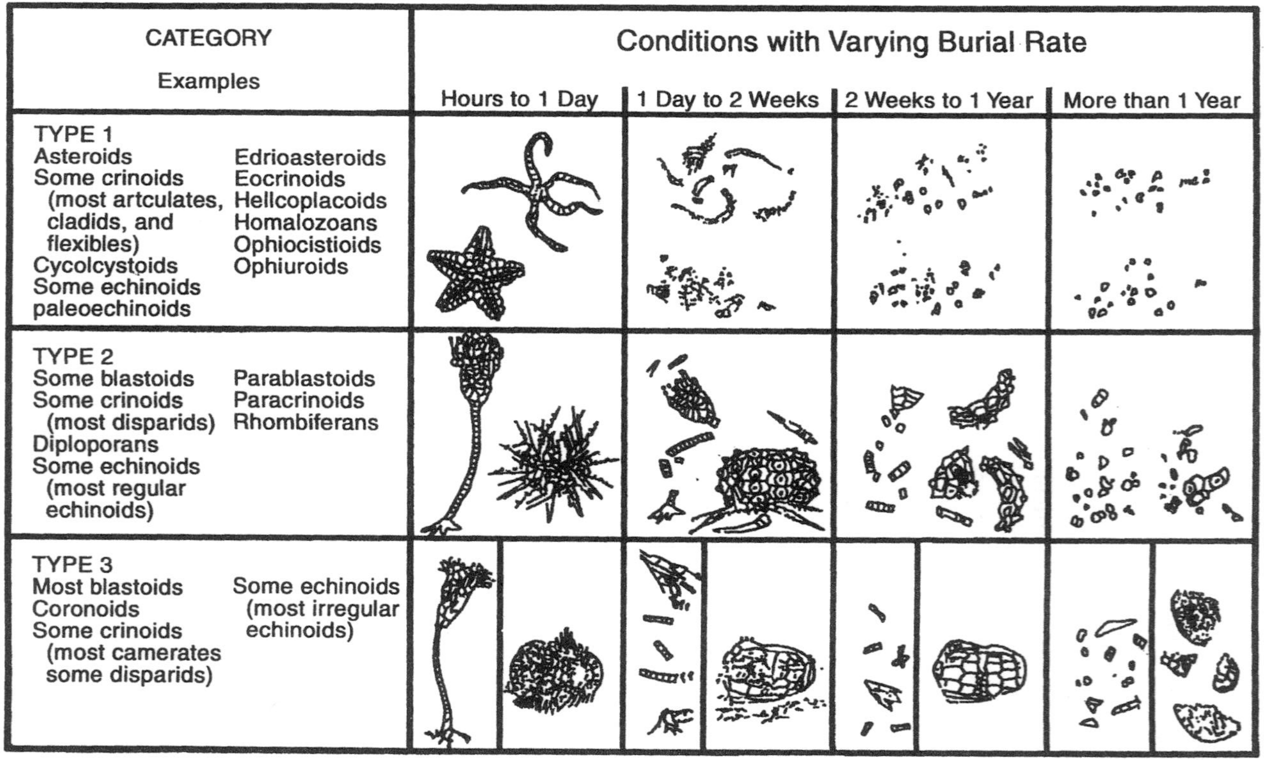

CATEGORY	Conditions with Varying Burial Rate			
Examples	Hours to 1 Day	1 Day to 2 Weeks	2 Weeks to 1 Year	More than 1 Year
TYPE 1 Asteroids Edrioasteroids Some crinoids Eocrinoids (most artculates, Helicoplacoids cladids, and Homalozoans flexibles) Ophiocistioids Cycolcystoids Ophiuroids Some echinoids paleoechinoids				
TYPE 2 Some blastoids Parablastoids Some crinoids Paracrinoids (most disparids) Rhombiferans Diploporans Some echinoids (most regular echinoids)				
TYPE 3 Most blastoids Some echinoids Coronoids (most irregular Some crinoids echinoids) (most camerates some disparids)				

Figure 5. Three-part taphonomic grade scale for echinoderm preservation (modified from Brett et al. 1997).

198

Seilacher (1979) and Smith (1984) emphasized the robustness of the test for understanding echinoid preservational biases. Smith (1984) also emphasized the habit of the echinoid as primary factors for echinoid preservation. The habit is an important distinction between regular and irregular echinoids (Kier 1977). However, preservational potential among regular and among irregular echinoids also varies in relationship to the suture connection. Imbricate plates of many Paleozoic echinoids and of the living echinothurioids is very susceptible to disarticulation. Those echinoids with the highest probability of preservation have interlocking stereom across plate sutures. Among regular echinoids, echinoidids and temnopleurids have interlocking stereom, and among irregular echinoids cylpeasteroids have the most significant interlocking stereom (Smith 1984). Smith (1984) argued that deep-sea echinoids and echinoids with imbricate plating were poorly represented in the fossil record and that those with interlocking stereom were over represented. Rose (1984) also emphasized the role of taphonomy in the echinoid fossil record.

Greenstein (1991, 1992, 1993) reexamined the question of preservational bias in the geological record, using a variety of preservational experiments as the baseline. Greenstein's (1991) work is much more robust than that completed previously, but the basic results are the same: i.e. irregular echinoids are preferentially better preserved than regular echinoids, which results in a biased fossil record. Donovan and Gordon (1993) reported consistent preservation for Pliocene to Holocene echinoids of the Caribbean. Furthermore, Greenstein (1991) demonstrated that constructional distinctions among three regular echinoids (one representative each of the Cidaridae, Echinometridae, and Toxopneustidae) yielded demonstrable taphonomic differences in decay experiments (see above) that corresponded to preservational patterns through time (Greenstein 1991, 1992). The conclusions of Greenstein (1993) differ some from those of Kier (1977) in that constructional morphology may be more important than life mode. Further, Greenstein's (1993) work in the Caribbean suggests that a reasonably high paleoecological reliability for regular echinoid fragments and that differences among irregular echinoids may lead to taphonomic bias among irregulars. The paleoecological utility of fragments from Caribbean regular echinoids agreed with the results of Gordon and Donovan (1992) who examined echinoid distribution in the Pleistocene of Jamaica.

Another potential bias among Cenozoic echinoids, based on mineralogy, considered by Beu et al. (1972). Beu et al. (1992) found a relatively high abundance of spatangoid echinoids in Cenozoic paleocommunites, compared to living communities. They attributed this to the dissolution of aragonitic molluscs.

As described above and with all factors equal, echinoderms have predictable sequences of decay and disarticulation dictated by connective tissue types and cementation across plate sutures. Given this predictability and the fact that contrasting tissue types may occur in both the column and the arms, Baumiller and Ausich (1992) Baumiller et al. (1995) and Ausich and Baumiller (1993, 1998) tested hypotheses of tissue organization among fossil crinoids by examination of crinoids preserved during the taphonomic window immediately surrounding initial disarticulation.

7.1 *Column connective tissue*

Living isocrinids have two types of ligamentary tissue in the column. Short, intercolumnal ligamentary tissue binds all adjacent columnals and penetrates only a short distance into adjacent columnals; and long, through-going ligamentary tissue penetrates through a stack of columnals (Grimmer et al. 1985). Thus, the column is organized as a serial repetition of segments bound by both intercolumnar and through-going ligaments. Segments are bound to adjacent segments by an articulation with only intercolumnar ligaments, thus creating relatively weak articulations along the column between segments. Furthermore, in isocrinids, this repetitive ligament structure is paralleled by repetitive hard-part morphologies. The through-going ligaments penetrate through one noditaxis, consisting of a set of internodals plus its distal nodal. The distal nodal facet is a synostosis, specialized for autotomy, whereas other columnal-columnal articulations are symplexies.

Disarticulation experiments demonstrated that isocrinid columns disarticulated first into noditaxis segments of approximately equal length, defined by soft tissue (Baumiller et al. 1991, Baumiller & Ausich 1992, Baumiller et al. 1995). Disarticulation into individual columnals occurred after considerably further decay. This explains both the occurrence of column segments on the substratum of the Straits of Florida and the preservation of column segments in fossil isocrinid assemblages (Baumiller et al. 1995). A similar pattern of column breakage was demonstrated for both Lower Mississippian (Baumiller & Ausich 1992) and Middle Ordovician (Ausich & Baumiller 1998) crinoid columns. Despite the fact that these Paleozoic crinoids lacked the serially repeated hard-part differentiation of nodals with a distal synostosis articulation, they were inferred to have had a soft-tissue organization analogous to isocrinids. Presumably, the hard-part differentiation on isocrinids is an exaptation on a primitive soft-tissue arrangement (Ausich & Baumiller 1998).

Crinoid column functional morphology was examined by Baumiller and Ausich (1996) using the assumption that the preserved posture of a complete column records the flexibility of the column in a living animal. The surprising conclusion of that study was that, except for significant morphological deviations, there was no correlation between inferred flexibility and variations in columnal morphology. Furthermore among crinoids with normal columns, column flexibility does not exceed the limits of echinoderm ligament stretching. The only exceptions among the Lower Mississippian crinoids considered (Baumiller & Ausich 1996) were *Platycrinites* and *Camptocrinus*, with synarthrial articulations between opposing columnals, and *Gilbertsocrinus*, whose column articulations mimic first-degree levers of most crinoids (Riddle et al. 1988).

7.2 Arm connective tissues

Taphonomic data have been used to test the long-standing view that crinoids only evolved muscular articulations in arms in one clade, the advanced, pinnulate cladids (Poteriocrinida *sensu* Moore & Teichert, 1978), beginning during the Lower Devonian (Moore & Laudon 1943, Van Sant & Lane 1964, Ubaghs 1978). This hypothesis was tested with the assumption that muscular articulations should disarticulate more rapidly than ligamentary articulations and that arm ligamentary articulations should disarticulate at approximately the same rate as column articulations with only intercolumnar ligaments. Two different studies verified this hypothesis. Among Lower Mississippian crinoids, advanced, pinnulate clades were consistently preserved differently than other crinoids (ramulate cladids, disparids, camerates, and flexibles) and consistent with predictions for a crinoid having muscular articulations in the arms (Ausich & Baumiller 1993). In contrast, among Middle Ordovician crinoids considered by Ausich and Baumiller (1998), the preservational style of all crinoids was consistent with the prediction for crinoids with only ligamentary articulations (Ausich & Baumiller, 1993).

8 ECHINODERM TAPHONOMY AND MODE OF LIFE

Taphonomic evidence has provided key information for the interpretation of some echinoderms with unusual life habits, especially crinoids. This was one reason for the early consideration of echinoderm preservation. Pseudoplanktonic crinoids attached to logs are an example. These occurred both during the Devonian and the Jurassic (Wells 1941, McIntosh 1978, Seilacher et al. 1968, 1985, Simms 1986, Wignall & Simms 1990, Hess et al. 1999). The debate has been whether crinoids hung beneath floating logs or whether the logs sat on the seafloor and were simply the bottom substratum for typical erect crinoids. By studying the remarkable occurrences of *Seirocrinus subangularis* from the Jurassic Holzmaden of Germany, Seilacher et al. (1968) were able to determine that crinoids were preserved beneath the logs and that the preserved position of these fossils indicated that 1, crinoid crowns had been dragged across the substratum; 2, the crowns settled onto the substratum first; 3, the long *Seirocrinus subangularis* columns (as long as 16 m) were deposited above the crowns; and 4, the log was incorporated into the substratum last. Although this interpretation has been challenged (Abel 1927, Rasmussen 1977, Kauffman 1981), the pseudoplanktonic mode is presently considered the most probable explanation for the occurrence of crinoids due to taphonomic, morphologic, and paleoecologic evidence (Simms 1986, Haude 1980, 1992, 1994, Hall 1991, Hess et al. 1999).

The Silurian to Devonian Scyphocrinitidae with their lobolith "root" structures are another example. Again, these are interpreted to have been planktonic, based on a variety of functional morphologic, biogeographic, and taphonomic criteria (Schuchert 1904, Kirk 1911, Springer 1917, Haude 1972, Haude 1974, Yetschewa 1974, Brett 1984, Haude 1989, Haude 1992, Haude et al. 1994, Hess 1998, Hess et al. 1999). Arguments for and against other planktonic crinoids are also a combination of taphonomic and functional morphologic interpretations. These include, among others, the Cretaceous crinoids *Uintacrinus* and *Marsupites* and Jurassic and younger antedontids (Springer 1901, Kirk 1911, Seilacher et al. 1985, Barthel et al. 1990, Milson et al. 1994, Manni et al. 1997, Meyer & Maples 1997, Hess 1998, Webber & Meyer 1999, Hess et al. 1999, Meyer et al. 1999). Also, Simms (1994) argued that the lack of many preserved Jurassic comatulid

Table 1. Relative preservation potential among clades of Lower Mississippian crinoids, as outlined by Meyer et al. (1990) and Ausich and Baumiller (1994). Exceptions to clade-level sequence are noted in parentheses.

MOST RESISTANT TO DISARTICULATION

Thick-plated Camerates
(most Monobathrid Camerates)

Disparids
(flexible: some species of *Gaulocrinus*)

Cladids
(thin-plated camerates: *Camptocrinus, Dichocrinus, Paradichocrinus, Platycrinites*)

most Diplobathrid Camerates

Flexibles

LEAST RESISTANT TO DISARTICULATION

---

crinoids resulted from the fact that they lived in shallow-water reefal settings, where preservational conditions were poor.

Taphonomic evidence has also been used to help interpret unusual benthic crinoids, such as *Agassizocrinus,* the calceocrinids, and *Metacrinus fossilis. Agassizocrinus* is a Lower Carboniferous unstalked crinoid. Preserved positions of crowns demonstrated that this crinoid lived directly on the substratum with the elongate infrabasals stuck into the substratum like a golf tee (Ettensohn 1975). The calceocrinids are a long-ranging (Ordovician to Permian) family of crinoids that lived with the column recumbent along the substratum. The calceocrinid crown is preserved completely much more commonly than sympatric crinoids with similar construction. This unusual complete preservation is consistent with the hypothesis that the crowns of these crinoids lay directly on the bottom where they could be easily buried (Ausich 1986). *M. fossilis* is an upper Eocene isocrinid crinoid from Antarctica (Meyer & Oji 1993). It is thought to have lost all but the proximal-most stalk early during ontogeny to become functionally a free-living crinoid.

The paleoecology of unusual, extinct echinoderms, such as the class Helicoplacoidea, is problematic because reasonable modern analogs do not exist. Preservational circumstances place boundaries on the paleoecological interpretations of such organisms. Thus, Durham (1993) concluded that the preservation of helicoplacoids supported the hypothesis that these ellipsoidal echinoderms were suspension feeders.

Taphonomy must also be considered when sampling for paleocommunity reconstruction. Rapid burial by tempestites collapses an epifaunally tiered community on the former seafloor. Thus, microstratigraphic sampling is best for epifaunal communities commonly dominated by stalked echinoderms (Bottjer & Ausich 1982). In contrast, infaunal communities, such as infaunal echinoids, are not collapsed, so sampling through as much as a meter is necessary for paleocommunity reconstruction (Bottjer & Ausich 1982).

9 ECHINODERM FACIES, TAPHOFACIES, AND LARGER SCALE TAPHONOMIC PROCESSES

The majority of taphonomic research has centered on preservation of individual echinoderm specimens. However, in certain settings during the past, echinoderm abundance was so high that entire rock units are dominated by or composed largely of echinoderms. Lane (1971) recognized that the geometry of such rocks, principally crinoidal limestones, was dependant on three factors: 1) Basic sedimentological setting, carbonate or siliciclastic; 2) sedimentation rate; and 3) current turbulence. Crinoidal limestones vary from thin stringers of bioclastic debris to the extensive regional encrinite deposits typified by the Burlington Limestone of Illinois, Iowa, and Missouri (Ausich 1997).

Speyer and Brett (1986, 1988, 1991) and Brett and Baird (1986) used taphonomy to better understand echinoderm-dominated facies and introduced the concept of taphofacies, which are sedimentary deposits characterized by similar preservational attributes of the contained fossils. In other words, various depositional, biostratinomic, and diagenetic processes are correlated so that various suites of fossils, differentiated on preservational characteristics, recur through the geological record. Echinoderms and other organisms that readily disarticulate are particularly susceptible to differentiation into taphofacies (Brett et al. 1997). Environmental turbulence, sedimentation rates, and oxygenation are critical extrinsic factors determining taphofacies, as are intrinsic constructional criteria, both between and within taxonomic groups.

Brett et al. (1997) and Hess et al. (1999) identified nine echinoderm taphofacies in carbonate-dominated settings and five in siliciclastic-dominated settings (Table 2) that represent repeated echinoderm occurrences in the geologic record. Taphofacies in carbonate-dominated settings include lagoons, skeletal sands, shoal-top hardgrounds, bioherm/reef, storm-dominated shelf/shoal margin/ramp, shoal margin hardground, deeper, storm-influenced shelf, distal shelf/ramp, and dysoxic/anoxic basin. These taphofacies preserve specific Konservat-Lagerstätten types (Brett et al. 1997, Hess et al. 1999). Examples of well-known echinoderm Konservat-Lagerstätten are accommodated within this model as follows (see Brett et al. 1997). The Lower Carboniferous Burlington Limestone of the east-central United States is an example of a Carbonate Shoal Taphofacies and Burlington-Type Kon-

Table 2. Echinoderm taphofacies and Konservat-Lagerstätten types for carbonate-dominated settings (from Brett et al. 1997, Hess et al. 1999).

Taphofacies	Konservat-Lagerstätten
Lagoon	Litchfield type
Carbonate shoal	Burlington type
Shoal-top hardground	Muschelkalk type
Bioherm/ reef	Racine type
Storm dominated shelf/ shoal margin/ramp	Cincinnatian type
Shoal margin hardground	Gmeund type
Deeper, storm-influenced shelf	Chalky Limestone type
Distal shelf/ramp	Rust Quarry type
Dysoxic/anoxic sasin	Solnhofen type

servat-Lagerstätte (Lane 1971, Ausich 1997). The Triassic Muschelkalk of Germany (Hagdorn 1985) is a Shoal-Top Hardground and Muschelkalk-Type Konservat-Lagerstätte. Examples of the Storm-Dominated Shelf/Shoal Margin/Ramp Taphofacies (Cincinnatian-Type Konservat-Lagerstätte) include the Late Ordovician of the Cincinnati, Ohio area (Meyer 1990, Schumacher & Ausich 1983); the Hampton Formation of Le Grande, Iowa (Laudon & Beane 1937); and other Muschelkalk Limestone facies (Linck 1965, Hagdorn 1985, Aigner 1985): and the Solnhofen Limestone represents the Solnhofen-Type Konservat-Lagerstätte and the Dysoxic/Anoxic Basin Taphofacies (Barthel et al. 1990, Hemleben & Swinburne 1991, Seilacher et al. 1985).

Taphofacies in siliciclastic dominated settings are shoreface and proximal sandy shelf, storm-dominated siliciclastic shelf/delta platform, storm-influenced deeper shelf/outer prodelta, distal shelf/basin/, and dysoxic/anoxic muddy basin (Table 3) (Brett et al. 1997). The Storm-Dominated Siliciclastic Shelf/Delta Platform Taphofacies yields Crawfordsville-Type Konservat-Lagerstätte, with examples being the Middle Devonian Hamilton Formation of New York (Goldring 1923) and the Lower Mississippian Edwardsville Formation at Crawfordsville, Indiana (Van Sant & Lane 1964, Lane 1963, 1973). The Middle Ordovician Bromide Formation of Oklahoma (Sprinkle 1982b), Middle Silurian Rochester Shale of New York (Taylor & Brett 1996), and Middle Silurian Waldron Shale of Indiana and Tennessee (Liddell & Brett 1982) represented the Storm-Influenced Deeper Shelf/Outer Prodelta Taphofacies and Waldron-Type Konservat-Lagerstätte. Hunsrück-Type Konservat-Lagerstätte occur in the Distal Shelf/Basin Taphofacies, as exemplified by the Early Devonian Hunsrück Slate of Germany (Seilacher et al. 1985); and the Jurassic Holzmaden of Germany (Seilacher 1985) records

Table 3. Echinoderm taphofacies and Konservat-Lagerstätten types for siliciclastic-dominated settings (from Brett et al. 1997, Hess et al. 1999).

Taphofacies	Konservat-Lagerstätten
Shoreface and proximal sandy shelf	Oriskany type
Storm-dominated siliciclastic shelf/delta platform	Crawfordsville type
Storm-influenced deeper shelf/outer prodelta	Waldron type
Distal shelf/basin	Hunsrück type
Dysoxic/anoxic muddy basin	Holzmaden type

the Dysoxic/Anoxic Muddy Basin Taphofacies and Holzmaden-Type Konservat-Lagerstätte (See Brett et al. 1997).

Brett et al. (1997) listed many other examples and discussed each taphofacies and Konservat-Lagerstätte type in detail. Furthermore, Hess et al. (1999) treated many crinoid Konservat-Lagerstätten in considerable detail. Specific studies that have utilized the taphofacies approach for echinoderms are Meyer et al. (1989) and Ausich and Sevastopulo (1994).

In the Lower Carboniferous of County Wexford, Ireland (Ausich & Sevastopulo, 1994), echinoderm faunas are preserved in facies that record contrasting water depths along a depth gradient of a carbonate ramp. Both the intrinsic resistance to disarticulation as discussed in Meyer et al. (1989) (Table 1) and the depth gradient accounted for crinoid preservation in this setting. The physical processes that imposed a primary control on preservation were the frequency and intensity of storm events that disturbed the substratum. Both of these processes were inversely correlated with water depth. In shallow water, high intensity storms frequently buried organisms. However, even if complete crinoids were buried in shallow-water settings, subsequent storms had sufficient turbulence to exhume and disarticulate previously buried and decayed individuals. The result in shallower facies was that the only articulated remains were the disarticulation-resistant camerate crinoid calyxes (lacking arms and column). Cladids lived in the shallower depth, but they were only preserved as isolated plates (Fig. 6). In the deeper ramp facies, nearly complete cladids (arms dorsal cup-column) were preserved rarely (apparently camerates were absent from these deep-water assemblages), because burial events were not frequent. Middle ramp facies had the best crinoid preservation with relatively common, complete (arms-calyx/dorsal cup-column-holdfast) preservation of cladids, flexibles, and camerates. It was inferred that burial events were sufficiently frequent to commonly bury individuals but that the intensity of storm turbulence did not routinely resuspend entire previous tempestites (Ausich & Sevastopulo 1994) (Fig. 6).

Nebelsick (1999b) developed a taphofacies model for preservation of *Clypeaster* in the Red Sea. Four species of *Clypeaster* live in the Red Sea, *C. humilis, C. fervens, C. reticulatus,* and *C. rarispinus,* with only *C. humilis* common. Four distinct taphofacies were recognized and correlated to topography, bottom facies, and environmental parameters (Nebelsick 1999b). Processes that control taphofacies distribution were identified as turbulence (especially as it affects abrasion rates), sedimentation rates, and biological processes such as predation and bioerosion. Taphofacies A has well-preserved fragments in deeper muddy sand and mud facies and in shallow-water seagrass facies. Taphofacies B has highly encrusted *Clypeaster* fragments but, otherwise, has well-preserved echinoids. This preservation occurs in the deeper muddy sand facies. Taphofacies C has an intermediate level of

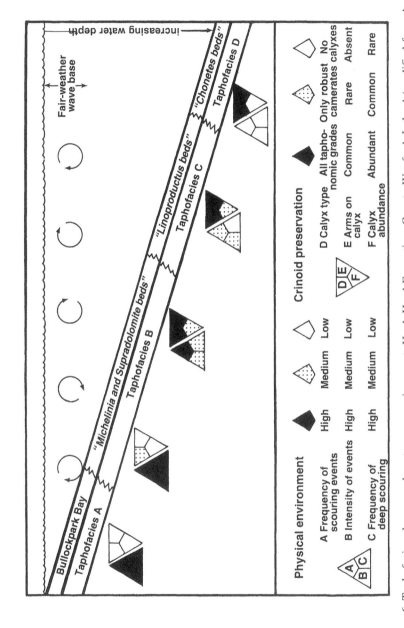

Figure 6. Taphofacies along a carbonate ramp environment, Hook Head Formation, County Wexford, Ireland (modified from Ausich & Sevastopulo 1994).

preservation and a wide distribution across many sedimentary facies. Tapho-facies D has the poorest preservation in the Red Sea. *Clypeaster* remains are highly fractured, have a high degree of surface detail loss, and are bioeroded to a considerable degree. These more poorly preserved specimens occur both in shallow water and in deeper water muddy sand and coral carpet facies (Nebelsick 1999b). Taphofacies C is apparently the "expected" pres-ervational mode of *Clypeaster* in the Red Sea. Exposure or shelter from wave turbulence changes preservation to Taphofacies D or Taphofacies A, respectively. Differences in sedimentation rate are reflected by the rate of encrustation. Taphofacies B has a high degree of encrustation and is present in a sediment-starved facies, whereas sedimentation is more rapid in facies supporting Taphofacies A (Nebelsick 1999b)

The concept of taphonomic feedback was introduced by Kidwell and Jablonski (1983) in their studies of shell beds. Taphonomic feedback recog-nizes that the remains of dead organisms may have an impact on the habitat where they lived, and this impact may be either positive or negative. In the case of echinoderms, positive taphonomic feedback may have been an important factor in maintaining extensive stands of crinoids. The very large accumulations of crinoidal remains are termed regional encrinites (Ausich 1997). Crinoids in this facies had holdfasts adapted to the coarse, poorly-sorted, episodically mobile crinoid column substratum, and when organisms died they produced more of this sediment type (Ausich 1997), thus a positive taphonomic feedback.

Carter and McKinney (1992) considered the preservational aspects of facies and its impact on biogeographic interpretations. They concluded in a study of Cenozoic echinoids that removal of sediment to unconformity-related erosion and environmental misinterpretations can lead to false bio-geographic patterns.

10 CONCLUSIONS AND THE OTHER CLASSES?

Many factors are important for echinoderm preservation, and at least some of these factors have a temporal or other aspect that should yield macroevo-lutionary or biogeographic trends in preservation. Echinoderm preservation improves in the absence of scavengers, with resistant constructional morphol-ogies and with an infaunal habit. Temperature also plays an important role, with echinoderm decay and disarticulation much more rapid in warm water. Constructional morphology and infaunal habit are important preservational biases. The evolution of scavengers do not appear to have had an impact, and there does not appear to have been any latitudinal controls (i.e., temperature) on preservation. Above all, even considering constructional morphology and infaunal habit, rapid, deep burial is essential for whole echinoderm preservation. Therefore, depositional circumstances that promote appropriate burial are the primary control on echinoderm preservation.

Echinoderm taphonomic studies have been largely restricted to living classes where taphonomic patterns and processes could be calibrated with actualistic and experimental approaches. Despite considerable advances in this area, it is limited to relative few "model" species considered under only a few environmental conditions. More comprehensive studies are needed. Also, relatively little is known about the taphonomy of the fifteen or more extinct echinoderm classes, and the taphonomic grades into which extinct echinoderms are placed is only an approximation (Fig. 5). However, sufficient detail is known about echinoid and crinoid taphonomy to use these two groups as models to understand the taphonomy of extinct classes and to test the model in Figure 5. Echinoids and crinoids can be used both as constructional analogs and as a taphonomic calibration where echinoids and crinoids are sympatric with other classes (see Meyer et al. 1989 for brief mention of blastoids with crinoids). This surely is an area of echinoderm taphonomy with much potential for future research. For example, are the imbricated tests of helicoplacoids and lepidocystids taphonomically analogous to paleoechinoids (Fig. 5)? Because thicker-walled blastoids commonly occur in crinoidal grainstones and packstones with camerate crinoid calyxes lacking arms and column, are they taphonomically analogous? At what point do thinner-walled blastoids become taphonomic analogs to cladid crinoids? Questions similar to these for extinct echinoderm classes should be answerable; and if combined with additional actualistic and experimental questions on the five living classes, it is clear that the research agenda for echinoderm taphonomy has a rich potential for significant advances.

ACKNOWLEDGMENTS

R. Brown and P. Dittoe helped with acquiring references; E. Ausich, A. Glass, and M. Jangoux helped with translations, and B. Heath typed the manuscript. This manuscript was greatly improved by careful readings and suggestions by G. Breton and M. Roux.

REFERENCES

At the end of each echinoderm taphonomy citation, the echinoderm class considered is indicated by A, asteroids; B, blastoids; C, crinoids; Co, coronoids; Cy, cyclocystoids; D, diploporans; E, echinoids; Ed, edrioasteroids; Eo, eocrinoids; H, holothurians; He, helicoplacoids; Hz, homalozoans; O, ophiuroids; Op, ophiocistioids; P, paracrinoid; R, rhombiferans. Abstracts are only included where they are considered important and a subsequent paper was not published.

Abel, O. 1927. *Ledensbilder aus deer Tierweit der Vorzeit*, G. Fisher, Jena, 714 pp.

Aigner, T. 1985. Storm depositional systems: dynamic stratigraphy in modern and ancient shallow-marine sequences. *Lecture Notes in the Earth Sciences 3*, Springer-Verlag, Berlin, 174 pp. [C]

Allison, P.A. 1988. The role of anoxia in the decay and mineralization of proteinaceous macro-fossils. *Paleobiology* 14: 139-154.

Allison, P.A. 1990. Variation in rates of decay and disarticulation of Echinodermata: Implications for the application of actualistic data. *Palaios* 5: 432-440. [A,E,H,O]

Améziane-Cominardi, N., J.P. Bourseau & M. Roux 1991. Les crinoïdes pédonculés de l'ouest Pacifique: un modele zoobathymétrique pour l'analyse des calcaires a entroques et du tectono-eustatisme au Jurassique. In: B. Lambert & M. Roux (eds), *L'environnement Çarbonaté Bathyal en Nouvelle-Caledonie (Programme ENVIMARGES). Documents et Travaux Institut Géologique Albert-de-Lapparent*, Paris, 15, pp. 182-198. [C]

Améziane-Cominardi, N. & M. Roux 1987. Biocorrosion et micritisation des ossicules d'Echinodermes en milieu bathyal au large de la Nouvelle-Calédonie. *Comptes-Rendus de l'Académie des Sciences, Paris, Serie II* 305: 701-705. [C]

Aronson, R.B. 1987. Predation on fossil and Recent ophiuroids. *Paleobiology* 13: 187-192. [O]

Aronson, R.B. 1989. A community-level test of the Mesozoic marine revolution theory. *Paleobiology* 15: 20-25. [O]

Aslin, C.J. 1968. Echinoid preservation in the Upper Estuarine Limestone of Blisworth, Northhamptonshire. *Geological Magazine* 105: 506-518. [E]

Ausich, W.I. 1986. Palaeoecology and history of the Calceocrinidae (Palaeozoic Crinoidea). *Palaeontology* 29: 85-99. [C]

Ausich, W.I. 1997. Regional encrinites: a vanished lithofacies. In: C.E. Brett & G.C. Baird (eds), *Paleontological Events: Stratigraphic, Ecological, and Evolutionary Implications,* pp. 509-519. Columbia University Press, New York. [C]

Ausich, W.I. & L.E. Babcock 1998. The phylogenetic position of *Echmatocrinus brachiatus*, a probable octocoral from the Burgess Shale. *Palaeontology* 41: 193-202.

Ausich, W.I. & T.K. Baumiller 1993. Taphonomic method for determining muscular articulations in fossil crinoids. *Palaios* 8: 477-484. [C]

Ausich, W.I. & T.K. Baumiller 1998. Disarticulation patterns in Ordovician crinoids: Implications for the evolutionary history of connective tissue in the Crinoidea. *Lethaia* 31: 113-123. [C]

Ausich, W.I. & C.P. Hart 1990. Cryogenic taphonomy of crinoids (Echinodermata) on the McMurdo Ice Shelf, Antarctica. *Geological Society of America Abstracts with Programs* 22: 1-2. [C]

Ausich, W.I., T. W. Kammer & N.G. Lane 1979. Fossil communities of the Borden (Mississippian) delta in Indiana and northern Kentucky. *Journal of Paleontology* 53: 1182-1196. [C]

Ausich, W.I. & N.G. Lane 1980. Platform communities and rocks of the Borden Siltstone delta (Mississippian) along the south shore of Monroe Reservoir, Indiana. In: R.H. Shaver (ed.), *Field Trips 1980 from the Indiana University Campus, Bloomington*, pp. 36-67. [C]

Ausich, W.I. & D.L. Meyer 1990. Origin and composition of carbonate buildups and associated facies in the Fort Payne Formation (Lower Mississippian, south-central Kentucky): an integrated sedimentologic and paleoecologic analysis. *Geological Society of America Bulletin* 102: 129-146. [C]

210

Ausich, W.I. & G.D. Sevastopulo 1994. Taphonomy of Lower Carboniferous crinoids from the Hook Head Formation, Ireland. *Lethaia* 27: 245-256. [C]

Ausich, W.I., G.D. Sevastopulo & H. Torrens in press. Middle Nineteenth Century crinoid studies of Thomas Austin, Sr. and Thomas Austin Jr.: Newly discovered unpublished materials. *Earth Sciences History.* [C]

Austin, J.J., A.B. Smith & R.H. Thomas 1977. Palaeontology in a molecular world: The search for authentic ancient DNA. *Tree* 12: 303-306.

Baird, G.C., S.D. Sroka, C.W. Shabica & G.J. Kuecher 1986. Taphonomy of Middle Pennsylvanian Mazon Creek area fossil localities, northeast Illinois: Significance of exceptional fossil preservation in syngenetic concretions. *Palaios* 1: 271-285. [H]

Bantz, H.-U. 1969. Echinoidea aus Plattenkalken der Altmühlab ihre Biostratinomie. *Erlanger Geologische Abhandlunge* 78: 1-35. [E]

Bartels, C. & W. Blind 1995. Röntgenuntersuchung pyritisch vererzter Fossilien aus dem Hunsrückschiefer (Unter-Devon, Rheinisches Schiefergebirge). *Metalla (Bochum)* 2: 79-100. [A,C,H]

Bartels, C., D.E.G. Briggs & G. Brassel 1998. *The Fossils of the Hunsrück slate, marine life in the Devonian.* Cambridge University Press, Cambridge, 309 pp. [A,C]

Barthel, K.W., N.H.M. Swinburne & S. Conway Morris 1990. *Solnhofen, a study in Mesozoic Paleontology.* Cambridge University Press, 236 pp. [A,C,E,H,O].

Bathurst, R.G.C. 1971. *Carbonate sediments and their diagensis.* Elsevier, Amsterdam, 658 pp.

Baumiller, T.K. 1994. Implications of stress induced shedding body parts in crinoids. *Geological Society of America Abstracts with Programs* 26: A429. [C].

Baumiller, T.K. & W.I. Ausich 1992. The broken-stick model as a null hypothesis for fossil crinoid stalk taphonomy and as a guide for distribution of connective tissue in fossils. *Paleobiology* 18: 288-298. [C]

Baumiller, T.K. & W.I. Ausich 1996. Crinoid stalk flexibility: Theoretical predictions and fossil stalk postures. *Lethaia* 29: 47-59. [C]

Baumiller, T.K. & H. Hagdorn 1995. Taphonomy as a guide to functional morphology of *Holocrinus*, the first post-Paleozoic crinoid. *Lethaia* 28: 221-228. [C]

Baumiller, T.K., M. LaBarbera & J.D. Woodley 1991. Ecology and functional morphology of the isocrinid *Cenocrinus asterius* (Echinodermata): In situ and laboratory experiments and observations. *Bull. Mar. Sci.* 48: 731-748. [C]

Baumiller, T.K., G. Llewellyn, C.G. Messing & W.I. Ausich 1995. Taphonomy of isocrinid stalks: Influence of decay and autotomy. *Palaios* 10: 87-95. [C]

Behrensmeyer, A.K. & S.M. Kidwell 1985. Taphonomy's contribution to paleobiology. *Paleobiology* 11: 105-119.

Bell, B.M. 1976. A study of North American Edrioasteroidea. *New York State Museum and Science Service Memoir* 21, 447 pp. [Ed]

Benton, M.J. & D.I. Gray 1981. Lower Silurian distal shelf storm-induced turbidites in the Welsh borders: sediments, tool marks and trace fossils. *J. Geol. Soc.* 138: 675-694. [C]

Bergström, J. 1990. Taphonomy of fossil-Lagerstatten: Hunsruck Slate. In: D.E. G. Briggs & P.R. Crowther (eds), *Palaeobiology: a synthesis*, pp. 277-279. Blackwell Scientific Publications, Oxford. [A]

Beu, A.G., R.A. Henderson & C.S. Nelson 1972. Notes on the taphonomy and paleoecology of New Zealand Tertiary spatangoida. *N. Z. J. Zool.* 15: 275-286. [E]

Blake, D.B. 1967. Pre-burial abrasion of articulated asteroid skeletons. *Paleobios* 2, 4 p. [A]

Blake, D.B. 1975. A new west American Miocene species of the modern Australian ophiuroid *Ophiocrassota. J. Paleontol.* 49: 501-507. [A]

Blake, D.B. & W.J. Zinsmeister 1979. Two early Cenozoic sea stars (Class Asteroidea) from Seymour Island, Antarctic peninsula. *J. Paleontol.* 53: 1145-1154. [A]

Blake, D.B. & W.J. Zinsmeister 1988. Eocene asteroids (Echinodermata) from Seymour Island, Antarctic Peninsula. *Geol. Soc. Amer. Mem.* 169: 489-498. [A]

Bloos, G. 1973. Ein Fund von Seeigeln der Gattung *Diademopsis* aus dem Hettangium Württembergs und ihr Lebensraum. *Stuttgarter Beiträge zur Naturkunde* B 5: 1-25. [E]

Blumer, M. 1951. Fossile Kohlenwasserstoffe und Farbstoffe in Kalksteinen. *Mikrochemie* 36/37: 1048-1055. [C]

Blumer, M. 1960. Pigments of a fossil echinoderm. *Nature* 188: 1100-1101. [C]

Blumer, M. 1962a. The organic chemistry of a fossil – I. The structure of the fringelite pigments. *Geochimica et Cosmochimica Acta* 26: 225-227. [C]

Blumer, M. 1962b. The organic chemistry of a fossil – II. Some rare polynuclear hydrocarbons. *Geochimica et Cosmochimica Acta* 26: 228-230. [C]

Blumer, M. 1965. Organic pigments: their long-term fate. *Science* 149:722-726. [C]

Blyth Cain, J.D. 1968. Aspects of the depositional environments and palaeoecology of crinoidal limestones. *Scottish J. Geol.* 4: 191-208. [C]

Bourseau, J.-P., J. David, M. Roux, D. Bertrand & V. Clochard 1997. *Balanocrinus maritimis* nov. sp., crinoïde pédonculé nouveau du Kimméridgien inférieur de La Rochelle (Charente-Maritime, France). *Geobios* 31: 215-227. [C]

Bourseau, J.-P., H. Hess, P. Bernier, G. Barale, E. Buffetaut, C. Gaillard, J.-C. Gall & S. Wenz 1991. Découverte d'ophiures dans les calcaires lithographiques de Cerin (Kimméridgien supérieur, Ain, France). Systématique et implications taphonomiques. *Comptes Rendus de l'Academie des Sciences, Serie 2,* 312: 793-799. [O]

Bottjer, D.J. & W.I. Ausich 1982. Tiering and sampling requirements in paleocommunity reconstruction. *Third North American Paleontological Convention Proceedings* 1: 57-59. [C,E]

Breton, G. 1992. Les Goniasteridae (Asteroidea, Echinodermata) jurassiques et crétacés de France, taphonomie, systématique, biostratigraphie, paléogéographie, évolution. Thèse de Doctorat d'Etats Sciences, Université de Caen (20.10.1990). Fasc. hors série du *Bulletin Trimestriel de la Société Géologique de Normandie et des Amis du Museum du Havre* 78: 592 pp. [A]

Breton, G. 1997. Deux étoiles de mer du Bajocien du nord-est du bassin de Paris (France): leur alliés actuels sont des fossiles vivants. *Bulletin trimestriel de la Société Géologique de Normandie et des Amis du Museum du Havre* 84: 23-34. [A]

Breton, G., J.P. Bourseau, P. Bernier, G. Barale, E. Buffetaut, C. Gaillard, J.C. Gall & S. Wenz 1994. Les astérides (Asteroidea, Echinodermata) des calcaires lithographiques Kimméridgiens de Cerin (Ain, France). *Geobios* 16: 49-60. [A]

Brett, C.E. 1984 Autecology of Silurian pelmatozoan echinoderms. *Special Papers in Paleontology* 32: 87-120.

Brett, C.E. 1985. Pelmatozoan echinoderms on Silurian bioherms, in western New York and Ontario. *J. Paleontol.* 59: 820-838. [C,Co,D,R]

Brett, C.E. 1990. Taphonomy of fossil-Lagerstatten: obrution deposits. In: D.E.G. Briggs & P.R. Crowther (eds), *Palaeobiology: A Synthesis*, pp. 239-243. Blackwell Scientific Publications, Oxford. [C,D,Ed,R]

Brett, C.E. 1995. Sequence stratigraphy, biostratigraphy, and taphonomy in shallow marine environments. *Palaios* 10: 597-616.

Brett, C.E. & G. Baird 1986. Comparative taphonomy: a key to paleoenvironmental interpretation based on fossil preservation. *Palaios* 1: 207-227. [B,C,E,O]

Brett, C.E. & G. Baird 1993. Taphonomic approaches to time resolution in stratigraphy: examples from Paleozoic marine mudrocks. In: S.M. Kidwell & A.K. Behrensmeyer (eds), *Taphonomic Approaches to Time Resolution in Fossil Assemblages*, pp. 250-274. Paleontological Society Short Courses in Paleontology 6. University of Tennessee Press, Knoxville, Tennessee.

Brett, C.E., V.B. Dick & G.C. Baird 1991 Comparative taphonomy and paleoecology of Middle Devonian dark gray and black shale facies from Western New York. In: E. Landing & C.E. Brett (eds), *Dynamic Stratigraphy and Depositional Environments of the Hamilton Group (Middle Devonian) in New York State, Part II*, pp. 5-36. *New York State Museum Bulletin* 469. [C]

Brett, C.E. & J.D. Eckert 1982 Palaeoecology of a well-preserved crinoid colony from the Silurian Rochester Shale in Ontario. *Royal Ontario Museum Life Sciences Contributions* 131, 20 pp. [C]

Brett, C.E. & W.D. Liddell 1978. Preservation and paleoecology of a Middle Ordovician hardground community. *Paleobiology* 4: 329-348. [C,Ed,P]

Brett, C.E., H.A. Moffat & W.L. Taylor 1997. Echinoderm taphonomy taphofacies, and Lagerstatten. *Paleontological Society Papers* 3: 147-190. [A,B,C,D,E,Ed, Eo,H,Hz,O,R]

Brett, C.E. & A. Seilacher 1991 Fossil-Lagerstätten: A taphonomic consequence of event sedimentation. In: G. Einsele, W. Ricken & A. Seilacher (eds), *Cycles and Events in Stratigraphy*, pp. 283-297. Springer-Verlag, New York, Berlin, Heidelberg. [A,C,E,Ed,O]

Brett, C.E. & W.L. Taylor 1997. The *Homocrinus* beds: Silurian crinoid Lagerstätten of western New York and southern Ontario. In: C.E. Brett & G.C. Baird (eds), *Paleontological Events: Stratigraphic, Ecological, and Evolutionary Implications*, pp. 181-223. Columbia University Press, New York. [C,Ed,O,R]

Briggs, D.E.G., R. Raiswell, S.H. Bottrell, D. Hatfield & C. Bartels 1996. Controls on the pyritization of exceptionally preserved fossils: An analysis of the Lower Devonian Hunsrück Slate of Germany. *Amer. J. Sci.* 296: 633-663. [O]

Broadhurst, F.M. & I.M. Simpson 1973. Bathymetry on a Carboniferous reef. *Lethaia* 6: 367-381. [C]

Brower, J.C. 1974. Crinoids from the Girardeau Limestone (Ordovician). *Palaeontographica Americana* 7: 259-499. [C]

Brower, J.C. & K.M. Kile 1994. Paleoautecology and ontogeny of *Cupulocrinus levorsoni* Kolata, a Middle Ordovician crinoid from the Guttenberg Formation of Wisconsin. *New York State Museum Bulletin* 481: 28-44. [C]

Brower, J.C. & J. Veinus 1978. Middle Ordovician crinoids from the Twin Cities area of Minnesota. *Bulletins of American Paleontology* 74: 369-506. [C]

Cadée, G.C. 1991. The history of taphonomy. In: S.K. Donovan (ed.), *The processes of fossilization*, pp. 3-21. Belham Press, London.

Carter, B.D. & M.L. McKinney 1992. Eocene echinoids, the Suwannee Strait and biogeographic taphonomy. *Paleobiology* 18: 299-325. [E]

Chave, K.E. 1964. Skeletal durability and preservation. In: J. Imbrie & N. Newell (eds), *Approaches to Paleoecology*, pp. 337-387. John Wiley & Sons, Inc. New York. [A,E]

Chestnut, D.R., Jr. & F.R. Ettensohn 1988. Hombergian (Chesterian) echinoderm paleontology and paleoecology, south-central Kentucky. *Bull. Amer. Paleontol.* 95(330), 102 pp. [B,C]

Clark, F.W. & W.C. Wheeler 1922. The inorganic constituents of the marine invertebrates. *U.S. Geological Survey Professional Papers* 124, 108 pp.

Conan, G., M. Roux & M. Sibueta 1981. A photographic survey of a population of a stalked crinoid *Diplocrinus (Annacrinus) wyvillethomsoni* (Echinodermata) from the bathyal slope of the Bay of Viscay. *Deep-Sea Research* 281A: 441-453. [C]

Conway, Morris, S. 1993. The fossil record and early evolution of the Metazoa. *Nature* 361: 219-225.

Courville, P, E. Vincent, J. Thierry & B. David 1989. La barre á scutelles du Burdigalien (Miocene) de Montbrison-Fontbonau (Bassin de Valréas, Vaucluse): du milieu de vie au milieu de dépôt. *Vie Marine, Hors Series* 10:3-16. [E]

Daley, P.E.J. 1996. The first solute which is attached as an adult: a Mid-Cambrian fossil from Utah with echinoderm and chordate affinities. *Zool. J. Linn. Soc.* 117: 405-440. [H]

Dietl, G. & R. Mundlos 1972. Ökologie und Biostratinomie von *Ophiopinna elegans* (Ophiuroidea) aus dem Untercallovium von La Voulte (Südfrankreich). *Neues Jahrbuch für Geologie und Paläontologie, Monatshafte* 8: 449-464. [O]

Dodd, J.R., R.R. Alexander & R.J. Stanton, Jr. 1985. Population dynamics in *Dendraster*, *Merriamaster*, and *Anadara* from the Neogene of the Kettleman Hills, California. *Palaeogeography, Palaeoclimatology, Palaeoecology* 52: 61-76. [E]

Donovan, S.K. 1991. The taphonomy of echinoderms: Calcareous multi-element skeletons in the marine environment. In: S.K. Donovan (ed.), *The processes of fossilization*, pp. 241-269. Belhaven Press, London. [A,C,E,H,O]

Donovan, S.K. & C.M. Gordon 1993. Echinoid taphonomy and the fossil record: Supporting evidence from the Plio-Pleistocene of the Caribbean. *Palaios* 8: 304-306. [E]

Donovan, S.K. & R.K. Pickerill 1995. Crinoid columns preserved in life position in the Silurian of Arisaig, Nova Scotia. *Palaios* 10:362-370. [C]

Durham, J.W. 1978. Polymorphism in the Pliocene sand dollar *Merriamaster* (Echinoidea). *J. Paleontol.* 52: 275-286. [E]

Durham, J.W. 1993. Observations on the Early Cambrian helicoplacoid echinoderms. *J. Paleontol.* 67: 590-604. [He]

Echols, C.S. & R.D. Lewis 1995. Epibionts and their effects on the taphonomy of Recent crinoid ossicles: In situ experiments at San Salvador, Bahamas. *Geological Society of America Abstract with Programs* 27: A136. [C]

Eckert, J.D. 1987. *Pycnocrinus altilis*, a new Late Ordovician channel-dwelling crinoid from southern Ontario. *Can. J. Earth Sciences* 24: 851-859. [C]

Efremov, I.A. 1940. Taphonomy, a new branch of paleontology. *Pan-American Geologist* 74: 81-93.

Erdtmann, B.D. & D.R. Prezbindowski 1974. Niagaran (Middle Silurian) interreef fossil burial environments. *Neus Jarhbuch für Geologie und Paläontologie Abhandlungen* 144: 342-372. [C]

Ernst, G., W. Hähnel & E. Seibertz 1973. Aktuopaläntologie und Merkmalsvariabilität bei mediterranen Echiniden und Rückschlüsse auf die Ökologie und Artumgrenzung fossiler Formen. *Paläontologische Zeitschrift* 47: 188-216. [E]

Ettensohn, F.R. 1975. The autecology of *Agassizocrinus lobatus*. *J. Paleontol.* 49: 1044-1061. [C]

Evamy, B.D. & D.J. Shearman 1965. The developments of overgrowths from echinoderm fragments. *Sedimentology* 5: 211-233. [C,E]

Evamy, B.D. & D.J. Shearman 1969. Early stages in development of overgrowths from echinoderm fragments in limestone. *Sedimentology* 12: 317-322. [E]

214

Feldman, H.R. 1989. Taphonomic processes in the Waldron Shale, Silurian, southern Indiana. *Palaios* 4: 144-156. [C]

Fisher, A.G. 1952. Echinoids. In: R.C. Moore, C.G. Lalicker & A.G. Fisher, *Invertebrate fossils*, pp. 675-714. McGraw-Hill Book Co., New York. [E]

Flessa, K.W. & T.J. Brown 1983. Selective solution of macroinvertebrate calcareous hard parts: a laboratory study. *Lethaia* 16: 193-205. [E]

Frankel, E. 1978. Evidence from the Great Barrier Reef of ancient *Acanthaster* aggregations. In: S.V. Smith (ed.), *Coral Reef Ecosystems*, pp. 75-93. *Atoll Res. Bull.* 220. [A]

Franzén, C. 1977. Crinoid holdfasts from the Silurian of Gotland. *Lethaia*, 10: 219-234. [C]

Franzén, C. 1982. A Silurian crinoid thanatope from Gotland. *Geologiska Föreningens i Stockholm Förhandlingar* 103: 439-490. [C]

Fujita, T., S. Ohta & T. Oji. 1987. Photographic observations of the stalked crinoid *Metacrinus rotundus* Carpenter in Suruga Bay, central Japan. *Journal of the Oceanographic Society of Japan* 43: 333-343. [C]

Gale, A.S. 1986. Goniasteridae (Asteroidea, Echinodermata) from the Late Cretaceous of north-west Europe. I. Introduction. The genera *Metopaster* and *Recurvaster*. *Mesozoic Research* 1: 1-69. [A]

Gale, A.S. 1987. Phylogeny and classification of the Asteroidea (Echinodermata). *Zool. J. Linn. Soc.* 89: 107-132. [A]

Gaspard, D. & M. Roux 1974. Quelques aspects de la fossilisation des tests chez les brachiopodes et les crinoïdes. Relation entre la présence de matière organique et le développement d'agrégats ferrifères. *Geobios* 7:81-89. [C]

Gluchowski, E. 1987. Jurassic and Early Cretaceous Crinoidea from the Plieniny Klippen Belt and the Tatra Mountains, Poland. *Studia Geologica Polonica* 94, 102 pp. [C]

Glynn, P.W. 1984. An amphinomid worm predator of the crown-of-thorns sea star and general predation on asteroids in eastern and western Pacific coral reefs. *Bull. Mar. Sci.* 35: 54-71. [A]

Goldring, R. & F. Laggenstrassen 1979. Open shelf and near-shore clastic facies in the Devonian. *Special Papers in Palaeontology* 23: 81-97. [C]

Goldring, R. & D.C. Stephenson 1972. The depositional environment of three starfish beds. *Neues Jahrbuch für Geologie und Paläontologie, Monatshelfte* 1972 (10): 611-624. [A,E]

Goldring, W. 1923. Devonian crinoids of New York. *New York State Museum Memoir* 16, 670 pp.

Gordon, C.M. & S.K. Donovan 1992. Disarticulated echinoid ossicles in paleoecology and taphonomy: the later interglacial Falmoouth Formation of Jamaica. *Palaios* 7: 157-166. [E]

Greenstein, B.J. 1989. Mass mortality of the West-Indian echinoid *Diadema antillarum* (Echinodermata: Echinoidea): A natural experiment in taphonomy. *Palaios* 4: 487-492. [E]

Greenstein, B.J. 1990. Taphonomic biasing of subfossil echinoid populations adjacent to St. Croix, U.S.V.I. In: D.K. Larve (ed.), *Transactions of the 12th International Caribbean Congress*, pp. 290-300. Miami Geological Society. [E]

Greenstein, B.J. 1991. An integrated study of echinoid taphonomy: Predictions for the fossil record of four echinoid families. *Palaios* 6: 519-540. [E]

Greenstein, B.J. 1992. Taphonomic bias and the evolutionary history of the family Cidaridae (Echinodermata: Echinoidea). *Paleobiology* 18 :50-79. [E]

Greenstein, B.J. 1993. Is the fossil record of regular echinoids really so poor? A comparison of living and subfossil assemblages. *Palaios* 8: 587-601. [E]

Greenstein, B.J. & D.L. Meyer 1988. Mass mortality of the West Indian echinoid *Diadema antillarum* adjacent to Andros Island, Bahamas: A natural experiment in taphonomy. In: J. Mylroie and D. Gerace (eds), *Fourth Symposium on the Geology of the Bahamas*, pp. 159-168. [E]

Greenstein, B.J., J.M. Pandolfi & P.J. Moran 1995. Taphonomy of crown-of-thorns starfish: implications for recognizing ancient population outbreaks. *Coral Reefs* 14: 91-97. [A]

Grimmer, J.C., N.D. Holland & I. Hayami 1985. Fine structure of an isocrinid sea lily (*Metacrinus rotundus*) (Echinodermata, Crinoidea). *Zoomorphology* 105:39-250.

Guensburg, T.E. 1984. Echinoderms of the Lebanon Limestone, central Tennessee. *Bull. Amer. Paleontol.* 86 (319), 100 pp. [C, P]

Hagdorn, H. 1985. Immigrations of crinoids into the German Muschelkalk basin. In: U. Bayer & A. Seilacher (eds), *Sedimentary and evolutionary cycles*, pp. 237-254. Lecture Notes in Earth Sciences, Springer Verlag, Berlin 1. [C]

Hagdorn, H. & T.K. Baumiller 1998. Distribution, morphology and taphonomy of *Holocrinus*. The earliest post-Palaeozoic crinoid. In: R. Mooi & M. Telford (eds), *Echinoderms: San Francisco*, pp. 163-168. Balkema, Rotterdam. [C]

Hall, R.L. 1991. *Seirocrinus subangularis* (Miller, 1821), A Pliensbachian (Lower Jurassic) crinoid from the Fernie Formation, Alberta, Canada. *J. Paleontol.* 65: 300-307. [C]

Häntzschel, W. 1936. Seeigel-Spülsäume. *Natur und Volk* 66: 293-298.

Haude, R. 1972. Bau und Funktion der *Scyphocrinites*-Lobolithen. *Lethaia* 5: 95-125.

Haude, R. 1980. Constructional morphology of the stems of Pentacrinitidae, mode of life of *Seirocrinus*. In: M. Jangoux (ed.), *Echinoderms Present and Past*: 17-23. Balkema Press; Rotterdam.

Haude, R. 1989. The Scyphocrinoids *Carolicrinus* and *Camarocrinus*. In: H. Jahnke and T. Shi (eds), *The Silurian-Devonian boundary strata and the early Devonian of the Shidian-Baoshan area (W. Yunnan, China)*, pp. 170-178. *Courier Forschungsinstitut Senckenberg*. [C]

Haude, R. 1992. Scyphocrinoiden, die Bojen-Seelilien im Hohen Silur-Tiefen Devon. *Palaeontographica, Pal.A* 222: 141-187. [C]

Haude, R., H. Jahnke & O.H. Walliser 1994. Scyphocrinoiden an der Wende Silur/ Devon. *Aufschluss* 45: 49-55. [C]

Haugh, B.N. 1979. Late Ordovician channel-dwelling crinoid from southern Ontario, Canada. *American Museum of Natural History Novitates* 2665: 1-25. [C]

Hawkins, H.L. & S.M. Hampton 1927. On the occurrence, structure, and affinities of *Echinocystis* and *Palaeodiscus*. *Quaternary J. Geol. Soc. London* 83: 574-603. [A,C,E,O]

Hemleben, C. & N.H.M. Swinburne 1991. Cyclical deposition of the plattenkalk facies. In: G. Einsele, W. Ricken & A. Seilacher (eds), *Cycles and events in stratigraphy*, pp. 572-591. Springer Verlag, New York, Berlin, Heidelberg.

Hess, H. 1972a. *Chariocrinus* n. gen. für *Isocrinus andraea* Desor aus dem unteren Hauptrogenstein (Bajocien) des Basler Juras. *Eclogae Geoloigcae Helvetiae* 65: 197-210. [C]

Hess, H. 1972b. The fringelites of the Jurassic sea. *CIBA-GEIGY Journal* 2: 14-17. [C]

Hess, H. 1972c. Eine Echinodermen-Fauna aus dem mittleren Dogger des Aargauer Juras. Schweiz. *Paläontologische Abhandlungen* 92. [A]

Hess, H. 1985. Schlangensterne und Seelilien aus dem unteren Lias von Hallau

(Kanton Schaffhausen). *Sonderdruck aus den Mitteilungen der Naturforschenden Gesellschaft Schaffhausen* 33: 1-15. [C,O]

Hess, H. 1991a. Neue Seesterne (Goniasteridea) aus mittleren Oxford von Reuchenette (Berner Jura). *Eclogoe Geologicae Helvetiae* 84:873-8791. [O]

Hess, H. 1991b. Neue Schlangensterne aud dem Toarcium und Aalenium des Schwäbischen Jura (Baden-Württemberg. *Stuttgarter Beiträge Zur Naturkunde Series B* 180: 1-11. [O]

Hess, H. 1994. New specimen of the sea star *Testudinaster peregrinus* Hess from the Middle Jurassic of northern Switzerland. *Ecologae geologicae Helvetiae* 87: 987-993. [A]

Hess, H. 1998. Habitat and lifestyle of crinoids from some fossil assemblages *Fifth European Conference on Echinoderms, Program and Abstracts,* p. 64. [C]

Hess, H., W.I. Ausich, C.E. Brett, M.J. Simms & R. Kindlimann 1999. *Fossil Crinoids.* Cambridge University Press, Cambridge, 316 pp. [C]

Hess, H. 1972b. The fringelites of the Jurassic sea. *CIBA-GEIGY Journal* 2: 14-17. [C]

Hess, H. & D.B. Blake 1995. *Coulonia platyspina* n. sp. A new astropectinid sea star from the Lower Cretaceous of Morocco. *Ecologae geologicae Helvetiae* 88: 777-788. [A,E]

Hess, H. & L. Pugin 1983. *Balanocrinus berchteni* n. sp., un nouveau crinoïde bajocien des Préalpes médianes fribourgeoises. *Eclogae Geologicae Helvetiae* 76: 691-700. [C]

Holterhoff, P.F. 1996. Crinoid biofacies in Upper Carboniferous cyclothems, midcontinental North America: faunal tracking and the role of regional processes in biofacies recurrence. *Paleogeography, Palaeoclimatology, Palaeoecology* 127: 47-81. [C]

Holterhoff, P.F. 1997a. Filtration models, guilds, and biofacies: Crinoid paleoecology of the Stanton Formation (Upper Pennsylvanian), midcontinent, North America. *Paleogeography, Palaeoclimatology, Palaeoecology* 130: 177-208. [C]

Holterhoff, P.F. 1997b. Paleocommunity and evolutionary ecology of Paleozoic crinoids. In: J.A. Waters & C.G. Maples (eds), *Geobiology of echinoderms,* pp. 69-106. *The paleontological Society Papers* 3. [C]

Hudson, R.G.S., M.J. Clark & G.D. Sevastopulo 1966. The palaeoecology of a Lower Viséan crinoid fauna from Feltrim, Co., Dublin. *Scientific Proc. Royal Dublin Society, Series A* 2: 273-286. [B,C,E]

Hughes, T.P., B.D. Keller, J.B.C. Jackson & M.J. Boyle 1985. Mass mortality of the echinoid *Diadema antillarum* Phillipi in Jamaica. *Bull. Mar. Sci.* 36: 377-384. [E]

Jacobson, S.R., J.R. Hatch, S.C. Teerman & R.A. Askin 1988. Middle Ordovician organic matter assemblage and their effect on Ordovician-derived oils. *Amer. Assoc. Petroleum Geologists Bull.* 72: 1090-1100.

Janicke, V. 1970. *Lumbricaria* – ein Cephalopoden – Koprolith. *Neues Jarhbuch für Geologie und Paläontologie, Montashelfte* 1970: 50-60. [C]

Jell, P.A. 1983. Early Devonian echinoderms from Victoria (Rhombifera, Blastoidea, and Ophiocistioidea). *Memoirs of the Association of Australasian Paleontologists* 1: 209-235. [Op]

Jensen, M. & E. Thomsen 1987. Ultrastructure, dissolution and "pyritization" of Late Quaternary and Recent echinoderms. *Geol. Soc. Denmark Bull.* 36: 275-287. [A,O]

Kammer, T.W. 1984. Crinoids from the New Providence Shale Member of the

Borden Formation (Mississippian) in Kentucky and Indiana. *J. Paleontol.* 58: 115-130. [C]

Kauffman, E.G. 1981. Ecological reappraisal of the German Posidonienschiefer (Toarcian) and the stagnant basin model. In: J. Gray & A.J. Boucot (eds), *Communities of the past*, pp. 311-381. Hutchinson Ross; Stroudsburg, Penn. [C]

Kesling, R.V. 1969. A new brittle-star from the Middle Devonian Arkona Shale of Ontario. *University of Michigan Museum of Paleontology Contributions* 23: 37-51. [O]

Kesling, R.V. 1971. *Michiganaster inexpectans*, a new many-armed starfish from the Middle Devonian Rogers City Limestone of Michigan. *University of Michigan Museum of Paleontology Contributions* 23: 247-262. [A]

Kesling, R.V. & D. Le Vasseur 1971. *Strataster ohioensis*, a new Early Mississippian brittle-star, and the paleoecology of its community. *University of Michigan Museum of Paleontology Contributions* 23: 305-341. [O]

Kidwell, S.M. & T.K. Baumiller 1989. Post-mortem disintegration of echinoids: Effects of temperature, oxygenation, tumbling, and algal coats. *Abstracts of the 28th International Geological Congress (Washington, D.C.)* 2: 188-189. [E]

Kidwell, S.M. & T.K. Baumiller 1990. Experimental disintegration of regular echinoids: Roles of temperature, oxygen, and decay thresholds. *Paleobiology* 16: 247-271. [E]

Kidwell, S.M. & D.W.J. Bosence 1991. Taphonomy and time-averaging of marine shelly faunas. In: P.A. Allison & D.E.G. Briggs (eds), *Taphonomy; releasing the data locked in the fossil record*, pp. 115-191. Plenum, New York and London. [C,E]

Kidwell, S.M. & D. Jablonski 1983. Taphonomic feedback; ecological consequences of shell accumulation. In: M.J.S. Tevesz & P.L. McCall (eds), *Biotic interactions in recent and fossil benthic communities*, pp. 195-248. *Plenum Press, New York.*

Kier, P.M. 1968. Triassic echinoids of North American. *J. Paleontol.* 42: 1000-1006. [E]

Kier, P.M. 1977. The poor fossil record of the regular echinoid. *Paleobiology* 3: 168-174. [E]

Kirk, E. 1911. The structure and relationships of certain eleutherozoic Pelmatozoa *U.S. National Museum Proceedings* 41: 1-137. [C]

Kobluck, D.R. & M.A. Lysenko 1984. *Carbonate rocks and coral reefs, Bonaire, Netherlands Antilles.* Geological Association of Canada, Mineralogical Association of Canada Joint Annual Meeting Field Trip 13, 67 pp. [E]

Kobluck, D.R. & W. Mielczarek 1984. Sediment. In: D.R. Kobluck & M.A. Lysenko (eds), *Carbonate rocks and coral reefs, Bonaire, Netherlands Antilles*, pp. 48-57. Geological Association of Canada, Mineralogical Association of Canada Joint Annual Meeting Field Trip 13, 67 [E]

Koch, D.L. & H.L. Strimple 1968. A new Upper Devonian cystoid attached to a discontinuity surface. *Iowa Geological Survey Report of Investigations* 5: 1-49. [Ed,R]

Kolata, D.R. & M. Jollie 1982. Anomalocystitid mitrates (Stylophora – Echinodermata) from the Champlanian (Middle Ordovician) Guttenberg Formation of the Upper Mississippi River Valley region. *J. Paleontol.* 56: 631-653. [H$_z$]

Kudrewicz, R. 1992. The endemic echinoids *Micraster (Micraster) maleckii* Mącyńska, 1979, from the Santonian deposits of Korzkiew near Cracow (southern Poland); their ecology, taphonomy and evolutionary position. *Acta Geologica Polonica* 42: 123-134. [E]

218

Land, L.S. 1967. Diagenesis of skeletal carbonates. *J. Sedimentary Petrology* 37: 914-930.

Lane, N.G. 1971. Crinoids and reefs. *Proceedings of the First North American Paleontological Convention* 1: 1430-1443. [C]

Lane, N.G. 1973. Paleontology and paleoecology of the Crawfordsville fossil site (Upper Osagian, Indiana). *University of California Special Publications in Geological Sciences* 99, 141 pp. [C]

Lane, N.G. & W.I. Ausich 1995. Interreef crinoid faunas from the Mississinewa Shale Member of the Wabash Formation (northern Indiana: Silurian; Echinodermata). *J. Paleontol.* 69:1090-1106. [C]

Lane, N.G. & G.D. Webster 1980. Crinoidea. In: T.W. Broadhead & J.A. Waters (eds), *Echinoderms, notes for a short course*, pp. 144-157. *University of Tennessee Studies in Geology* 3. [C]

Lasker, H. 1976. Effects of differential preservation on the measurement of taxonomic diversity. *Paleobiology* 2: 84-93. [E]

Laudon, L.R. & B.H. Beane 1937. The crinoid fauna of the Hampton Formation at LeGrand, Iowa. *University of Iowa Studies in Natural History* 17: 227-272. [B,C]

Lawrence, D.R. 1968. Taphonomy and information losses in fossil communities. *Geol. Soc. Amer. Bull.* 79: 1315-1330.

Lawrence, J.M. 1996. Mass mortality of echinoderms from abiotic factors. In: *Echinoderm Studies 5*, pp. 103-137. Balkema Press, Rotterdam. [A, C, E, H, O]

LeClair, E.E. 1993. Effects of anatomy and environment on the relative preservability of asteroids: a biomechanical comparison. *Palaios* 8: 233-243. [A]

Lehmann, D. & J.K. Pope 1989. Upper Ordovician tempesites from Swatara Gap, Pennsylvania: depositional processes affecting sediments and paleoecology of fossil faunas. *Palaios* 4: 553-564. [C,O,Hz]

LeMenn, J. & N. Spjeldnaes 1996. Un nouveau crinoïde Dimerocrinitidae (Camerata, Diplobathrida) de l'Ordovicien supérieur du Maroc, *Rosfacrinus robustus* nov. gen., nov. sp. *Geobios* 29: 341-351. [C]

Lessios, H.A. 1988. Mass mortality of *Diadema antillarum* in the Caribbean: What have we learned? *Ann. Rev. Ecol. Syst.* 19: 371-393. [E]

Lessios, H.A., J.D. Cubit, D.R. Robertson, M.J. Shulman, M.R. Parker, S.D. Garrity & S.C. Levings 1984. Mass mortality of *Diadema antillarum* on the Caribbean coast of Panama. *Coral Reefs* 3: 168-182. [E]

Lessios, H.A., D.R. Robertson & J.D. Cubit 1984. Spread of *Diadema* mass mortality through the Caribbean. *Science* 226: 335-337. [E]

Lewis, R.D. 1980. Taphonomy. In: T.W. Broadhead & J.A. Waters (eds), Echinoderms: Notes for a Short Course, pp. 40-58. . *University of Tennessee Department of Geological Sciences, Studies in Geology* 3. [A,C,E,H,O]

Lewis, R.D. 1986. Relative rates of skeletal disarticulation in modern ophiuroids and Paleozoic crinoids. *Geological Society of America Abstracts with Programs* 18: 672. [C,O]

Lewis, R.D. 1987. Post-mortem decomposition of ophiuroids from the Mississippi Sound. *Geological Society of America Abstracts with Programs* 19(2): 94-95. [O]

Lewis, R.D. 1997. Degradation and transport of modern crinoid ossicles, San Salvador, Bahamas: How good is the analog? *Geological Society of America Abstract with Programs* 29: 31. [C]

Lewis, R.D., C.R. Chambers & M.W. Peebles 1990. Grain morphologies and surface textures of Recent and Pleistocene crinoid ossicles, San Salvador, Bahamas. *Palaios* 5: 570-579. [C]

Lewis R.D. & C.M.H.M.S. Echols 1994. Attachment studies on plastic panels and calcium-carbonate skeletal substrata in Fernandez Bay, San Salvador Island, Bahamas. *Proceedings of the 26th Meeting of the Association of Marine Laboratories of the Caribbean, June 11-16, 1994.* Bahamian Field Station, Ltd, pp. 126-146. [C]

Lewis, R.D. & M.W. Peebles 1988. Surface textures of *Nemaster rubiginosa* (Crinoidea: Echinodermata) San Salvador, Bahamas. In: J. Mylroie (ed.), *Proceedings of the Fourth Symposium on the Geology of the Bahamas,* pp. 203-207. Bahamas Field Station, San Salvador, Bahamas. [C]

Liddell, W.D. 1975. Recent crinoid biostratinomy. *Geological Society of America Abstracts and Programs* 7: 1169. [C]

Liddell, W.D. & C.E. Brett 1982. Skeletal overgrowths among epizoans from the Silurian (Wenlockian) Waldron Shale. *Paleobiology* 8:67-78. [C]

Linck, O. 1965. Stratigraphische, stratinomische und ökologisches Betrachtung zu *Encrinus lilliformis* Lamarck. *Jahrbuch Geologie Landsesamt, Bad-Wurtemburg* 7: 123-148. [C]

Llewellyn, G. & C.G. Messing 1993. Compositional and taphonomic variations in modern crinoid-rich sediments from the deep-water margin of a carbonate bank. *Palaios* 8: 554-573. [C, E]

LoDuca, S.K. & C.E. Brett 1997. The *Medusaegraptus* epibole and lower Ludlovian Konservat-Lagerstätten of eastern North America. In: C.E. Brett & G.C. Baird (eds), *Paleontological Events: Stratigraphic, ecological, and evolutionary implications,* pp. 369-406. Columbia University Press, New York. [C]

MacCurdy, E. 1938. *The Notebooks of Leonardo da Vinci* (two volumes). Jonathan Cape, London, 455 pp.

MacQueen, R.W., E.D. Ghent & G.R. Davies 1974. Magnesium distribution in living and fossil specimens of the echinoid *Peronella lesueuri* Agassiz, Shark Bay, Western Australia. *J. Sediment. Petrology* 44: 60-69. [E]

Malzahn, E. 1968. Uber neue Funde von *Janassa bituminosa* (Schloth.) Im niederrheinischen Zechstein. *Geologische Jahresberichte* 85: 67-96.

Manni, R., U. Nicosia & L. Tagliacozzo 1997. *Saccocoma* as a normal benthonic stemless crinoid: an opportunistic reply within mud-dominated facies. *Palaeopelagos* 7: 121-132. [C]

Manten, A.A. 1971. *Silurian Reefs in Gotland.* Developments in Sedimentology 13. Elsevier, Amsterdam, 539 pp. [C]

Maples, C.G. & A.W. Archer 1989. Paleoecological and sedimentological significance of bioturbated crinoid calyces. *Palaios* 4: 379-383. [C]

Marcher, M.V. 1962. Petrography of Mississippian limestones and cherts from the northwestern Highland Rim, Tennessee. *J. Sediment. Petrology* 32: 819-832. [C]

Martin, R.R. 1999. *Taphonomy: a process approach.* Cambridge University Press, Cambridge, U.K., 508 pp. [A,C,E,O]

Mather, K.F. & S.L. Mason 1939. *A Source Book in Geology.* McGraw-Hill, New York, 702 pp.

McIntosh, G. 1978. Pseudoplanktonic crinoid colonies attached to Upper Devonian logs. *Geological Society of America Abstracts with Programs* 10(7): 453. [C]

Messing, C.G. 1997. Living comatulids. In: J.A. Waters & C.G. Maples (eds), *Geobiology of Echinoderms,* pp. 3-30. *Paleontological Society Papers* 3. [C]

Messing, C.G. & G. Llewellyn 1992. Variations in post-mortem disarticulation and sediment production in two species of Recent stalked crinoids. *Geological Society of America Abstracts with Programs* 27: A136. [C]

Messing, C.G., A.C. Neumann & D.J.C. Lang 1990. Biozonation of deep-water limestone lithoherms and associated hardgrounds in the northeastern Straits of Florida. *Palaios* 5: 15-33. [C]

Messing, C.G. & D. Rankin 1995. Local variations in skeletal contribution to sediment by a modern stalked crinoid (*Chladocrinus decorus*) (Echinodermata) relative to distribution of a living population. *Geological Society of America Abstracts with Programs* 27: A136. [C]

Messing, C.G., M.C. Rosesmyth, S.R. Mailer & J.E. Miller 1988. Relocation movement in a stalked crinoid (Echinodermata). *Bull. Mar. Sci.* 42 :480-487. [C]

Meyer, C.A. 1984. Palökologie und Sedimentologie der Echinodermen Lagerstätte Scholgraben (mittleres Oxfordian, Weissenstein, Kt. Solothurn). *Eclogae Geologicae Helvetiae* 77: 649-673. [A, O]

Meyer, C.A. 1988a. Paléoécologie d'une communauté d'ophiures du Kimméridgien supérieur de la region Havraise (Seine-Maritime). *Bulletin Trimestriel de la Société Géologique de Normandie et des Amis du Museum Havre* 75: 25-35. [O]

Meyer, C.A. 1988b. Palökologie, Biofazies und Sedimentologie von Seeliliengemeinschaften aus dem unteren Hauptrogenstein des Nordwestschweizer Jura. *Revista Paleobiologica* 7: 359-433. [C]

Meyer, D.L. 1971. Post-mortem disintegration of Recent crinoids and ophiuroids under natural conditions. *Geological Society of America Abstracts with Programs* 3: 645-646. [C,O]

Meyer, D.L. 1990. Population paleoecology and comparative taphonomy of two edrioasteroid (Echinodermata) pavements: Upper Ordovician of Kentucky and Ohio. *Hist. Biol.* 4: 155-178. [Ed]

Meyer, D.L. 1997. Implications of research on living stalked crinoids for paleobiology. *Paleontological Society Papers* 3: 31-43. [C]

Meyer, D.L. & W.I. Ausich 1983. Biotic interactions among recent and among fossil crinoids. In: M.J.S. Tevesz & P. L. McCall (eds), *Biotic Interactions in Recent and Fossil Benthic Communities*, pp. 377-427. Plenum Publishing Corp. [B,C]

Meyer, D.L., W.I. Ausich & R.E. Terry 1989. Comparative taphonomy of echinoderms in carbonate facies: Fort Payne Formation (Lower Mississippian) of Kentucky and Tennessee. *Palaios* 4: 533-552. [B,C]

Meyer, D.L., C.A. LaHaye, N.D. Holland, A.C. Arneson & J.R. Strickler 1984. Time-lapse cinematography of feather stars (Echinodermata: Crinoidea) on the Great Barrier Reef, Australia: Demonstrations of posture changes, locomotion, spawning and possible predation by fish. *Mar. Biol.* 78: 179-184. [C]

Meyer, D.L. & C.G. Maples 1997. Radial orientation pattern of *Uintacrinus* aggregations: New evidence for life at the air-water interface using surface tension. *Geological Society of America Abstracts with Programs* 29: A106. [C]

Meyer, D.L. & K.B. Meyer 1986. Biostratinomy of Recent crinoids (Echinodermata) at Lizard Island, Great Barrier Reef, Australia. *Palaios* 1: 294-302. [C]

Meyer, D.L., C.V. Milson & A.J. Webber 1999. *Uintacrinus*: A riddle wrapped in an enigma. *Geotimes*, August:14-16. [C]

Meyer, D.L. & T. Oji 1992. Experimental taphonomy of a Recent stalked crinoid: implications for the crinoid fossil record. *Geological Society of America Abstract with Programs* 24: 55. [C]

Meyer, D.L. & T. Oji 1993. Eocene crinoids from Seymour Island, Antarctica Peninsula: Paleobiogeographic and paleoecologic implications. *J. Paleontol.* 67: 250-257. [C]

Meyer, D.L., R.C. Tobin, W.A. Pryor, W.B. Harrison, R.G. Osgood, G.D. Hinterlong, B.J. Krumpolz & T.K. Mahan 1981. Stratigraphy, sedimentology, and paleoecology of the Cincinnatian Series (Upper Ordovician) in the vicinity of Cincinnati, Ohio. In: T.G. Roberts (ed.), *Geological Society of America Cincinnati 1981 Field Trip Guidebook, vol. 1*, pp. 31-72. [C,Ed]

Meyer, D.L. & T.R. Weaver 1980. Biostratinomy of crinoid-dominated communities in the lower Bull Fork Formation (Upper Ordovician) of southwestern Ohio. *Geological Society of America Abstracts with Programs* 12: 251. [C]

Miller, K.B., C.E. Brett & K.M. Parsons 1988. The paleoecologic significance of storm-generated disturbance within a Middle Devonian muddy epeiric sea. *Palaios* 3: 35-52. [C]

Milson, C.V., M.J. Simms & A.S. Gale 1994. Phylogeny and palaeobiology of *Marsupites* and *Uintacrinus*. *Palaeontology* 37: 595-607. [C]

Moore, R.C. & L.R. Laudon 1943. Evolution and classification of Paleozoic crinoids. *Geological Society of America Special Papers* 46, 153 pp.

Moore, R.C. & K. Teichert (eds) 1978. *Treatise on Invertebrate Paleontology Part T, Echinodermata 2. The Geological Society of America and University of Kansas Press; Boulder, Colorado and Lawrence, Kansas*, 1027 pp.

Moran, P.J. 1992. Preliminary observations of the decomposition of crown-of-thorns starfish, *Acanthaster planci* (L.). *Coral Reefs* 11: 115-118. [A]

Moran, P.J., R.E. Reichelt & R.H. Bradbury 1986. An assessment of the geological evidence for previous *Acanthaster* outbreaks. *Coral Reefs* 4: 235-238. [A]

Müller, A.H. 1953. Bemerkungen zur Stratigraphie und Stratonomie der ober-senone Schreibkreide von Rügen. *Geologie* 2:2 5-34 [C,E]

Müller, A.H. 1963. *Lehrbuch der Paläozoologie I. Allgemeine Grundlagen,* Second edition, G. Fisher, Jena, 405 pp. [O]

Müller, A.H. 1979. Fossilization (taphonomy). In: R.A. Robison & C. Teichert (eds), *Treatise on Invertebrate Paleontology. Part A, Introduction*, pp. 2-78. Geological Society of America and University of Kansas Press, Boulder, Colorado and Lawrence, Kansas. [C,H,O]

Nagle, J.S. 1967. Wave and current orientation of shells. *J. Sediment. Petrology* 37: 1124-1138. [C]

Nebelsick, J.H. 1992a. The Northern Bay of Safaga (Red Sea, Egypt): An actuopaläontological approach, III Distribution of echinoids. *Beiträge zur Paläontologie von Österreich* 17: 5-79. [E]

Nebelsick, J.H. 1992b. Echinoid distribution by fragment identification in the northern Bay of Safaga, Red Sea, Egypt. *Palaios* 7: 316-328. [E]

Nebelsick, J.H. 1992c. Components analysis of sediment composition in Early Miocene temperate carbonates from the Austrian Paratethys. *Palaeogeography, Palaeoclimatology, Palaeoecology* 91: 59-69. [E]

Nebelsick, J.H. 1995a. Actuopalaeontological investigations on echinoids: The potential for taphonomic interpretation. In: R. Emson, A. Smith & A. Campbell (eds), *Echinoderm Research 1995*, pp. 209-214. Balkema, Rotterdam. [E]

Nebelsick, J.H. 1995b. Uses and limitations of autopalaeonotological investigations on echinoids. *Geobios* 18: 329-336. [E]

Nebelsick, J.H. 1995c. Comparative taphonomy of Clypeasteroids. *Ecologae Geologicae Helvetiae* 88: 685-693. [E]

Nebelsick, J.H. 1996. Biodiversity of shallow-water Red Sea echinoids: Implications for the fossil record. *Journal of the Marine Biological Association U.K.* 76: 185-194. [E]

Nebelsick, J.H. 1999a. Taphonomic legacy of predation on echinoids. In: M.D. Candia Carnevali & F. Bonasoro (eds), *Echinoderm Research 1998*, 347-352. [E]

Nebelsick, J.H. 1999b. Taphonomy of *Clypeaster* fragments: Preservation and tapho-facies. *Lethaia* 32: 241-252. [E]

Nebelsick, J.H. & S. Kampfer 1994. Taphonomy of *Clypeaster humilis* and *Echino-discus auritus* (Echinoidea, Clypasteroida) from the Red Sea. In: B. David, A. Guille, J.-P. Feral & M. Roux (eds), *Echinoderms through time*, pp. 803-808. Balkema, Rotterdam. [E]

Nebelsick, J.H. & A. Kroh 1999. Palaeoecology and taphonomy of *Parascutella* beds from the Lower Miocene of the Eastern Desert, Egypt. In: M.D. Candia Carnevali & F. Bonasoro (eds), *Echinoderm Research 1998*, p. 353. [E]

Nebelsick, J.H., B. Schmid & M. Stachowitsch 1997. The encrustation of fossil and recent sea-urchin tests: ecological and taphonomic significance. *Lethaia* 30: 271-284. [E]

Néraudeau, D. & G. Breton 1993. Un assemblage de *Macaster* c.f. *polygonus*, spatan-goïde primitif de l'Albien moyen de Saint-Jouin (Seine-maritime): Développe-ment ontogénique, démographie, paléoécologie. *Bulletin trimestriel de la Société Géologique de Normandie et des Amis du Muséum du Havre* 80: 53-62. [E]

Néraudeau, D., J. Thierry, G. Breton & P. Moreau 1998. Comparative variation in asteroid and echinoid diversity during the Late Cenomanian transgression in Charentes (France). In: R. Mooi & M. Telford (eds), *Echinoderms: San Fran-cisco*, pp. 65-70. Balkema, Rotterdam. [A,E]

O'Brien, N.J., C.E. Brett & W.L. Taylor 1994. The significance of microfabric and taphonomic analysis in determining sedimentary processes in marine mudrocks: Examples from the Silurian of New York. *J. Sediment. Res., Section A* A64: 847-852. [C]

Oji, T. & S. Amemiya 1998. Survival of crinoid stalk and its taphonomic implica-tions. *Paleontol. Res.* 2: 67-70. [C]

Okulitch, V.J. & W.M. Tovell 1941. A crinoidal marking in the Dundas Formation at Toronto. *J. Paleontol.* 15: 89. [C]

Olson, E.C. 1980. Taphonomy: Its history and role, in community evolution. In: A.K. Behrensymeyer & A.P. Hill (eds), *Fossils in the Making*, pp. 5-19. University of Chicago Press, Chicago.

Parsons, K.M., C.E. Brett & K.B. Miller 1988. Taphonomy and depositional dynam-ics of Devonian shell-rich mudstones. *Palaeogeography, Palaeoclimatology, Palaeoecology* 63: 109-141. [C]

Parsons, K.M., E.N. Powell, C.E. Brett, S.E. Walker, A. Raymond, R. Callender & G. Staff 1997. Experimental taphonomy on the continental shelf and slope. *Proceed-ings of the 8th International Coral Reef Symposium, Panama City, Panama.*

Pawson, D.L. 1980. Holothuroidea. In: T.W. Broadhead & J.A. Waters (eds), *Echi-noderms, notes for a short course*, pp. 175-189. *University of Tennessee Studies in Geology* 3. [H]

Peterson, C.H. 1976. Relative abundance of living and dead molluscs in two Cali-fornia lagoons. *Lethaia* 9: 137-148. [E]

Pinna, G. 1985. Exceptional preservation in the Jurassic of Osteno. *Philosophical Trans. Roy. Soc. London B* 311:171-180. [O]

Plotnick, R.E. 1986. Taphonomy of modern shrimp: Implications for the arthropod fossil record. *Palaios* 1: 286-293.

Poinar, G., Jr. 1999. Ancient DNA. *Amer. Sci.* 87: 446-457.

Powers, B.G. & W.I. Ausich 1990. Epizoan associations in a Lower Mississippian paleocommunity (Borden Group, Indiana, U.S.A.). *Historical Biology* 4: 245-265. [C]

Rasmussen, H.W. 1950. Cretaceous Asteroidea and Ophuroidea with special reference to the species found in Denmark. *Danmarks Geologiske Undersøgelse. II. række* 77: 1-134. [A]

Rasmussen, H.W. 1977. Function and attachment of the stem in Isocrinidae and Pentacrinitidae: Review and interpretation. *Lethaia* 10: 51-57. [C]

Régis, M.B. 1977. Organisation microstructurale du stéréome de l'Échinoïde *Paracentrotus lividus* Lamarck et ses éventuelles incidences physiologiques. *Comptes Rendus de l'Académie des Sciences Paris, Séries D* 285: 189-192. [E]

Reyment, R.A. 1986. Necroplanktonic dispersal of echinoid tests. *Palaeogeography, Palaeoclimatology, Palaeoecology* 52: 347-349. [E]

Riddle, S.W., J.I. Wulff & W.I. Ausich 1988. Biomechanics and stereomic microstructure of the *Gilbertsocrinus tuberosus* column. In: R.D. Burke, P.V. Mladenov, P. Lambert & R.L. Parsley (eds), *Echinoderm Biology*, pp. 641-648. Balkema, Rotterdam. [C]

Roman, J. & J. Fabre 1986. Un rivage à échinoïdes régulier de la base du Crétacé à Canjuers (Aiguines, Var). *Actes du IIIe congrès national des Sociétés Savantes* 111: 147-158. [E]

Roman, J. & A. Strougo 1987. *Fayoumaster pharaonum* n. gen., n. sp. (Asteroidea, Goniasteridae) et son cadre stratigraphique et paléoécologique (Éocène d'Egypte). *Annales de Paléontologie* 73: 29-50. [A]

Rose, E.P.F. 1984. Problems and principles of neogene echinoid biostratigraphy. *Annales Géologiques des Pays Helléniques* 32: 171-181. [E]

Rosenkranz, D. 1971. Zur Sedimentologie und Okölogie von Echinodermen-Lagerstätten. *Neues Jahrbuch für Geologie und Paläontologie, Abhandlungen* 138: 221-258. [A,C,E,O]

Roux, M. 1975. Microstructural analysis of the crinoid stem. *University of Kansas Paleontological Contributions* 75, 7 pp. [C]

Ruhrmann, G. 1971a. Riffe-nahe Sedimentation paläozoischer Krinoiden-Fragmente. *Neues Jahrbuch für Geologie und Paläontologie Abhandlungen* 138: 56-100. [C]

Ruhrmann, G. 1971b. Riffe-nahe Sedimentation unterdevonischer krinoidenkalk im Kantabrischen Gebirge (Spanien). *Neues Jahrbuch für Geologie und Paläontologie Monatshafte 1971* H4: 231-248. [C]

Sadler, M. & R.D. Lewis 1996. Actualistic studies of the taphonomy and ichnology of the irregular echinoid *Meoma ventricosa* at San Salvador, Bahamas. *Geological Society of America Abstracts with Programs* 28: 293-294. [E]

Sass, D. & R.A. Chondrate 1985. Destruction of a Late Devonian ophiuroid assemblage: a victim of changing ecology at the Catskill delta front. In: D.L. Woodrow & W.D. Sevon (eds), *The Catskill Delta*, pp. 237-263. *Geol. Soc. Amer. Special Paper* 201. [O]

Savarese, M., J.R. Dodd & N.G. Lane 1997. Taphonomic and sedimentologic implications of crinoid intraskeletal porosity. *Lethaia* 29: 141-156. [C]

Schäfer, W. 1962. *Aktuo-Paläöntologie nach Studien in der Nordsee.* Verlag W. Kramer, Frankfurt am Main, 666 pp. [A,E,H,O]

Schäfer, W. 1972. *Ecology and palaeoecology of marine environments.* University of Chicago Press, Chicago, 568 pp. [A,E,H,O]

Schopf, T.J.M. 1978. Fossilization potential of an intertidal fauna: Friday Harbor, Washington. *Paleobiology* 4: 261-270. [A,E,H,O]

Schubert, J.K., D.J. Bottjer & M.J. Simms 1992. Paleobiology of the oldest known articulate crinoid. *Lethaia* 25: 97-110. [C]

Schuchert, C. 1904. On siluric and devonic cystidea and *Camarocrinus. Smithsonian Miscellaneous Contributions* 47: 201-272. [C]

Schumacher, G.A. 1986. Storm processes and crinoid preservation. *Fourth North American Paleontological Convention, Boulder, 12-15 August, Abstracts*, A41. [C]

Schumacher, G.A. & W.I. Ausich 1983. New Upper Ordovician echinoderm site: Bull Fork Formation, Caesar Creek Reservoir (Warren County, Ohio). *Ohio J. Sci.* 83: 60-64. [C]

Schumacher, G.A. & W.I. Ausich 1985. Catastrophic sedimentation: Impact on a Late Ordovician crinoid assemblage. *Geological Society of America Abstracts with Program* 17: 278. [C]

Schwarzacher, W. 1961. Petrology and structure of some Lower Carboniferous reefs in northwestern Ireland. *American Association of Petroleum Geologists Bulletin*, 45: 481-1503. [C]

Schwarzacher, W. 1963. Petrology and structure of some Lower Carboniferous reefs in northwestern Ireland. *J. Sediment. Petrology* 33: 580-586. [C]

Seilacher, A. 1960. Strömungsanzeichen in Hunsrückshiefer Notizblatt des Hessischen *Landesant für Bodenforschung zu Wiesbaden*. 88: 88-106. [A,C,O]

Seilacher, A. 1968. Origin and diagenesis of the Oriskany Sandstone (Lower Devonian, Appalachians) as reflected in its fossil shells. In: G. Müller & G.M. Friedman (eds), *Recent Developments in Sedimentology in Central Europe*, pp. 175-185. Springer Verlag, New York, 33: 580-586. [C]

Seilacher, A. 1970. Begriff und Bedeutung der Fossil-Lagerstätten. *Neus Jahrbuch für Geolologie und Paläontologie Monatshafte* 1970: 34-39.

Seilacher, A. 1973. Biostratinomy: The sedimentology of biologically standardized particles. In: R.N. Ginsburg (ed.), *Evolving concepts in sedimentology*, pp. 159-177. Johns Hopkins Univ. Press, Baltimore. [A,C,E,O]

Seilacher, A. 1979. Constructional morphology of sand dollars. *Paleobiology* 5: 191-221. [E]

Seilacher, A., G. Drozdzewski & R. Haude 1968. Form and function of the stem in a pseudoplanktonic crinoid (*Seirocrinus*). *Palaeontology* 11: 275-282. [C]

Seilacher, A. & C. Hemleben 1966. Spürenfauna und Bildungstiefe der Hunstrückschiefer (Unterdevon). *Notizblatt des Hessichen Landesamt für Bodenforschung zu Wiesbaden* 94: 40-53.

Seilacher, A., W.E. Reif & F. Westphal 1985. Sedimentological, ecological and temporal patterns of fossil-Lagerstätten, In: H.B. Whittington and S. Conway Morris (eds), *Extraordinary biotas: their ecological and evolutionary significance*, pp. 5-23. Phil. Trans. R. Soc. London, B, 311. [A,C,E,O]

Sevastopulo, G.D. & N.G. Lane 1988. Ontogeny and phylogeny of disparid crinoids. In: C.R.C. Paul & A.B. Smith (eds), *Echinoderm phylogeny and evolutionary biology*, p. 245-253. Clarendon Press, Oxford, England.

Simms, M.J. 1986. Contrasting lifestyles in Lower Jurassic crinoids: A comparison of benthic and pseudopelagic Isocrinida. *Palaeontology* 29: 475-493. [C]

Simms, M.J. 1994. Crinoids from the Chambara Formation, Pucará Group, central Peru. *Palaeontographica Abteilung A* 223: 169-175. [C]

Simon, A. & M. Poulicek 1990. Biodégradation anaérobique des structures squelettiques en milieu marin: I – Approche morphologique. *Cah. Biol. Mar.* 31:95-105. [E]

Simon, A., M. Poulicek, R. Machiroux & J. Thorez 1990. Biodégradation anaérobique des structures squelettiques en milieu marin: II. Approche chimique. *Cah. Biol. Mar.* 31: 365-384. [E]

Smith, A.B. 1980. Stereom microstructure of the echinoid tests. *Special Papers in Palaeontology* 25, 81 pp.

Smith, A.B. 1984. *Echinoid paleobiology*. George Allen and Unwin, London, 190 pp. [E]

Smith, A.B. & J. Gallemí 1991. Middle Triassic holothurians from northern Spain. *Palaeontology* 34: 49-76 [H]

Smith, A.B. & C.R.C. Paul 1982. Revision of the class Cyclocystoidea (Echinodermata). *Phil. Trans. Royal Soc. London, B. Biological Series* 296: 577-684. [Cy]

Spencer, W.K. & C.W. Wright 1966. Asterozoans. In: R.C. Moore (ed), *Treatise on invertebrate paleontology. Part V, Echinodermata 3*, pp. U4-U107. Geological Society of America and University of Kansas Press, Boulder, Colorado and Lawrence, Kansas. [A]

Speyer, S.E. & C.E. Brett 1986. Trilobite taphonomy and Middle Devonian taphofacies. *Palaios* 1: 312-327. [B,C,O]

Speyer, S.E. & C.E. Brett 1988. Taphofacies models for epeiric sea environments: Middle Paleozoic examples. *Palaeogeography, Palaeoclimatology, Palaeoecology* 63: 225-262. [C]

Speyer, S.E. & C.E. Brett. 1991 Taphonomic controls: background and episodic processes in fossil assemblage preservation. In: P.A. Allison & D.E.G. Briggs (eds), *Taphonomy: releasing the data locked in the fossil record*, pp. 502-546. Plenum Press, New York. [C]

Springer, F. 1901. *Uintacrinus*, its structure and relations. *Harvard College Museum of Comparative Zool. Mem.* 25, 89 pp. [C]

Springer, F. 1917. On the crinoid genus *Scyphocrinus* and its bulbous root, *Camarocrinus*. *Smithsonian Institution Publication* 2440, 74 pp.

Sprinkle, J. 1973a. Morphology and evolution of blastozoan echinoderms. *Harvard Museum of Comparative Zoology Special Publication*, 283 pp. [Eo]

Sprinkle, J. 1973b. *Tripatocrinus*, a new hybocrinid crinoid based on disarticulated plates from the Antelope Valley Limestone of Nevada and California. *J. Paleontol.* 47: 861-882. [C]

Sprinkle, J. 1982a. Echinoderm zones and faunas. In: J. Sprinkle (ed.), *Echinoderm faunas of the Bromide Formation (Middle Ordovican) of Oklahoma*, p. 46-56. *University of Kansas Paleontological Contribution Monograph 1*. [C]

Sprinkle, J. 1982b. Echinoderm faunas of the Bromide Formation (Middle Ordovician) of Oklahoma. *University of Kansas Paleontological Contributions Monograph* 1, 369 pp.

Sprinkle, J. & D. Collins 1999. Revision of *Echmatocrinus* from the Middle Cambrian Burgess Shale of British Columbia. *Lethaia* 31: 269-282.

Sprinkle, J. & R.C. Gutschick 1967. *Costatoblastus*, a channel fill blastoid from the Sappington Formation of Montana. *J. Paleontol.* 41: 385-402. [B]

Sprinkle, J. & M.W. Longman 1982. Echinoderm paleoecology. In: J. Sprinkle (ed), *Echinoderm faunas of the Bromide Formation (Middle Ordovician) of Oklahoma*, p. 68-75. *University of Kansas Paleotological Contributions Monograph 1*. [C]

Sroka, S.D. 1988. Preliminary studies on a complete fossil holothurian from the Middle Pennsylvanian Francis Creek Shale of Illinois. In: R.D. Burke, P.V. Mladenov, P. Lambert & R.L. Parsley (eds), *Echinoderm biology*: 159-160. Balkema, Rotterdam. [H]

Stilwell, J.D., R.E. Fordyce & P.J. Rolfe 1994. Paleocene isocrinids (Echinodermata: Crinoidea) from the Kauru Formation, South Island, New Zealand. *J. Paleontol.* 68: 135-141.

Strathmann, R.R. 1981. The role of spines in preventing structural damage to echinoid tests. *Paleobiology* 7: 400-406. [E]

226

Taylor, W. & C.E. Brett 1996. Taphonomy and paleoecology of echinoderm Lager-stätten from the Silurian (Wenlockian) Rochester Shale. *Palaios* 11: 118-140. [C]

Tetreault, D.K. 1995. An unusual Silurian arthropod/echinoderm dominated soft-bodies fauna from the Eramosa Member (Ludlow) of the Guelph Formation, southern Bruce Peninsula, Ontario, Canada. *Geological Society of America Abstracts with Programs* 27(6): A-114. [E,O]

Thomas, D.W. & M. Blumer 1964. The organic chemistry of a fossil--III. The hydro-carbons and their geochemistry. *Geochim. et Cosmochim. Acta* 28: 1467-1477. [C]

Ubaghs, G. 1978. Origin of crinoids. In: R.C. Moore & K. Teichert (eds), *Treatise on invertebrate paleontology, Part T, Echinodermata 2(2)*, pp. T275-T281 *University of Kansas and Geological Society of America, Lawrence, Kansas, and Boulder, Colorado.*

Van Sant, J.F. & N.G. Lane 1964. Crawfordsville (Indiana) crinoid studies. *University of Kansas Paleontological Contributions Echinoderm Article 7*, 136 pp.

Watkins, R. 1991. Guild structure and tiering in a high-diversity Silurian community, Milwaukee County, Wisconsin. *Palaios* 6: 465-478. [C]

Watkins R. & J.M. Hurst 1977. Community relations of Silurian crinoids at Dudley, England. *Paleobiology* 3: 207-217. [C]

Webber, A.J. & D.L. Meyer 1999. *Uintacrinus socialis* Grinnell (Cretaceous, San-tonian): New observations and re-evaluation of its life habit. *Geological Society of America Abstracts with Programs* 31: A79. [C]

Weber, J.N. 1969. The incorporation of magnesium into the skeletal calcite of echinoderms. *Amer. J. Sci.* 267: 537-566.

Welch, J.R. 1984. The asteroid *Lepidasterella montanaensis*, n. sp., from the upper Mississippian Bear Gulch Limestone of Montana. *J. Paleontol.* 58: 843-851. [A]

Wells, J.W. 1941. Crinoids and *Callixylon*. *Amer. J. Sci.* 239: 454-456.

Whiteley, T.E., C.E. Brett & D.M. Lehmann 1993. The Walcott-Rust quarry: a unique Ordovician trilobite Konservatte-Lagerstätte. *Geological Society of America Abstracts with Programs* 25(2): 89. [C]

Wigley, R.L. & F.C. Stinton 1973. Distribution of macroscopic remains of Recent animals from marine sediments off Massachusetts. *Fish. Bull.* 71: 1-40. [E]

Wignall, P.B. & M.J. Simms 1990. Pseudoplankton. *Palaeontology* 33: 359-378. [C]

Wilson, M.V.H. 1988. Taphonomic processes: Information loss and information gain. *Geoscience Canada* 15: 131-148.

Wright, C.W. & E.V. Wright 1941. Notes on Cretaceous Asteroidea. *Quarterly J. Geol. Soc. London* 96: 231-248. [A]

Zangrel, R. & E.S. Richardson, Jr. 1963. The paleoecological history of two Pennsylvanian black shales. *Fieldiana Geol. Mem.* 4: 352 pp. [C]

An index of names of recent Asteroidea – Part 4: Forcipulatida and Brisingida

AILSA M. CLARK[1] AND CHRISTOPHER MAH[2]

[1]*Formerly of Department of Zoology, The Natural History Museum, (London, UK). Present address: Gyllyngdune, Wivelsfield Green, Sussex, UK.*
[2]*Department of Invertebrate Zoology, California Academy of Sciences, Golden Gate Park, San Francisco, USA*

KEYWORDS: Taxonomy, geographical range, Bathymetry, Asteroidea, Echinodermata.

INTRODUCTION

Treatment of nearly all the Forcipulatida is by Ailsa Clark, as before, but that of the Labidiasteridae and the Brisingida is by Christopher Mah, whose timely researches have enabled him to take over this specialized order.

Explanation of the procedure followed in this index was given in part 1. However, the type conventions followed are briefly repeated here: Valid names for genera and species are given in bold type when in their definitive position alphabetically but in italics in cross references where either genus-group names have been altered in rank or species-group names have been transferred to other genera; names in ordinary type are synonyms or otherwise invalid. Asterisks before names signify doubtful or threatened names needing further attention, while asterisks under 'Range' indicate the type localities where noted during compilation.

The classification of the Forcipulatida followed here is largely that initiated by Downey in Clark and Downey (1992: 401). Apart from the long-established Zoroasteridae, Fisher's subfamilies Neomorphasteridae, Pedicellasterinae and Labidiasterinae (1928)'s were raised to the rank of families apart from the Asteriidae, with which A.M.C. had previously (1962a) merged the Coscinasteriinae for want of a character to distinguish all the genera, despite considerable divergence in general facies. The long-established family Heliasteridae and the small subfamily Pycnopodiinae of the Asteriidae being absent from the Atlantic were not covered in 1992. Pending further assessment, it seems best to treat the Pycnopodiinae in this index on a par with the similarly multibrachiate Labidiasteridae as another family. (C.M. concurs with this ranking). [A.M.C.]

229

As for the Brisingida, the classification is based on preliminary data published in Mah (1998a). A full revision is in preparation. Phylogenetic data support a valid Freyellidae Downey, 1986 but show the Brisingidae (sensu Downey in Clark & Downey, 1992) to be a paraphyletic assemblage. Some genera formerly included in the Brisingidae have been assigned to three families: the Odinellidae, the Brisingidae and the Hymenodiscidae. Range and bathymetric data given here for brisingidans differ slightly from the rest of the list in that they show full bathymetric ranges as currently known rather than type localities. [C.M.]

Order FORCIPULATIDA Perrier

Family ZOROASTERIDAE Sladen

Zoroasteridae Sladen 1889:416; Fisher 1919a:470; 1919b:387; H.L.
    Clark 1920:94; Fisher 1928a:32; Downey 1970a:1-18; McKnight
    1977b:159; Maluf 1988:124; Downey *in* Clark & Downey 1992:401.
Genus-group names: **Bythiolophus, Cnemidaster, Doraster,
Mammaster, Myxoderma, Pholidaster**, Prognaster, **Zoroaster**.

BYTHIOLOPHUS Fisher, 1916
    Fisher 1916b:31; 1919a:484; 1919b:389.
    Type species: *Bythiolophus acanthinus* Fisher, 1916.
acanthinus Fisher, 1916
    Fisher 1916b:31; 1919a:485.
    Range: Celebes (Sulawesi), Indonesia, 1020 m.
macracanthus (H.L. Clark, 1916)
    H.L. Clark 1916:68; 1920:101 (in key) (as *Zoroaster*).
    Rowe *in* Rowe & Gates 1995:115 (as *Bythiolophus*).
    Range: South Australia, 457-820 m.
    Possibly a synonym of *Zoroaster spinulosus* Fisher, 1906 (type locality off the Hawaiian Is but also recorded from New Zealand) according to Fell (1958) but having a very different facies according to Rowe (1995).
CNEMIDASTER Sladen, 1889
    Sladen 1889:423; Fisher 1919a:480; 1919b:389.
    Type species: *Cnemidaster wyvilli* Sladen, 1889.
nudus (Ludwig, 1905)
    Ludwig 1905:164 (as *Zoroaster*).
    Downey 1970a:14; Maluf 1988:43,124 (as *Cnemidaster*).
    Range: Off Lower California, 1366-2600 m.

squameus (Alcock, 1893)

Alcock 1893a:109 (as *Zoroaster*).

Fisher 1919a:481; Macan 1938:415 (as *Cnemidaster*).

Range: Laccadive Sea and off Aden, 1900-2000 m.

wyvilli Sladen, 1889

Sladen 1889:424; Fisher 1919a:480.

Range: Arafura Sea to Borneo, 1390-1990 m.

DORASTER Downey, 1970

Downey 1970a:5; Moyana & Larrain Prat 1976:103.

Type species: *Doroaster constellatus* Downey, 1970.

cancellatus Jangoux 1978:98. Lapsus for *D. constellatus*.

constellatus Downey, 1970

Downey 1970a:5; Moyana & Larrain Prat 1976:103; Downey *in* Clark & Downey 1992:402.

Range: Gulf of Mexico, atlantic Panama, Surinam, 350-640 m.

qawashqari Moyana & Larrain Prat, 1976

Moyana & Larrain Prat 1976:105.

Range: S Chile, 300 m.

MAMMASTER Perrier, 1894

Perrier 1894:114 (in key); Fisher 1919b:389; Downey 1970a:12; Downey *in* Clark & Downey 1992:403.

Type species: *Zoroaster sigsbeei* Perrier, 1880.

sigsbeci Perrier, 1880:436. Lapsus for *sigsbeei* (Perrier, 1881)

Perrier 1881a:5; 1884:195 (as *Zoroaster*).

Perrier 1894:125; H.L. Clark 1920:73; 1941:67; Downey 1970a:12; Downey *in* Clark & Downey 1992:403 (as *Mammaster*).

Range: Leeward Is, S Florida, Trinidad, 310-640 m.

MYXODERMA Fisher, 1905

Fisher 1905:316 (as subgenus of *Zoroaster*).

Fisher 1919b:389,391; 1928:44; Aziz & Jangoux 1984b:193 (as genus).

Type species: *Zoroaster (Myxoderma) sacculatus* Fisher, 1905.

acutibrachia Aziz & Jangoux, 1984

Aziz & Jangoux 1984b:192.

Range: Macassar Strait, 715-800 m.

derjugini Djakonov, 1950

Djakonov 1950a:104; 1950b:32.

Range: Okhotsk Sea, 590-665 m.

ectenes Fisher, 1919b:392. A subspecies of *M. sacculatum* Fisher, 1905.

longispinum: Fisher, 1928:51, see *Zoroaster*

platycanthum (H.L. Clark, 1913) (with subsp. *rhomaleum* Fisher, 1919)

platycanthum platycanthum H.L. Clark, 1913

H.L. Clark 1913:199 (as *Zoroaster*).

Fisher 1919b:392; 1928:52; Aziz & Jangoux 1984b:193; Maluf

1988:44,124 (as *Myxoderma*).

Range: Southern to Lower California, 256-768 m.

platycanthum rhomaleum Fisher, 1919

Fisher 1919b:393.

Range: Oregon and mid-California, 500-540 m.

rhomaleum Fisher, 1919. A subspecies of *M. platycanthum* H.L. Clark, 1913

sacculatum (Fisher, 1905) (with subspecies *ectenes* Fisher, 1919)

sacculatum sacculatum Fisher, 1905

Fisher 1905:316 (as *Zoroaster (Myxoderma)*).

Fisher 1919b:392; 1928:45; Baranova 1957:175; Maluf 1988:44,125 (as *Myxoderma*).

Range: Bering Sea and mid-California, 1000-1400 m.

sacculatum ectenes Fisher, 1919

Fisher 1919b:392; 1928:49.

Range: Southern California, 520-1940 m.

PHOLIDASTER Sladen, 1889 [?1885:616]

Sladen 1889:426; Fisher 1919a:471,484; 1919b:388.

Type species: *Pholadaster squamatus* Sladen, 1889 (designated by Fisher, 1919).

distinctus Sladen, 1889

Sladen 1889:429.

Range: Indonesia, 256 m.

squamatus Sladen, 1889

Sladen 1889:427; Fisher 1919a:484; H.L. Clark 1920:98; Marsh 1976:221; Jangoux 1981:475.

Range: Philippines, W and NW Australia, 164-246 m.

PROGNASTER Perrier, 1891

Perrier 1891c:1226; 1896:22 (Probably not Perrier, 1894 according to Fisher (1928a:32); Fisher 1919b:388.

Type species: *Prognaster grimaldii* Perrier, 1891.

Listed as a doubtful genus by Downey (1970) and type species cited as a synonym of type species of *Zoroaster* Thomson, 1873 by Downey *in* Clark & Downey (1992).

grimaldii Perrier, 1891c:1226; 1896:22; Fisher 1928:32. Probably a synonym of *Zoroaster fulgens* Thomson, 1873 according to Downey (1970a) and cited as a synonym by Downey *in* Clark & Downey (1992).

*longicauda Perrier, 1894:120 (cited as sp. nov. of *Prognaster*
gen. nov., so probably non *Zoroaster longicauda* Perrier, 1885 according to Fisher 1928a:32). Presumably also referable to *Zoroaster*.

ZOROASTER Thomson, 1873 (with synonym *Prognaster* Perrier, 1891)

Thomson 1873:154; Sladen 1889:416; Fisher 1919:472; H.L. Clark 1920:100; Fisher 1928:33; Downey 1970a:14; Downey *in* Clark &

Downey 1992:403.

Type species: *Zoroaster fulgens* Thomson, 1873.

*ackleyi Perrier, 1880:436 (but barely validated); 1881:6; 1884: 191; 1894:117; H.L. Clark 1920:102; 1941:66. A synonym of *Z. fulgens* Thomson, 1873 according to Downey (1970) but inferred as having varietal status following Farran (1913) by Harvey et al. (1988).

actinocles Fisher, 1919
Fisher 1919b:390; 1928:37.
Range: Aleutian Is, 2220 m.

adami Koehler, 1909
Koehler 1909b:108; H.L. Clark 1920:101 (in key).
Range: Andaman Sea, 1040 m.

alfredi Alcock, 1893
Alcock 1893a:102; 1893b:173; H.L. Clark 1920:101 (in key); Macan 1938:415.
Range: Bay of Bengal and Arabian Sea, 2380-3350 m.

angulatus Alcock, 1893
Alcock 1893a:105; H.L. Clark 1920:101 (in key); Macan 1938:415.
Range: Gulf of Mannar, Maldive-Laccidive area, 910-1460 m.

barathri Alcock, 1893
Alcock 1893a:103; 1893b:173[?]; H.L. Clark 1920:101 (in key).
Range: Bay of Bengal, 2780 m.

bispinosus Koehler, 1909a:316. Lapsus for *Z. trispinosus* Koehler, 1895. A synonym of *Z. fulgens* Thomson, 1873 according to Downey (1970a).

carinatus Alcock, 1893 (with synonym *philippinensis* Fisher, 1916 (as subsp.))
Alcock 1893a:107; H.L. Clark 1920:102 (in key); Jangoux 1981:459; Jangoux & Aziz 1988: ; Rowe *in* Rowe & Gates 1995:116;
Liao & A.M. Clark 1995 [1996]:140 (as *Z. carinatus philippinensis*).
Range: Andaman Sea, off Reunion I., S China Sea, Borneo, 180-1120 m.

diomedeae Verrill, 1884:217; H.L. Clark 1920:101 (in key). A synonym of *Z. fulgens* Thomson, 1873 according to Downey (1970a).

evermanni Fisher, 1905 (with subsp. *mordax* Fisher, 1919)

evermanni evermanni Fisher, 1905
Fisher 1905:317 (as Zoroaster (Myxoderma)).
Fisher 1919b:390; 1928:40; Maluf 1988:44,125.
Range: S California, 395-930 m.

evermanni mordax Fisher, 1919
Fisher 1919b:391.
Range: Washington to S California, 437-1390 m.

evermanni: H.L. Clark, 1913:198, non *Z. evermanni* Fisher, 1905, = *Myxoderma sacculatum* (Fisher, 1905) according to Fisher (1928).

fulgens Thomson, 1873 (with synonyms *Zoroaster ackleyi* Perrier, 1880, *Z. bispinosus* (lapsus) Koehler, 1909, *Z. diomedeae* Verrill, 1884, *Prognaster grimaldii* Perrier, 1891, *Zoroaster longicauda* Perrier, 1885 and *Z. trispinosus* Koehler, 1895 but *ackleyi* possibly recognisable as a long-armed form or variety)

Thomson 1873:154; Sladen 1883b:160; Perrier 1885c:16; Sladen 1889:418; Perrier 1894:116; Koehler 1895:442; Grieg [1921]1932:24; Downey 1970a:15; Gage et al. 1983:286; Harvey et al. 1988:167; Downey *in* Clark & Downey 1992:403.

Range: Faeroe Channel and entire Atlantic Ocean, 900-4810 m.

gilesi Alcock, 1893

Alcock, 1893a:108; H.L. Clark 1920:101 (in key).

Range: Andaman Sea, c.900 m.

hirsutus Ludwig, 1905

Ludwig 1905:172; H.L. Clark 1920:101 (in key); Fisher 1928:37.

Range: Off Acapulco, Mexico, 3440 m.

longicauda Perrier, 1885b:885; 1885c:19. A probable synonym of *Z. fulgens* Thomson, 1873 according to Downey (1970a) and cited as a synonym by Downey *in* Clark & Downey (1992).

longispinus Ludwig, 1905

Ludwig 1905:180; Downey 1970:13 (listed without comment on generic position); Maluf 1988:45,255.

Fisher 1928:51 (as *Myxoderma*).

Range: Gulf of California S to Panama and the Galapagos, 1430-2420 m.

macracantha H.L. Clark, 1916, see *Bythiolophus*

magnificus Ludwig, 1905

Ludwig 1905:159; H.L. Clark 1920:104; Fisher 1928:37; Maluf 1988:44,125.

Range: Panama to N Peru, 3060-3670 m.

microporus Fisher, 1916

Fisher 1916b:30; 1919a:475; H.L.Clark 1920:101 (in key).

Range: Molucca Is, 1280 m.

nudus Ludwig, 1905, see *Cnemidaster*

ophiactis Fisher, 1916

Fisher 1916b:29; 1919a:473; H.L. Clark 1920:101 (in key); Hayashi 1073a:94 (R only 18 mm).

Range: Philippines and Celebes (Suluwesi), also Sagami Bay, Japan (?), 1020-1630 (?500) m.

ophiurus Fisher, 1905

Fisher 1905:315; H.L. Clark 1920:102 (in key); Fisher 1928:34; Baranova 1957:175; Maluf 1988:45, 125.

Range: Bering Sea S to N Peru, 695-2230 m.

orientalis Hayashi, 1950
Hayashi 1950:
Range: Japan, 192-870 m.
perarmatus H.L. Clark, 1920
H.L. Clark 1920:102; Maluf 1988:45, 125.
Range: Off N Peru, 980 m.
philippinensis Fisher, 1916b:30 (as subsp. of *Z. carinatus* Alcock, 1893). A synonym of *Z. carinatus* according to Jangoux & Aziz (1988) and Rowe *in* Rowe & Gates (1995). Retained as subspecies by Liao *in* Liao & Clark (1996).
planus Alcock, 1893
Alcock 1893:104; H.L. Clark 1920:101 (in key).
Range: Laccadive Sea, 2200 m.
platyacanthus H.L. Clark, 1913, see *Myxoderma*
sacculatus Fisher, 1905, see *Myxoderma*
sigsbeci (lapsus) Perrier, 1880:436, = *sigbeei* Perrier, 1881:5, see *Mammaster*
spinulosus Fisher, 1906
Fisher 1906a:1102; H.L. Clark 1920:101 (in key); Fell 1958:19; 1960:65; McKnight 1967a:302; Rowe *in* Rowe & Gates 1995:116.
Range: Hawaiian Is, S Australia and N of Chatham Is, New Zealand, 371-1020 m.
squameus Alcock, 1893, see *Cnemidaster*
tenuis Sladen, 1889
Sladen 1889:421; Koehler 1907a:141; 1908a:566.
Range: N of New Guinea and Southern Ocean between Gough and Bouvet Islands, 3190 and 1960 m.
trispinosus Koehler, 1895b:442; 1896:42; 1909a:108. A synonym of *Z. fulgens* Thomson, 1873 according to Downey (1970).
zea Alcock, 1893
Alcock 1893:110.
Range: S India, 1090 m.

Family NEOMORPHASTERIDAE Fisher

Asteriidae: Neomorphasterinae Fisher 1923:250; 1928a:56; 1930:211; Spencer & Wright 1966:U77.
Neomorphasteridae: Downey *in* A.M. Clark & Downey 1992:401,404.
Genus-group names: Calycaster, Glyptaster, Gastraster, **Neomorphaster**

CALYCASTER Perrier, 1891
Perrier 1891c:1226; 1891b:258,262; 1896:27.

Type species: *Calycaster monoecus* Perrier, 1891

A synonym of *Neomorphaster* Sladen, 1889 according to Fisher (1930).

monoecus Perrier, 1891b:262. A synonym of *Neomorphaster talismani* Sladen, 1889 according Mortensen (1927), Fisher (1930).

monoecus: Grieg 1921[1932]:44 (from S of Newfoundland). More likely = *N. forcipatus* Verrill, 1894 from locality but R only 4 mm. [New observation.]

GASTRASTER Perrier, 1894

Perrier 1894:103; Fisher 1918:103; Mortensen 1927:103; Fisher 1930:206.

Type species: *Pedicellaster margaritaceus* Perrier *in* Milne Edwards, 1882. A synonym of *Neomorphaster* Sladen, 1889 according to Downey *in* Clark & Downey (1992).

margaritaceus Perrier *in* Milne Edwards, 1882, see *Neomorphaster*

*studeri de Loriol, 1904

de Loriol 1904:34.

Range: Port San Antonio, Argentina.

Not congeneric with *Gastraster margaritaceus Perrier, 1882, according to Downey in* Clark & Downey (1992) but generic position not clarified.

GLYPTASTER Sladen *in* Thomson & Murray, 1885:612. A NOMEN NUDUM, no type species named; also invalidated by *Glyptaster* Hall, 1852. Replaced by

NEOMORPHASTER Sladen, 1889 (with synonyms *Glyptaster* Sladen, 1885, *Calycaster* Perrier, 1891 and *Gastraster* Perrier, 1894)

Sladen 1889:436; Fisher 1923:596; 1930:211; Downey *in* Clark & Downey 1992:405.

Type species: *Neomorphaster eustichus* Sladen, 1889, a synonym of *Pedicellaster margaritaceus* Perrier *in* Milne Edwards, 1882.

eustichus Sladen, 1889:438. A synonym of *Pedicellaster*

margaritaceus Perrier *in* Milne Edwards, 1882, according to Perrier (1894), Mortensen (1927).

forcipatus Verrill, 1894

Verrill 1894:269; 1895:206; Downey *in* Clark & Downey 1992:405.

Range: S of Nantucket to Hudson Canyon, USA, 1400-2000 m.

margaritaceus (Perrier *in* Milne Edwards, 1882) (with synonyms *Stichaster talismani* Perrier, 1885, *N. eustichus* Sladen, 1889, *Calycaster monoecus* Perrier, 1891 and *N. parfaiti* Koehler, 1896 [lapsus]).

Perrier *in* Milne Edwards 1882:46 (as *Pedicellaster*).

Perrier 1894:103; Mortensen 1927:137; Fisher 1930:207 (as *Gastraster*).

Downey *in* Clark & Downey 1992:406 (as *Neomorphaster*).

Range: Bay of Biscay*, Azores, Rockall Trough, SW Ireland and Morocco, 400-5410 m.

parfaiti Koehler 1895:443; 1896:44. Misprint [surely lapsus since repeated] for *N. talismani* according to Mortensen (1927).

talismani: Perrier 1894:134; 1896:30; Koehler 1909a:107; Mortensen 1927:134; Harvey et al. 1988:167.

A synonym of *Pedicellaster margaritaceus* Perrier *in* Milne Edwards, 1882 according to Downey *in* Clark & Downey (1992) (though merited retention by submission to ICZN because of greater familiarity and the immaturity of holotype of *margaritaceus* A.M.C.)

Family PEDICELLASTERIDAE Perrier

Pedicellasteridae Perrier 1884:154,167,194; Sladen 1889:556; Perrier 1891c:76; 1894:90; Mortensen 1927:129; Djakonov 1950a:102 [88], 105[91]; Downey *in* Clark & Downey 1992:406.

Asteriidae: Pedicellasterinae: Fisher 1918:108; 1923:249; 1928a:57; Spencer & Wright 1966:U76.

Genus-group names: **Ampheraster, Anteliaster, Hydrasterias, Pedicellaster, Peranaster, Tarsaster.** [*Plazaster* see family Labidiasteridae]

AMPHERASTER Fisher, 1923

Fisher 1923:253; Downey 1971c:51; 1973:90; Downey *in* Clark & Downey 1992:407.

Type species: *Sporasterias mariana* Ludwig, 1905.

alaminos Downey, 1971

Downey 1971c:52; Downey *in* Clark & Downey 1992:408.

Range: Gulf of Mexico and Great Bahama Bank, 256-2650 (3200) m.

atactus Fisher, 1928

Fisher 1928a:86.

Range: Southern California, 490-590 m.

chiroplus Fisher, 1928

Fisher 1928a:86.

Range: Southern California, 820-930 m.

distichopus (Fisher, 1917)

Fisher 1917b:92; 1919a:420 (as *Tarsaster*).

Fisher 1928a:81 (as *Ampheraster*; in key).

Range: Macassar Strait, 730 m.

hyperonchus (H.L. Clark, 1913)

H.L. Clark 1913:201 (as *Pedicellaster*).

Fisher 1928:81,84 (as *Ampheraster*, in key).

Range: Lower California, 520 m.

marianus (Ludwig, 1905)

Ludwig 1905:231 (as Sporasterias).

Fisher 1923:253; 1928a:81 (as *Ampheraster*).

Range: W Mexico N to Washington, 510-1240 m.

ANTELIASTER Fisher, 1923

Fisher 1923:252; 1928a:69.

Type species: *Anteliaster coscinactis* Fisher, 1923.

australis Fisher, 1940

Fisher 1940:215; A.M. Clark 1962:72.

Range: Falkland Is, Shag Rocks, South Georgia, ? Marion I, 79-341 m.

coscinactis Fisher, 1923 (with subspecies *megatretus* Fisher, 1928)

coscinactis coscinactis Fisher, 1923

Fisher 1923:252; 1928a:70.

Range: Southern California, 820-930 m.

coscinactis megatretus Fisher, 1928

Fisher 1928a:71.

Range: Lower California, 520 m.

microgenis Fisher, 1928 (with subspecies *nannodes* Fisher, 1928)

microgenys microgenys Fisher, 1928

Fisher 1928a:73.

Range: Southern California, 88 m.

microgenys nannodes Fisher, 1928

Fisher 1928a:74; Baranova 1957:176.

Range: Bering Sea, 450 m.

megatretus Fisher, 1928. A subspecies of *A. coscinactis* Fisher, 1923.

nannodes Fisher, 1928. A subspecies of *A. microgenys* Fisher, 1928.

scaber (E.A. Smith, 1876)

E.A. Smith 1876:107: 1879:274; Sladen 1889:558; ?Perrier 1894:100; Doederlein 1928:295 (as *Pedicellaster*).

Fisher 1940:217 (as *Anteliaster* but see below); A.M. Clark 1962a:72.

Range: Kerguelen, ?Tierra del Fuego (Perrier), ?Marion Island (Fisher), 0-46 (?-113) m.

scaber: Fisher, 1940:217, non *Pedicellaster scaber* E.A. Smith, 1876,

?= *Anteliaster australis* Fisher, 1940 according to A.M. Clark (1962a).

HYDRASTERIAS Sladen, 1889

Sladen 1889:563 (in key), 581 (as subgenus of *Asterias*.

Fisher 1923:251; 1928a:68 (as genus).

Type species: *Asterias (Hydrasterias) ophidion* Sladen, 1889.

diomediae Ludwig, 1905:242. A juvenile and synonym of *Sclerasterias alexandri* (Ludwig, 1905) according to Fisher (1928).

improvisus (Ludwig, 1905)

Ludwig 1905:216 (as *Pedicellaster*).

Fisher 1928a:69 (as *Hydrasterias*).

Range: Cocos Island, SW of Panama and Galapagos Is, 1620-2420 m.

ophidion (Sladen, 1889)

Sladen 1889:581 (as *Asterias (Hydrasterias)*).

Verrill 1894:279; 1895:211; Perrier 1896:32; Fisher 1928a:69; A.H.
Clark 1949:375; Downey *in* Clark & Downey 1992:409.

Range: S of Nova Scotia and mid-Atlantic seamounts N of Azores,
2290-3190 m.

ophidion: A.M.C. *in* Herring 1974:405; non *H. ophidion* (Sladen, 1889),
= *H. sexradiata* (Perrier, 1882) according to Downey *in* Clark &
Downey (1992).

*richardi (Perrier *in* Milne Edwards, 1882:20) 1894:109. See A.M.C.
in Clark & Downey 1992:448,451,511. Proposed for suppression by the
ICZN.

sexradiata (Perrier, 1882)

Perrier *in* Milne-Edwards 1882:46,50; Perrier 1885c:15; 1894:100
Koehler 1909a:110 (as *Pedicellaster*).

Fisher 1928a:69 (possibly only racially distinct from *H. ophidion*
(Sladen, 1889)); Gage et al. 1983:285; Downey *in* Clark & Downey
1992:410 (as *Hydrasterias*).

Range: Azores*, Rockall Trough S to Cape Verde Is, 600-4269 m.

verrilli Fisher, 1906:1106, see *Tarsastrocles* (Asteriidae).

HYDRASTERIAS: Perrier 1893:848; 1896:32, non *Hydrasterias* Sladen,
1889,
= *Sclerasterias* Perrier, 1891.

PEDICELLASTER M. Sars, 1861

Sars 1861:77; Perrier 1884:194; Sladen 1889:557; Perrier 1894:99;
Fisher 1923:251; 1928a:58; Downey *in* Clark & Downey 1992:410.

Type species: *Pedicellaster typicus* M. Sars, 1861.

antarcticus Ludwig, 1903:35. A subspecies of *P. hypernotius*.

antarcticus: Döderlein 1928: , non *antarcticus* Ludwig,
= *P. hypernotius formatus* Koehler, 1920, according to A.M. Clark
(1962a).

atratus Alcock, 1893

Alcock 1893:114; Fisher 1928a:59.

Range: Andaman Sea, Bay of Bengal, 400-530 m.

chirophorus Fisher 1917b, see *Peranaster*

eximius Djakonov, 1949 (cited as 1948 by Djakonov (1950a))

Djakonov 1949:33 (in key; short description, probably enough to validate
though not given as 'sp.nov.'); 1950a:109; 1950c:68.

Range: Okhotsk Sea, 120-180 m.

formatus Koehler, 1920. A subspecies of *P. hypernotius* Sladen, 1889.

hypernotius Sladen, 1889 (with subspecies *antarcticus* Ludwig, 1903
and *formatus* Koehler, 1920)

hypernotius hypernotius Sladen, 1889

Sladen 1889:558; Fisher 1940:214; A.M. Clark 1962a:70; H.E.S. Clark

1963b:66; McKnight 1976:29.

Range: Marion Island*, South Georgia and Shag Rocks, 37-450 m.

hypernotius antarcticus Ludwig, 1903

Ludwig 1903:35 (as species).

Fisher 1940:214 (as synonym of *P. hypernotius*).

A.M. Clark 1962a:71; Jangoux & Massin 1986:89 (as subspecies of *P. hypernotius*).

Range: Bellingshausen Sea*, Ross Sea[?], 93-450 m.

hypernotius formatus Koehler, 1920

Koehler 1920:106 (as species).

A.M. Clark 1962a:71 (as subspecies of *P. hypernotius*).

Range: Antarctica, Enderby Land - Queen Mary Land, 220 m.

hyperoncus H.L. Clark, 1913, see *Ampheraster*

improvisus Ludwig, 1905, see *Hydrasterias*

improvisus: H.L. Clark, 1913:202, non *P. improvisus* Ludwig, 1905, = *Anteliaster coscinactis megatretus* Fisher, 1928, according to Fisher.

indistinctus Djakonov, 1950

Djakonov 1950a:107; 1950c:66.

Range: Okhotsk Sea, 128 m.

magister Fisher, 1923 (with subspecies *megalabis* Fisher, 1928, *ochotensis* Djakonov, 1950a and *sagaminus* Hayashi, 1973a)

magister magister Fisher, 1923

Fisher 1923:251; 1928:59; Baranova 1957:176.

Range: S Bering Sea to Alaska, 77-220 m.

magister megalabis Fisher, 1928

Fisher 1928:64.

Range: S California to Washington, 520-970 m.

magister ochotensis Djakonov, 1950

Djakonov 1950a:107.

Range: Okhotsk Sea, 100-645 m.

magister sagaminus Hayashi, 1973

Hayashi 1973a:11; 1973b:96.

Range: SE Japan, 300-450 m.

margaritaceus Perrier *in* Milne Edwards, 1882, see *Neomorphaster* (Neomorphasteridae).

megalabis Fisher, 1928, a subspecies of *P. magister* Fisher, 1923, also *ochotensis* Djakonov, 1950

orientalis Fisher, 1928

Fisher 1928:66 (as subspecies of *P. magister*).

Djakonov 1950a:108; 1950c:64 (as species).

Range: Japan Sea, 112-670 m.

palaeocrystallus: Duncan & Sladen, 1881:455; Mortensen 1912:264.

Lectotype selected: Rowe 1974:216. A synonymof *P. typicus* M. Sars, 1861, according to Fisher (1928).

parvulus Perrier, 1891b:258; 1896:21. A young *Sclerasterias* according to Fisher (1928a:59) under *Pedicellaster* only. Not mentioned by Downey *in* Clark & Downey (1992).

pourtalesi Perrier, 1881

Perrier 1881:7; 1884:194; Fisher 1928:59 (`possibly a true *Pedicellaster*'; H.L. Clark 1941:67; Downey *in* Clark & Downey 1992:411.

Range: Grenada, St Kitts, Cayman Is, E Florida and Nicaragua, 232-560 m.

reticulatus H.L. Clark, 1916, see *Stylasterias* (Asteriidae).

sagaminus Hayashi, 1973a. A subspecies of *P. magister* Fisher, 1923.

*sarsi Studer, 1885:151. A *Pedicellaster* according to Koehler (1920) but four rows of tube feet so probably not a pedicellasterid according to A.M.C. (1962a:73). South Georgia.

scaber E.A. Smith, 1876, see *Anteliaster*

sexradiatus Perrier *in* Milne Edwards, 1882, see *Hydrasterias*

typicus M. Sars, 1861 (with synonym *Asteracanthion palaeocrystallus* Duncan & Sladen, 1877)

M. Sars 1861:77; Danielssen & Koren 1884b:37; Verrill 1895:205; Mortensen 1927:130; 1932:18; Djakonov 1933:60; 1950a:106; Rowe 1974:216; Downey *in* Clark & Downey 1992:411,

Range: Norway*, Arctic Atlantic, S to c. 40°N, New Jersey, in west, c.60°N, Bergen, in east, 100-223 m (Downey), 91-1130 m (Sladen).

PERANASTER Fisher, 1923

Fisher 1923:252; 1928a:67.

Type species: *Pedicellaster chirophorus* Fisher, 1917c.

PLAZASTER Fisher, 1941, see Labidiasteridae

chirophorus (Fisher, 1917)

Fisher 1917c:93; 1919a:499 (as *Pedicellaster*).

Fisher 1923:252; 1928a:67 (as *Peranaster*).

Range: Sulawesi, 885 m.

TARSASTER Sladen, 1889

Sladen 1889:439; Fisher 1928a:75.

Type species: *Tarsaster stoichodes* Sladen, 1889.

alaskanus Fisher, 1928

Fisher 1928a:78.

Range: S. Alaska, 198-440 m.

cocosanus (Ludwig, 1905)

Ludwig 1905:235 (as *Sporasterias*).

Fisher 1928a:77; Maluf 1988:48,128 (as *Tarsaster*).

Range: Cocos I (SW from Panama), S California, 245-412 m.

distichopus Fisher, 1917b:92, see *Ampheraster*

fascicularis (Perrier, 1881)

Perrier 1881:1; 1884:200 (as *Asterias*)

Verrill 1915:22 (as *Leptasterias*)

Fisher 1923:252; 1928a:77; Downey *in* Clark & Downey 1992:412 (as *Tarsaster*).

Range: Guadeloupe, Lesser Antilles, 0-100 m.

galapagensis (Ludwig, 1905)

Ludwig 1905:240 (as *Sporasterias*).

Fisher 1928:77; Maluf 1988:48,128 (as *Tarsaster*).

Range: Galapagos Ridge*, S. California, 457-700 m.

neozelanica Farquhar, 1894:207. A synonym of *Allostichaster polyplax* (Müller & Troschel, 1844) according to Koehler (1920), Mortensen (1925).

stoichodes Sladen, 1889

Sladen 1889:440; Fisher 1928a:77 (in key).

Range: N of Admiralty Is, Papua New Guinea, 274 m.

Family ASTERIIDAE Gray, 1840

Asteriadae Gray 1840:178.

Stichasteridae (part): Perrier 1885b:885; Sladen 1889:450; Mortensen 1927:133.

Asteriidae: Sladen 1889:560 (restricted); Perrier 1894:105,128; Verrill 1914a:27-45; Fisher 1923:247-250; 1928a:65 (emended); Djakonov 1950a: 109-112 [1968:95-98]; Spencer & Wright 1966:U75; A.M. Clark & Courtman-Stock 1976:91; A.M. Clark & Downey 1992:413 (restricted to former Asteriinae).

Asteriidae: Asteriinae: Verrill 1914a:42; Fisher 1923:250; 1930:2-5, 217-220.

Asteriidae: Coscinasteriinae: Fisher 1923:249; 1928a:93.

Asteriidae: Notasteriinae Fisher 1923:249; 1928a:93.

Genus group names: **Adelasterias**, Allasterias, **Allostichaster, Anasterias, Aphanasterias, Aphelasterias**,Asteracanthion, **Asterias**, Asteroderma, **Asteroderma, Astrometis, Astrostole, Australiaster**, Autasterias, Bathyasterias, **Caimanaster, Calasterias**, Calliasterias, Carlasterias, Coelasterias, Comasterias, **Coscinasterias, Cosmasterias, Cryptasterias**, Ctenasterias, **Diplasterias, Distolasterias**, Endogenasterias, **Eoleptasterias**, Eremasterias, Eustolasterias, Evasterias}, Gastraster, **Granaster**, Hemiasterias, **Icasterias**, Kalyptasterias, **Kenrickaster**, Koehleraster, **Leptasterias, Lethasterias, Lysasterias**, Lytaster, Margaraster, **Marthasterias, Meyenaster**, Mortensenia, Nanaster, **Nesasterias, Neosmilaster, Notasterias, Orthasterias**, Paedasterias, Parasterias,

Parastichaster, **Perissasterias, Pisaster**, Podasterias, **Psalidaster, Pseudechinaster**, Quadraster, **Saliasterias, Sclerasterias, Smilasterias**, Sporasterias, Stellonia, **Stenasterias, Stephanasterias, Stichaster**, Stichorella, **Stichastrella, Stolasterias, Stylasterias, ‡Taranuiaster, Tarsastrocles**, Tonia, Triplasterias, **Uniophora**, Uraster, **Urasterias**

ADELASTERIAS Verrill, 1914
Verrill 1914a:360; Fisher 1930:220.
Type species: *Diplasterias papillosa* Koehler, 1906.
papillosa (Koehler, 1906)
Koehler 1906:21 (as *Diplasterias*).
Verrill 1914a:360; Fisher 1930:220 (as *Adelasterias*).
Range: Graham Land (Antarctic Peninsula).
ALLASTERIAS Verrill, 1909
Verrill 1909:65.
Type species: *Allasterias rathbuni* Verrill, 1909.
A synonym of *Asterias* Linnaeus, 1758 according to Fisher (1923, 1930).
anomala Verrill, 1909:66 (as variety of *A. rathbuni*; 1914a:193 (as species).
A synonym of *Asterias rathbuni* (Verrill, 1909) according to Djakonov (1950a,c).
forficulosa Verrill, 1914a:194. A synonym of *Asterias rollestoni* Bell, 1881 according to Djakonov (1950a,c).
migrata (Sladen, 1879:432, as var. of *Asteracanthion rubens*) Verrill, 1914:373. A variety or forma of *Asterias amurensis* Lütken, 1871, or possibly young *A. versicolor* Sladen, 1899, according to Fisher (1930).
nortonensis Verrill, 1909 (variety of *A. rathbuni*), see *Asterias rathbuni*
Verrill, 1909, see *Asterias*
ALLOSTICHASTER Verrill, 1914
Verrill 1914a:363; Fisher 1930:220; H.L. Clark 1946:157; A.M. Clark & Downey 1992:415.
Type species: *Asteracanthion polyplax* Müller & Troschel, 1844.
capensis (Perrier, 1875) (with synonyms *Stephanasterias hebes* Verrill, 1915 and *Allostichaster inaequalis* Koehler, 1923)
Perrier 1875:73 [337]; [?*A. capensis* Bell, 1905]; H.L. Clark 1923a:306 (as *Asterias*).
Fisher 1926:198; Mortensen 1933a:276 (as *Cosmasterias*).
Mortensen 1941:4; Madsen 1956a:44; A.M. Clark & Courtman-Stock 1976:91; A.M. Clark *in* Clark & Downey 1992:417 (as *Allostichaster*).
Range: South Africa*, S Chile to 41.5øS, ?Juan Fernandez, Falkland-Magellan area, Argentina, Tristan da Cunha and the W coast of South Africa, intertidal - c.100 m.
hartti (Rathbun, 1879)

Rathbun 1879:145; Verrill 1915:23; Madsen 1956a:44 (as *Leptasterias*).

Fisher 1930:221; Tommasi 1966:244; 1970:20; [?Carrera-Rodriguez & Tommasi 1977:107] A.M. Clark *in* Clark & Downey 1992:417.

Range: Cape Frio/Rio de Janeiro area, Brazil, ?S to c.34°S off Rio Grande do Sul, 113 and ?166-338 m.

hebes (Verrill, 1915) Fisher 1930:221;Madsen 1950:44. A synonym of *A. capensis* (Perrier, 1875), possibly according to Madsen (1956a), confirmed A.M. Clark *in* Clark & Downey (1992).

inaequalis Koehler, 1923:50. A synonym of *A. capensis* (Perrier, 1875) according to Madsen (1956a).

insignis (Farquhar, 1895) (with variety *gymnoplax* Fell, 1953)

Farquhar 1895:203; 1898b:313; Benham 1909:97 (as *Stichaster*).

Koehler 1920:85; Mortensen 1925a:316; Farquhar 1927:238; Fell 1953:94 and 1958:19 (var. *gymnoplax*); Pawson 1965:254; McKnight 1967:303; H.E.S. Clark 1970:22.

Range: E and S from Cook Strait, New Zealand, to Chatham and Auckland Is, 0-222 m.

peleensis Marsh, 1974

Marsh 1974:96.

Range: Pitcairn I, Polynesia.

polyplax (Müller & Troschel, 1844)

Müller & Troschel 1844:178 (as *Asteracanthion*).

Perrier 1875:63 (as *Asterias*).

Sladen 1889:432 (as *Stichaster*).

Verrill 1914a:363; H.L. Clark 1916:70; Mortensen 1925a:315;

Farquhar 1927:238; H.L. Clark 1928:399; 1946:157; Shepherd 1968:752;

H.E.S. Clark 1970:23 (as *Allostichaster*).

Range: S Australia, New Zealand and Chatham Is, 0-238 m.

regularis H.L. Clark, 1928

H.L. Clark 1928:400; 1946:158.

Range: Spencer and St Vincent Gulfs, South Australia.

ANASTERIAS Perrier, 1875 (with synonyms *Asteroderma* Perrier, 1888 (nom. nud.), *Sporasterias* Perrier, 1894, *Parastichaster* Koehler, 1920, *Kalyptasterias* Koehler, 1923 and *Eremasterias* Fisher, 1930)

Perrier 1875:81 [345]; 1891c:K91; Fisher 1908a:52; 1923:592; 1930:221; 1940:231; A.M. Clark 1962a:93; A.M. Clark *in* Clark & Downey 1992:418.

Type species: *Anasterias minuta* Perrier, 1875, possibly a synonym of *Anasterias antarctica* (Lütken, 1857).

Anasterias: Studer, 1885, Ludwig, 1903, non Perrier, 1875, = *Lysasterias* Fisher, 1908.

244

adeliae Koehler, 1920, see *Lysasterias*

antarctica (Lütken, 1857) (with synonyms *Asterias rugispina* Stimpson, 1862, *A. cunninghami* Perrier, 1875 and *A. hyadesi* Perrier, 1886)

Lütken 1857a:108 (as *Asteracanthion*, possibly also *A. minuta* Perrier, 1875, see below).

Studer 1884:7; Meissner 1896:105 (as *Asterias*).

Ludwig 1903:39; Tortonese 1934b:4 (as *Sporasterias*).

Fisher 1940:233; Madsen 1956a:38; Bernasconi 1964b:270; 1973a:315; Codoceo & Andrade 1979:159; Hernandez & Tablado 1985:9; A.M. Clark *in* Clark & Downey 1992:419.

Range: Southern Chile and Argentina N to Buenos Aires Province, (?Falkland Is), 1-185 m.

belgicae Ludwig, 1903, see *Lysasterias*

chirophora Ludwig, 1903, see *Lysasterias*

conferta (Koehler, 1923) Fisher 1931:9; 1940:239. A synonym of *Anasterias spirabilis* (Bell, 1881) according to A.M. Clark (1962a).

cupulifera Koehler, 1908:566. A synonym of Lysasterias perrieri} (Studer, 1885) according to Fisher (1940).

directa (Koehler, 1920)

Koehler 1920:97 (as *Parastichaster*).

Fisher 1930:241 (as *Sporasterias*).

Fisher 1940:75 (by inference to *Anasterias*, with other species of *Parastichaster* though not named individually).

A.M. Clark 1962a:76,97 (as *Anasterias*).

Range: Macquarie I, Southern Ocean.

*haswelli Koehler, 1920:15. Listed; no other mention. ?A a mistake for *Notasterias haswelli* or a NOMEN NUDUM.

lactea Ludwig, 1903, see *Lysasterias*

lysasteria Verrill, 1914:354. An unnecessary substitute name and a synonym of *Lysasterias perrieri* Studer, 1885 according to Fisher (1922).

mawsoni (Koehler, 1920)

Koehler 1920:91 (as *Parastichaster*).

Fisher 1930:241 (as *Sporasterias*).

Fisher 1940:232 (by inference to *Anasterias*, being type species of *Parastichaster*, synonymized).

A.M. Clark 1962a:96 (as *Anasterias*).

Range: Macquarie I, Southern Ocean, 'littoral'.

*minuta Perrier, 1875:81. Possibly a synonym of *Anasterias antarctica* Lütken, 1857 according to A.M. Clark (1962a and *in* Clark & Downey 1992:419), though treated as valid by Hernandez & Tablado (1985).

Holotype with R only 12 mm.

multicostata (Hamann) Verrill, 1899:717. A mistake for *Brisinga multicostata* (Verrill, 1894) according to Ludwig (1903).

nuda Perrier, 1878:44,75. A NOMEN NUDUM, undescribed; see Ludwig (1903:42 footnote).

octoradiata Koehler. 1914, see *Diplasterias*

*pedicellaris (Koehler, 1923)

Koehler 1923:18 (as *Sporasterias*).

Fisher 1930:225; 1940:236; Bernasconi 1973:287 (as *Anasterias*).

Possibly a synonym of *Anasterias* now *(Lysasterias) studeri* Perrier, 1891, according to A.M. Clark (1962a).

Range: Rio de la Plata, Argentina to Falkland-Magellan area, 0-120 m.

perrieri (E.A. Smith, 1876) (with synonyms *Othilia sexradiata* Studer, 1876 and *Pisaster antarcticus* Koehler, 1917)

E.A. Smith 1876:106 (as *Asterias*).

Döderlein 1928:294; Fisher 1930:241 (as *Sporasterias*).

Fisher 1940:234 (by inference to *Anasterias* from synonymy of *Sporasterias*).

Madsen 1955:15; A.M. Clark 1962a:94; Guille 1974:39 (as *Anasterias*).

Liable to confusion with *Anasterias perrieri* Studer, 1885, transferred by Fisher (1930) to *Lysasterias*, but not contemporaneously in *Anasterias*.

Range: Kerguelen, 0-200 m.

perrieri Studer, 1885, see *Lysasterias*

rupicola (Verrill, 1876)

Verrill 1876:71; Koehler 1917:9 (as *Asterias*).

Ludwig 1903:40 (as var. of *Sporasterias antarctica*).

Döderlein 1928:294 (as *Sporasterias*).

Fisher 1940:234; Madsen 1955:15; A.M. Clark 1962:9; Guille 1974:40 (as *Anasterias*).

Range: Kerguelen, Marion and Crozet Is, Southern Ocean, 0-11 m.

sphoerulata (Koehler, 1920)

Koehler 1920:101 (as *Parastichaster*).

Fisher 1930:241 (as *Sporasterias*).

Fisher 1940:75 (by inference to *Anasterias* with synonymy of *Parastichaster*).

A.M. Clark 1962a:96 (as *Anasterias*).

Range: Macquarie I, Heard I, Southern Ocean.

spirabilis (Bell, 1881) (with synonym *Kalyptasterias conferta* (Koehler, 1923))

Bell 1881b:513 (as *Asterias*).

A.M. Clark 1962a:76 (a valid species, not a synonym of *Anasterias antarctica* Lütken, as suggested by Fisher (1940:233)).

Range: Falkland Is.

stolidota (Sladen, 1889:590 (as *Calvasterias*). 'May be the same species' as *conferta* (Koehler, 1923) according to Fisher (1940:239 but a synonym of *A. varium* (Philippi, 1870) according to Madsen (1956), probably so according to A.M. Clark (1962a).

studeri Perrier, 1891

Perrier 1891a:99; Fisher 1940:234; Bernasconi 1973:317.

Fisher 1930:236 (as *Lysasterias*).

Range: Magellan area, Santa Cruz Province, Argentina to Falkland Is, 51-320 m.

studeri: Koehler, 1914b:61, non *Anasterias studeri* Perrier, 1891,
? = *Lysasterias perrieri* (Studer, 1885) according to Fisher (1940).

tenera Koehler, 1906:12. A synonym of *Lysasterias perrieri* (Studer, 1885) according to Fisher (1940).

varia (Philippi, 1870) (with synonyms *Anasterias fulgens* (Philippi, 1870) and *A. stolidota* (Sladen, 1889)

Philippi, 1870:272 (as *Asteracanthion*).

Madsen 1956:40; A.M. Clark 1962a:76 (as *Anasterias*).

Range: Southern Chile.

verrilli: Koehler, 1920:14, non *Asterias verrilli* Bell, 1881, =
Anasterias antarctica (Lütken, 1857) according to Fisher (1940).

victoriae Koehler, 1920:17; 1923:12. A synonym of *Lysasterias perrieri* (Studer, 1885) according to Fisher (1940).

APHANASTERIAS Fisher, 1923

Fisher 1923:601; 1930:159.

Type species: *Aphanasterias pycnopodia* Fisher, 1923.

pycnopodia Fisher, 1923

Fisher 1923:601; 1930:160.

Range: Shumagin Is, Alaska.

pycnopodia: Caso, 1943, non Fisher, 1923, = *Astrometis sertulifera* (Xantus, 1860) according to Maluf (1988)

APHELASTERIAS Fisher, 1923

Fisher 1923:602 (in key); 1930:205.

Type species: *Asterias japonica* Bell, 1881.

changfengyingi Baranova & Wu, 1962

Baranova & Wu 1962:110 (Chinese), 114 (Russian); Chang & Liao 1964:7.

Range: Yellow Sea, China, 50-65 m.

japonica (Bell, 1881) (with synonym *Asterias torquata* Sladen, 1889)

Bell 1881b:515; Uchida 1928:799 (as *Asterias*).

Fisher 1930:205; Hayashi 1938a:117; Djakonov 1950a:120; Baranova & Wu 1962:109,114; Chang & Liao 1964:72; Baranova 1971:258 (as *Aphelasterias*).

Range: Tatar Strait, Yellow Sea, Japan Sea, 0-80 m.

ASTERACANTHION Müller & Troschel, 1840 Müller & Troschel 1840a:102; 1840b:320; 1842:14.

A synonym of *Asterias* Linnaeus, 1758 according to Perrier (1875).

africanus Müller & Troschel, 1842. A forma of *Marthasterias glacialis* (Linnaeus, 1758).

albulus Stimpson, 1853, see *Stephanasterias*

antarcticus Lütken, 1857, see *Anasterias*

aurantiacus: Müller & Troschel, 1842. An invalid homonym and = *Stichaster striatus* Müller & Troschel, 1840.

australis Valenciennes (MS) *in* Perrier, 1875 = *Coscinasterias calamaria* (Gray, 1840) [?or *muricata* Verrill if from Australia]

berylinus A. Agassiz, 1866:106; 1877:94. A synonym of *Asterias forbesi* (Desor, 1848) according to Perrier (1875).

*bootes Müller & Troschel, 1842. ? Identity; omitted from Clark & Downey (1992).

calamaria: Dujardin & Hupé, 1862, see *Coscinasterias*

clavatum Philippi, 1870:269. A synonym of *Cosmasterias lurida* (Philippi, 1858) according to Leipoldt (1895).

distichum Brandt, 1851. Could be identical with *Asterias* [i.e. *Leptasterias*] *groenlandica* (Steenstrup, 1857) according to Döderlein, 1900 but a synonym of *Asterias rubens* Linnaeus, 1758 according to Fisher (1930).

flaccida A. Agassiz, [1863] 1866. A synonym of *Leptasterias tenera* (Stimpson, 1862) according to Fisher (1930).

forbesi Desor, 1848, see *Asterias*

fulgens Philippi, 1870:274. A synonym of *Anasterias varia* (Philippi, 1870) according to Madsen (1956).

fulvum Philippi, 1870:270. A synonym of *Cosmasterias lurida* (Philippi, 1858) according to Leipoldt (1895).

gelatinosus: Müller & Troschel, 1842, see *Meyenaster*

*gemmifer Valenciennes (MS) *in* Perrier, 1869:237. A probable synonym of *Coscinasterias echinata* (Gray, 1840) according to Fisher (1928), after H.L. Clark (1916) but alternatively conspecific with *Astrostole* (or *Coscinasterias*) *platei* (Meissner, 1896) (from Pacific islands west of Chile), which would then be a junior synonym, according to Madsen (1956).

germaini Philippi, 1858:265. A synonym of *Cosmasterias lurida* (Philippi, 1858) according to Madsen (1956).

glacialis: Müller & Troschel, 1842, see *Marthasterias*

graniferus Müller & Troschel, 1842, see *Uniophora*

groenlandica Steenstrup, 1857, see *Leptasterias*

helianthus: Müller & Troschel, 1842, see *Heliaster*

katherinae: Müller & Troschel, 1842. A subspecies of *Leptasterias polaris* (Müller & Troschel, 1842, according to Fisher (1930).

lacazei Perrier, 1869:51. A synonym of *Asterias arenicola* Stimpson, 1862, according to Perrier (1875), which is a synonym of *Asterias forbesi* (Desor, 1848) according to Fisher (1930).

lincki Müller & Troschel, 1842, see *Urasterias*

littoralis Stimpson, 1853, see *Leptasterias*

luridum Philippi, 1858, see *Cosmasterias*

margaritifer Müller & Troschel, 1842:20. A synonym of *Pisaster ochraceus* (Brandt, 1835) according to Fisher (1930).

mexicanum LÜtken, 1859, see *Leptasterias*

migratum Sladen, 1878:432, as variety of *A*. [i.e. *Asterias*] *rubens* Linnaeus, 1758, but from Korean Strait so presumably rather *A. amurensis* Lütken, 1871.

mite Philippi, 1870:272. A synonym of *Cosmasterias lurida* (Philippi, 1858) according to Leipoldt (1895).

muelleri M. Sars, 1846, see *Leptasterias*

novaeboracensis Perrier, 1869:41[413]. A synonym of *Asterias forbesi* (Desor, 1848) according to Fisher (1930).

ochotense Brandt, 1851, see *Leptasterias*

pallidus A. Agassiz, [1863] 1866. A NOMEN NUDUM (undescribed?) and = *Asterias vulgaris* Verrill, 1866 [i.e. *A. rubens* Linnaeus, 1758] according to Fisher (1930).

polaris Müller & Troschel, 1842, see *Leptasterias*

polyplax Müller & Troschel, 1842, see *Allostichaster*

problema Steenstrup, 1854:240 [1855:240, acc. M.]. A synonym of *Stephanasterias albula* (Stimpson, 1853) according to Sladen (1889).

rosea: Müller & Troschel, 1842, see *Stichastrella*

rubens: Müller & Troschel, 1842, see *Asterias*

rubens: Müller & Troschel, 1843:113, non *Asterias rubens* Linnaeus, 1758. Probably = *Anasterias varium* (Philippi, 1870) according to Madsen (1956).

spectabile Philippi, 1870:271. A synonym of *Cosmasterias lurida* (Philippi, 1858) according to Leipoldt (1895).

stellionura Perrier, 1869:48[420]. A synonym of *Urasterias lincki* (Müller & Troschel, 1842) according to Fisher (1930)[?Perr. '75].

striatus: Müller & Troschel, 1842, see *Valvaster* (Asteropseidae, pt. 2).

sulcifer Perrier, 1869:235. A synonym of *Cosmasterias lurida* (Philippi, 1858) according to Fisher (1930).

tenuispinum: Müller & Troschel, 1842, see *Coscinasterias*

varium Philippi, 1870, see *Anasterias*

violaceus: Müller & Troschel, 1842:16. A synonym of *Asterias rubens* Linnaeus, 1758.

webbianum: Dujardin & Hupé, 1862:340. A synonym of *Asterias* [i.e. *Marthasterias*] *glacialis* Linnaeus, 1758, according to Sladen (1889).

ASTERIAS Linnaeus, 1758 (with synonyms *Stellonia* Nardo, 1834, *Uraster* L. Agassiz, 1836, *Asteracanthion* Müller & Troschel, 1840, [*Asteracanthium* Brandt, 1851], *Allasterias* Verrill, 1909 and *Parasterias* Verrill, 1914)

Linnaeus, 1758:661; Sladen 1889:560; Ludwig 1897:303; Verrill 1914a:101; Fisher 1930:5; Djakonov 1950a:188; 1950c:92; Tortonese 1963a:212; Downey *in* A.M. Clark & Downey 1992:420.

Type species: *Asterias rubens* Linnaeus, 1758.

acanthostoma Verrill, 1909. A forma of *Evasterias troscheli* (Stimpson, 1862) according to Djakonov (1950c).

acervata Stimpson, 1862. A subspecies of *Leptasterias polaris* (Müller & Troschel, 1842) according to Fisher (1930).

acervispinis Djakonov, 1950a. A forma of *A. amurensis* Lütken, 1871.

aculeata O.F. Müller, 1776. An OPHIUROID (*Ophiopholis aculeata*).

acutispina Stimpson, 1862, see *Coscinasterias*

aequalis Stimpson, 1862, see *Leptasterias*

affinis Brandt, 1835:71. A synonym of *Crossaster papposus* (Linnaeus, 1758) according to Fisher (1911c) (Solasteridae, see part 3).

africana: Perrier, 1875. A variety of *Marthasterias glacialis* (Linnaeus, 1758) according to Mortensen (1933a).

alba Bell, 1881a:92. A synonym of *Cosmasterias lurida* (Philippi, 1858) according to Madsen (1956).

alboverrucosa Brandt, 1835:71. A synonym of *Solaster endeca* (Linnaeus, 1771) according to Fisher (1911c) (Solasteridae, see part 3).

albula: Stimpson, 1864, see {Stephanasterias

alta Quijada, 1911. A NOMEN NUDUM according to Madsen (1956).

alternata Say, 1825, see *Luidia* (Luidiidae, see part 1).

alta Philippi (MS) *in* Quijada, 1911 [Not seen]. A NOMEN NUDUM according to Madsen (1956).

alveolata Djakonov, 1950a, as a forma of *A. rathbuni* (Verrill, 1909)

amurensis Lütken, 1871 (with synonym *Parasterias albertensis* Verrill, 1914, probably also *Asterias migratum* Sladen, 1878 [new obs.] also forms acervispinis, flabellifera, gracilispinis, latissima and robusta, all of Djakonov, 1950)
Lütken 1871:296; Fisher 1930:6; Hayashi 1936:1-20; 1938a:117; Djakonov 1950a:125; 1950c:107; 1958:324; Baranova & Wu 1962:112,118; Rho & Kim 1966:281; Baranova 1971:259; Byrne, Morrice & Wolf 1997:673.
Range: Yellow Sea, Japan Sea, Tatar Strait S to Sagami Bay, also Tasmania, 0-106 m.

angulosa Abildgaard *in* O.F. Müller, 1789:1. A synonym of *Asterias* [i.e. *Marthasterias glacialis* Linnaeus, 1758, according to Sladen (1889).

angulosa: Perrier, 1881:3; 1884:202, non *A. angulosa* Abildgaard *in* O.F. Müller, 1789, = *Sclerasterias contorta* Perrier, 1881 according to Downey (1973).

anomala Verrill, 1909:66. A forma of *Asterias rathbuni* (Verrill, 1909), according to Djakonov (1950a).

250

*antarctica: Studer, 1884:7; de Loriol 1904:36. A synonym [and homonym]
of *Anasterias antarctica* Lütken, 1857 according to Downey *in* Clark &
Downey (1992).

antarctica (Sporasterias): Meissner, 1896, see *Anasterias*

arancia Dewhurst, 1834:283. A synonym of *Ctenodiscus corniculatus*
[i.e. *C. crispatus* (Retzius, 1805)] according to Sladen (1889)
(see Goniopectinidae: Ctenodiscinae, see part 1).

aranciaca Linnaeus, 1758, see *Astropecten*

aranciaca: O.F. Müller, 1776:234, non *aranciaca* Linnaeus, 1758,
= *Astropecten irregularis* (Pennant, 1777) according to Sladen (1889).

aranciaca: Audouin, 1826:10, non *aranciaca* Linnaeus, 1758,
= *Astropecten polyacanthus* Müller & Troschel, 1842, according to
Döderlein (1917).

aranciaca: Johnston, 1836:289, non *aranciaca* Linnaeus, 1758,
= *Astropecten pentacanthus* [i.e. *A. irregularis pentacanthus*]
(Delle Chiaje, 1827) according to Perrier (1875).

aranciaca: Gould, 1841, non *aranciaca* Linnaeus, 1758, = *Astropecten articulatus* (Say, 1825) according to Perrier (1875).

aranciata Parelius, 1768. ?lapsus for *aranciaca* Linnaeus, 1758,
= *Psilaster andromeda* (Müller & Troschel, 1842) according to Sladen
(1889).

arctica Murdoch, 1885, see *Leptasterias*

*arenata Lamarck, 1816:566; Dujardin & Hupé, 1862 (as *Ophidiaster*), unidentifiable according to H.L. Clark (1921).

arenicola Stimpson, 1862:268. A synonym of *Asterias forbesi* (Desor, 1848)
according to Verrill (1895).

argonauta Djakonov, 1950
Djakonov 1950a:124; 1950c:103; Baranova & Wu 1962:111,117; Chang
& Liao 1964:69.
Range: Peter the Great Bay, Japan Sea, Yellow Sea, 1-32 m.

articulata Say, 1825, see *Astropecten (pt. 1)*.

articulata Linnaeus, 1753:114. Invalid (pre-1758).

aspera O.F. Müller 1776:234. A synonym of Solaster endeca (Linnaeus,
1771) according to Sladen (1889).

*aster Gray, 1840
Gray 1840:178.
Range: Unknown.
Types with 12,13 rays; lost. ? *Coscinasterias* according to Verrill
1870:248; 1914:46. Should be suppressed by ICZN.

atlantica Verrill, 1868:368; Rathbun 1879:145. A synonym of *Asterias*
(i.e. *Coscinasterias*) *tenuispina* Lamarck, 1816 according to Sladen
(1889).

attenuata Bell, 1891, as var. of *A. rubens* Linnaeus, 1758, subsequently ignored or overlooked.

aurantiaca Tiedemann, 1816:33. A synonym of *Astropecten aranciaca* (Linnaeus, 1758) according to Döderlein (1917).

aurantiacus: Meyen, 1834:222, non aurantiaca Tiedemann, 1816, = *Stichaster striatus* Müller & Troschel, 1840 according to Fisher (1930).

austera Verrill, 1895, see *Leptasterias*

belli Studer, 1884:12. A synonym of *Diplasterias brandti* (Bell, 1881) according to Bell (1908).

*bicolor Lamarck, 1816
>Lamarck 1816:566. [Range unknown.]
>Referred to *Ophidiaster* (Ophidiasteridae, see part 2) by Dujardin et Hupé (1862) but unidentifiable according to H.L. Clark (1921).

bispinosa Otto, 1823, see *Astropecten* (part 1).

*bootes (Müller & Troschel, 1842). Cited by Bell 1881:493 (footnote).

*borbonica Perrier, 1875:61. Probably a young *Sporasterias i.e. (Anasterias)* according to Fisher (1930) but the locality Réunion Island (if correct) makes this unlikely [new obs.].

borealis Perrier, 1875:59 [323]. A synonym of *Leptasterias polaris* (Müller & Troschel, 1842) according to Verrill (1895), or a race according to Fisher (1930).

*brachiata Linnaeus (date ? but not 1758): Fisher, 1930:140, distinct from brachiata Perrier, 1875:329, which is a synonym of *Evasterias troscheli* (Stimpson, 1862) according to Verrill (1914a) and Fisher (1930).

brandti Bell, 1881a, see *Diplasterias*

brasiliensis Cuvier, 1829:485 (listed only). A NOMEN NUDUM.

brevispina Stimpson, 1857, see *Pisaster*

briareus Verrill, 1882, see *Coronaster*

brisingoides Perrier, 1885b:885. Barely enough indication to validate.
>Intermediate between *Coronaster* and *Asterias tenuispina* but no locality given. Presumably = *Coronaster brisingoides* Perrier 1884:272, no more adequately described, so better treated as a NOMEN NUDUM.

calamaria Gray, 1840, see *Coscinasterias*

calamaria var. japonica Döderlein, 1902:352, = *Coscinasterias acutispina* (Stimpson, 1862) according to Liao *in* Liao & Clark (1995).

calcar Lamarck, 1816, and var. c octagona, see *Patiriella* (Asterinidae, pt. 2).

calcar var. a quinqueangula Lamarck, 1816 = *Asterina* (i.e. *Patiriella*) *regularis* Verrill, 1870, according to Perrier (1875) but strictly has priority and needs rejection by the ICZN.

calcar var. b hexagona Lamarck, 1816 = *Asterina (i.e. (Patiriella) gunni* Gray, 1840, according to Perrier (1875) but strictly has priority and needs rejection by the ICZN.

calcar var. a :Audouin, 1826, non Lamarck, 1816, = *Asterina cephea*
(*A. burtoni cepheus* according to A.M.C., Asterinidae, part 2)
Müller & Troschel, 1842 according to Perrier (1875).

camtschatica Brandt, 1835, see *Leptasterias*

canariensis d'Orbigny, 1839, see *Narcissia* (Ophidiasteridae, part 2).

capensis Cuvier, 1829:485 (listed only). A NOMEN NUDUM.

capensis Perrier, 1875, see *Allostichaster*

capitata Stimpson, 1862:264. A subspecies of *Pisaster giganteus* (Stimpson,
1857) according to Fisher (1930).

caput medusae Linnaeus, 1758:663. An ophiuroid.

carinifera Lamarck, 1816, see *Asteropsis* (Asteropseidae, part 2).

cartilaginea Fleming, 1828:485. A synonym of *Palmipes membranaceus*
(Retzius, 1783), itself a synonym of *Anseropoda placenta*
(Pennant, 1777), Asterinidae, see part 2).

cayennensis Cuvier, 1829:485 (listed only). A NOMEN NUDUM.

ciliaris Philippi, 1837, see *Luidia* (Luidiidae, part 1).

ciliata O.F. Müller, 1776:235. AN OPHIUROID.

*cilicia: Compter, 1886:764.

*clathrata Pennant, 1777:61. A synonym of *Asterias rubens* Linnaeus, 1758
according to Sladen (1889) but proposed by A.M.C. (1982) for suppression by the ICZN as a senior homonym of

clathrata Say, 1825, see *Luidia* (Luidiidae, see part 1).

clavigera Lamarck, 1816:562, see *Mithrodia* (Mithrodiidae, part 2)

compta Stimpson, 1852, see *Leptasterias*

conceptionis Cuvier, 1829:485 (only listed). A NOMEN NUDUM.

conferta Stimpson, 1862:263, a forma of *Pisaster ochraceus* (Brandt, 1835)
according to Fisher (1930).

contorta Perrier, 1881a:1, see *Sclerasterias*

*coriacea Grube, 1840:22. Treated as a synonym of *Hacelia attenuata*
Gray, 1840 since Ludwig (1897) but if published earlier than December,
1840, then should have priority, so better suppressed.

crassispina H.L. Clark, 1941:68. A synonym of *A. forbesi* (Desor, 1848)
according to Downey *in* Clark & Downey (1992).

crassispinis Djakonov, 1950a as a subspecies of *A. rathbuni* (Verrill, 1909)

cribraria Stimpson, 1862:270. A synonym of *Leptasterias groenlandica*
(Steenstrup, 1857) according to Fisher (1930) and Djakonov (1950).

crispata Retzius, 1805, see *Ctenodiscus* (Goniopectinidae, part 1)

cumingii Gray, 1840, see *Heliaster* (Heliasteridae)

cumpullis Philippi (MS) *in* Quijada, 1911. A NOMEN NUDUM according
to Madsen (1956).

cunninghami Perrier, 1875:75; Bell 1881a:93. A synonym of *Anasterias
antarctica* (Lütken, 1857) according to Fisher (1940).

cuspidata Lamarck, 1816:553, Goniodiscus according to Sladen (1889)

(Oreasteridae, pt.2)

cylindricus Lamarck, 1816, see *Dactylosaster* (Ophidiasteridae, see part 2)

decemradiatus Brandt, 1835:271, undescribed. A NOMEN NUDUM according to Fisher (1911c) = *Solaster stimpsoni* (see part 3).

discoidea Lamarck, 1816. A synonym of *Culcita schmideliana* (Retzius, 1805) according to Sladen (1889).

disticha: Sladen, 1889:820. Could be identical with *A. [Leptasterias] groenlandica* (Steenstrup, 1857) according to Döderlein (1900) (from locality White Sea) but a synonym of *Asterias rubens* Linnaeus, 1758 according to Fisher (1930).

dorsatus Linnaeus, 1753:114. Invalid, = *A.* [i.e. *Protoreaster*] *nodosa* Linnaeus, 1758:661 (Oreasteridae, pt. 2). See A.M. Clark (1962b:175).

douglasi Perrier, 1875:69. A synonym of *Leptasterias polaris katherinae* (Gray, 1840) according to Fisher (1930).

dubia Verrill, 1909:545. A synonym of *L. polaris katherinae* (Gray, 1840) according to Fisher (1930).

echinata Gray, 1840:179. (Locality Valparaiso, Chile.) Probably *Coscinasterias* according to Fisher (1928) but might be *Meyenaster gelatinosus* (Meyen, 1834) or *Astrostole platei* (Meissner, 1896) according to Madsen (1956). Type not found.

echinata Philippi *in* Quijada, 1911. A NOMEN NUDUM according to Madsen (1956).

echinites Ellis & Solander, 1786, pls 60-62. A synonym of *Acanthaster planci* (Linnaeus, 1758) according to Verrill (1914b[AJS]). (Acanthasteridae, see part 2)

echinophora Lamarck, 1816, see *Othilia* and *Echinaster* (Echinasteridae, pt. 3)

echinophora: Delle Chiaje, 1825:356, non *echinophora* Lamarck, 1816,
    = *Asterias* [i.e. *Marthasterias*] *glacialis* Linnaeus, 1758 according to Sladen (1889).

'echinus Ellis' *in* Verrill 1914c:484, cited as synonym of *A. planci* [i.e. *Acanthaster*] but Ellis & Solander (1786) used '*echinites*' in plate captions.

edmondi Benham, 1911:151 (*Asterias (Stolasterias)*, presumably lapsus for *edmundi* Ludwig, 1897.

edmundi Ludwig, 1897:395, nom. nov. for *Asterias (Stolasterias) neglecta* Perrier, 1891, non *Asterias neglecta* Bell, 1881 but homonymy not contemporaneous since Bell's name already referred to synonymy of *A.* [Diplasterias] brandti} Bell, 1881 by Perrier (1891) so *edmundi* = *Sclerasterias neglecta* (Perrier, 1891).

endeca Linnaeus, 1771, see *Solaster* (Solasteridae, p. 3)

endeca var. decemradiata Brandt, 1835:71. A NOMEN NUDUM according to Fisher (1911c).

enopla Verrill, 1895, see *Urasterias*

epichlora Brandt, 1835:270. Has been used for *Leptasterias* but types were probably *Evasterias troscheli* forma *alveolata* (Verrill, 1914) according to Fisher (1930:91). *A. (Diplasterias) epichlora* : de Loriol, 1897:19 doubtfully = *E. troscheli* according to Djakonov (1950c).

epichlora: Verrill, 1909b:544, non epichlora Brandt, 1835, = *Leptasterias hexactis* (Stimpson, 1862) according to Fisher (1930).

epichlora var. *alaskensis* Verrill, 1909:549, raised to rank of species of *Leptasterias* by Fisher (1930).

*equestris Linnaeus, 1758:662, needs suppression by the ICZN as a threat to *Hippasteria phrygiana* (Parelius, 1768); see under *Stellaster* (Goniasteridae, part 2).

equestris Retzius, 1895:12. Invalid and = *Stellaster childreni* Gray, 1840, according to A.M.C. (1993, p.2, p. 285).

eustyla Sladen, 1889, see *Sclerasterias*

exigua Lamarck, 1816, see *Patiriella* (Asterinidae, see part 2)

exigua Delle Chiaje, 1825:353, non *A. exigua* Lamarck, 1816, = *Asterina gibbosa* Pennant, 1777 according to Ludwig (1897) (Asterinidae, see part 2).

exquisita de Loriol, 1888:403. A synonym of *Pisaster giganteus* (Stimpson, 1857) according to Fisher (1926a, 1930).

fabricii Perrier, 1875:56. A synonym of *Asterias vulgaris* [i.e. *A. rubens* Linnaeus, 1758] according to Verrill (1895).

fascicularis Perrier, 1881, see *Tarsaster* (Pedicellasteridae).

*fernandensis Meissner, 1896:104 (subgenus *Polyasterias*). 'Almost certainly the young of *Astrostole platei*' (Meissner, 1896) according to Fisher (1930).

fernandensis: de Loriol, 1904:41 (subgenus *Polyasterias*), non *A. fernandensis* Meissner, 1896, = *Allostichaster capensis* Perrier, 1875) according to Madsen (1956). ? *A.* fernandensis: Lieberkind, 1924:387 from Mas a Fuera, SE Pacific, also.

*fernandezianus Philippi *in* Quijada, 1911 [not seen] mentioned by Madsen (1956). ?MS.

filiformis O. F. Müller, 1776, an OPHIUROID.

fissispina Stimpson, 1862:264. A synonym of *Pisaster ochraceus* (Brandt, 1835) according to Fisher (1926a, 1930).

flabellifera Djakonov, 1950a, a forma of *Asterias amurensis* Lütken, 1871.

forbesi (Desor, 1848) (with synonyms *A. arenicola* Stimpson, 1862, *Asteracanthion berylinus* A. Agassiz, 1866, *A. lacazii* Perrier, 1869:51[243], *A. novae boracensis* Perrier, 1869:41[233] and *Asterias crassispina* H.L. Clark, 1941)

Desor 1848:67 (as *Asteracanthion*).

Verrill 1866:345; 1895:206; H.L. Clark 1904:552; Fisher 1930:205; Tortonese 1937:109; Galtsoff & Loosanov 1943:76; Engel & Schrovers

1960:8; Gray et al. 1968:156; Downey 1973:93; Franz et al. 1981:397; Downey *in* Clark & Downey 1992:421 (as *Asterias*).

Range: Maine to Cuba [to Gulf of Mexico according to Gray et al. 1968], 0-92 [?613] m.

forcipulata Verrill, 1909:67. A synonym of *Stylasterias forreri* (de Loriol, 1887) according to Fisher (1928).

forreri de Loriol, 1887, see *Stylasterias*

forreri: H.L. Clark 1913:203, non *forreri* de Loriol, 1887, = *Astrometis sertulifera* (Xantus, 1860) according to Fisher (1928).

*fragilis Studer, 1884:11 (E of New Zealand). Types juvenile. See Fisher (1930:217).

fulgens: Sladen, 1889:566. A synonym of *Anasterias varia* (Philippi, 1870 according to Madsen (1956).

fulva: Sladen, 1889:566. A synonym of *Cosmasterias lurida* (Philippi, 1858) according to Leipoldt (1895) but not endorsed by Fisher (1940).

fungifera Perrier, 1875:73[338]. A synonym of *Uniophora granifera* (Lamarck, 1816) according to Shepherd (1967).

gelatinosa Meyen, 1834, see *Meyenaster*

georgiana Studer, 1885, see *Neosmilaster*

gemmifera (Perrier, 1869) Perrier, 1875:46. A possible synonym of *Coscinasterias[?] echinata* (Gray, 1840) according to Fisher (1928) since also from Chile.

germaini: Perrier, 1875. Probably a forma of *Cosmasterias lurida* (Philippi, 1858) according to Fisher (1940).

gibbosa Pennant, 1777, see *Asterina* (Asterinidae, part 2)

gibbosa Leach, 1817 *in* Gray, 1840:288 (cited as type species of *Porania*). ?MS name. A NOMEN NUDUM according to Sladen (1889) and = *Porania pulvillus* (O.F. Müller, 1776) (Poraniidae, part 2).

gigantea Stimpson, 1857, see *Pisaster*, with homonym

gigantea Bell, 1891, as var. of *A. rubens* Linnaeus, 1758.

gigas Linnaeus, 1753 [invalidated by date], see *Oreaster gigas* Lütken, 1858 (Oreasteridae, part 2)

glacialis Linnaeus, 1758, see *Marthasterias*

glacialis: Pennant, 1777, non *A. glacialis* Linnaeus, 1758, = *A. rubens* Linnaeus, 1758 according to Sladen (1889).

glacialis: Grube, 1840, non *A. glacialis* Linnaeus, 1758, = *A.* [*i.e. Coscinasterias*] *tenuispina* Lamarck, 1816 according to Sladen (1889).

glomerata Sladen, 1889:571. A synonym of *Diplasterias brandti* forma *neglecta* (Bell) according to Fisher (1940).

gracilis Perrier, 1881, see *Stephanasterias*

*grandis Brandt, 1835. [Not found]

granifera Lamarck, 1816, see *Uniophora*

gracilispinis Djakonov, 1950. A forma of *A. amurensis* Lütken, 1871.

granularis Retzius, 1783, see *Ceramaster* (Goniasteridae, see part 2).

groenlandica: Stimpson, 1862, see *Leptasterias*

gunneri Danielssen & Koren, 1884:7. A synonym of *Urasterias lincki* (Müller & Troschel, 1842) according to Döderlein (1900).

hartii Rathbun, 1879, see *Allostichaster*

helianthemoides Pennant, 1777:56. A synonym of *Crossaster papposus* (Linnaeus, 1766) according to Sladen (1889).

helianthoides Brandt, 1835, see *Pycnopodia*

helianthus Lamarck, 1816, see *Heliaster* (Heliasteridae)

hexactis Stimpson, 1862, see *Leptasterias*

hirtum Philippi *in* Quijada, 1911. A NOMEN NUDUM, undescribed, according to Madsen (1956).

hispida Pennant, 1777: pl. 30, fig. 28. Said to be synonymous with *Leptasterias muelleri* (M. Sars, 1844 but proposed for suppression by Brun, 1970:238; Opinion 984, 1972:115, suppressed under the plenary powers.

holsatica Retzius, 1805:22. A synonym of *A. rubens* Linneaus, 1758 according to Sladen (1889).

hyadesi Perrier, 1886:1146. A synonym of *Anasterias antarctica* Lütken, 1857 according to Fisher (1940) but a NOMEN NUDUM in 1886, only the habit of incubation described.

hyperborea Danielssen & Koren, 1882, see *Leptasterias*

hyperborea: Kalischewsky, 1907, non *hyperborea* Danielssen & Koren, 1882, = *Leptasterias groenlandica* (Steenstrup, 1857) according to Djakonov *in* Fisher (1930).

hyperborea: Verrill, 1909:553, non *hyperborea* Danielssen & Koren, 1882, = *Leptasterias arctica* (Murdoch, 1885) according to Fisher (1930).

ianthina Brandt, 1835:774[?], alternative spelling of janthina Brandt, 1835:69 and a synonym of *Pisaster ochraceus* (Brandt, 1835) according to Fisher (1930).

imperati Delle Chiaje, 1841:57. A synonym of *Luidia ciliaris* (Philippi, 1837) (Luidiidae, see part 1) according to Sladen (1889).

inermis Bell, 1881:512. A synonym of *Leptasterias groenlandica* (Steenstrup, 1857 according to Fisher (1930).

irregularis Pennant, 1777, see *Astropecten* (Astropectinidae, part 1)

jacksonensis Cuvier, 1929:485 (listed). A NOMEN NUDUM, undescribed.

janthina Brandt, 1835:69. A synonym of *Pisaster ochraceus* (Brandt, 1835) according to Fisher (1930).

japonica Bell, 1881, see *Aphelasterias*

japonica Stimpson, 1862, non *A. japonica* Bell, 1881. A NOMEN NUDUM according to Fisher (1930).

jehennesii Cuvier *in* Perrier, 1875:47 [311]. By inference, probably con-

specific with *Coscinasterias calamaria* (Gray, 1840) according to Perrier.

johnstoni Gray *in* Johnston, 1836:146. A synonym of *Hippasteria plana* [i.e.
 H. phrygiana (Parelius, 1768)] according to Sladen (1889) (Goniasteridae, part 2).

jonstoni Delle Chiaje, 1827, see *Astropecten* (Astropectinidae, part 1).

katherinae Gray, 1840. A subspecies of *Leptasterias polaris* (Müller & Troschel, 1842) according to Djakonov (1950a).

katherinae: Perrier, 1875:67[331], non *A. katherinae* Gray, 1840, = *Pisaster
 grayi* Verrill, 1914, nom. nov., but that is also synonymous with *Leptasterias polaris katherinae* (Gray, 1840) according to Fisher (1930).

koehleri de Loriol, 1897, see *Orthasterias*

lacazei: Perrier, 1875:52, a variety of *Asterias arenicola* Stimpson, 1862,
 which is a synonym of *Asterias forbesi* (Desor, 1848) according to Fisher
 (1930).

laevigata Linnaeus, 1758, see *Linckia* (Ophidiasteridae, part 2)

latissima Djakonov, 1950a. A forma of *A. amurensis* Lütken, 1871.

lincki de Blainville, 1830, see *Protoreaster* (Oreasteridae, part 2)

linckii: Danielssen & Koren, 1884b [a secondary but transitory homonym
 of *{lincki* de Blainville], see *Urasterias*

linearis Perrier, 1881 ? *Coscinasterias*
 Perrier, 1881:2; A.M. Clark *in* Clark & Downey 1992:425.
 Range: SE Gulf of Mexico, *c.* 180 m.
 To *Coscinasterias* according to Verrill (1915) but only five arms in the
 holotype [Paris Museum], which needs reexamination.

littoralis Verrill: 1866:349, see *Leptasterias*

longipes Retzius, 1805, see *Chaetaster* (Chaetasteridae, part 2)

longstaffi Bell, 1908:7. A NOMEN NUDUM according to Fisher (1940).

luetkeni Stimpson, 1862:265. A synonym of *Pisaster giganteus* Stimpson,
 1857, according to Fisher (1926, 1930).

luna Linnaeus, 1758:661 [subdivision *integra* as opposed to *stellatae*,
 which includes all other asteroids]. '= *Anseropoda rosacea* (Lamarck,
 1816) without any reasonable doubt' according to Verrill (1914c:484).
 However, *luna* has priority. (See part 2, p.206.)

macropora Verrill, 1909, see *Stenasterias*

madeirensis Stimpson, 1862:263. A synonym of *Asterias* [i.e. *Marthasterias*] *glacialis* (Linnaeus, 1758) according to Sladen (1889).

mammillatus Audouin, 1827, see *Pentaceraster* (Oreasteridae, see part 2)

margaritifera: Brandt, cited by Fisher (1930:167) from label so possibly MS
 comb. *Asteracanthion margaritifer* Müller & Troschel, 1842, also from
 Sitka, Alasca, a synonym of *Pisaster ochraceus* (Brandt, 1835), according to Fisher (1930).

mazophorus Wood-Mason & Alcock, 1891, see *Sclerasterias*

membranacea Retzius, 1783:238. A synonym of *Anseropoda placenta* (Pennant, 1777) according to Ludwig (1897)

membranacea: Risso, 1826, non *A. membranacea* Retzius, 1783, = *Asterina gibbosa* Pennant, 1777, according to Ludwig (1897).

meridionalis Perrier, 1875, see *Diplasterias*

mexicana: Verrill, 1870:344, see *Leptasterias*

‡microdiscus} Djakonov, 1950a
    Djakonov 1950a:127; 1950c:120; Baranova 1957:178.
    Range: Kamtchatka, Bering Sea, 3-24 m.
    Forma brandti Djakonov, 1950 a homonym of *Asterias brandti* Bell, 1881 (*Diplasterias*).

*migratum Sladen, 1878:432 (as variety of *A. rubens* Linnaeus, 1758), now referred to synonymy of *A. amurensis* Lütken, 1871 on account of 'Korean Seas' locality.

milleporella Lamarck, 1816, see *Fromia* (Ophidiasteridae, part 2)

*migratum Sladen, 1878:432, as variety of *Asterias rubens* Linnaeus, 1758, but from Korean Strait so presumably = *A. amurensis* Lütken, 1871.

militaris O.F. Müller, 1776, see *Pteraster* (Pterasteridae, part 3)

miniata Brandt, 1835, see *Asterina* (Asterinidae, part 2)

minuta Linnaeus, 1761:512. A synonym of *A. rubens* Linnaeus, 1758 according to Madsen (1987).

minuta: Fabricius, 1780:375, non *A. minuta* Linnaeus, 1761, = *Leptasterias polaris* (Müller & Troschel, 1842) according to Madsen (1987).

minuta: Gmelin, 1788:3164, non *A. minuta* Linnaeus, 1761, = *Asterina gibbosa* (Pennant, 1777).

minuta: de Blainville, 1834:288, non *A. minuta* Linnaeus, 1761, = *Asterina* [i.e. *Patiriella*] *exigua* (Lamarck, 1816) according to Sladen (1889).

mollis Hutton, 1872, see *Sclerasterias*

mollis Studer, 1877:457, non *A. mollis* Hutton, 1872, = *Diplasterias meridionalis* (Perrier, 1875) according to Koehler (1917)

muelleri: Norman, 1865:127, see *Leptasterias*

muelleri: Ludwig, 1900, non *Asteracanthion muelleri* M. Sars, 1866, = *Leptasterias arctica* (Murdoch, 1885)

multiclava Verrill, 1914a:114. A synonym of *Leptasterias camtschatica* (Brandt, 1835) according to Fisher (1930).

multiradiata Linnaeus, 1758:663. A comasterid crinoid

multiradiata Gray, 1840:180. A homonym of *A. multiradiata* Linnaeus, 1758, = *Heliaster solaris* A.H. Clark, 1920, nom. nov.

murrayi Bell, 189a1:478. A synonym of *A. rubens* Linnaeus, 1758 according to Mortensen (1927).

nanimensis Verrill, 1914, see *Lethasterias*

*nautarum Bell, 1883:333; Sladen 1889:824 (not assignable to subgenus).

The locality 'Equador' was rejected by Fisher 1926:198, who found the type material to be conspecific with *Leptasterias arctica* (Murdoch, 1885), which it antedates; *nautarum* declared a NOMEN NUDUM by Fisher (1930) but Bell's description was fair so the name should perhaps be suppressed by the ICZN.

neglecta Bell, 1881a:94. A synonym of *A.* [i.e. *Diplasterias*] *brandti* Bell, 1881 according to Bell (1908) but treated as a forma by Fisher (1940).

nipon Döderlein, 1902, see *Distolasterias*

noctiluca Viviani, 1805:5. According to Millott *in* Boolootian (1966) AN OPHIUROID conspecific with *Amphipholis squamata* (Delle Chiaje, 1828) which should be saved by being put on the Official List.

nodosa Linnaeus, 1758:661, pt, see *Protoreaster* (Oreasteridae, part 2), pt = *Oreaster clavatus* Müller & Troschel, 1842, according to Madsen (1959). See A.M.C. (1962b:174).

normani Danielssen & Koren 1883:1; 1884b:25. A synonym of *Leptasterias hyperborea* (Danielssen & Koren, 1882) according to Fisher (1930).

nortonensis: A.H. Clark, 1920:8c. A synonym of *A. amurensis* Lütken, 1871 according to Fisher (1930) but a forma of *A. rathbuni* (Verrill, 1909) according to Djakonov (1950a).

NovaeZelandiae Cuvier, 1829:485 (listed). A NOMEN NUDUM, unde-scribed.

nuda Perrier, 1875, see *Uniophora*

obtusangula Lamarck, 1816, see *Pseudoreaster* (Oreasteridae, part. 2)

obtusata Bory de St Vincent, 1827, see *Pentaster* (Oreasteridae, part 2)

obtusispinosa Bell, 1881a:92. A synonym of *Cosmasterias lurida* (Philippi, 1858) according to Madsen (1956).

ochotensis: Perrier, 1869, see *Leptasterias*

ochracea Brandt, 1835, see *Pisaster*

octagona Lamarck, 1816:557, as variety of *A.* [i.e. *Patiriella*] *calcar* Lamarck, 1816. A synonym of *calcar* as restricted by Gray (1840).

oculata Pennant, 1777, see *Henricia* (Echinasteridae, part 2)

ophidiana Lamarck, 1816, see *Ophidiaster* (Ophidiasteridae, part 2)

ophidion Sladen, 1889, see *Hydrasterias* (Pedicellasteridae)

ophiura Linnaeus, 1758:662, an ophiuroid

pallida: Perrier, 1875:53, from *Asteracanthion pallidus* A. Agassiz, 1863, a NOMEN NUDUM according to Fisher (1930) and = *Asterias vulgaris* Verrill, 1866, i.e. *A. rubens* Linnaeus, 1758.

palmipes Olivi, 1792:66 (from Linck, 1733). A synonym of *A. membranacea* Retzius, 1783 according to Ludwig (1897), i.e. *Anseropoda placenta* (Pennant, 1777).

panopla Stuxberg, 1879, see *Icasterias*

papposus Linnaeus, 1776, see *Crossaster* (Solasteridae, part 3) papulosa Verrill, 1909:63. A synonym of *Pisaster brevispinus* (Stimpson, 1857) according to Fisher (1930).

papyracea Konrad, 1814:3. A synonym of *Asterina gibbosa* (Pennant, 1777) according to Ludwig (1897).

paucispina Stimpson, 1862:266. Reduced to a forma of *Pisaster brevispinus* (Stimpson, 1857) by Fisher (1930).

pectinata Linnaeus, 1758:663. A CRINOID.

pectinata: Brandt, 1835:270, non *A. pectinata* Linnaeus, 1758, ? = *A. amurensis* (Lütken, 1871) according to Fisher (1930).

pectinata: Couch, 1840:34, non *A. pectinata* Linnaeus, 1758, = *Luidia ciliaris* (Philippi, 1837) (Luidiidae, see part 1), according to Forbes (1841).

pedicellaris Koehler, 1907a, see *Notasterias*

pedicellaris: Bell, 1917, non *A. pedicellaris* Koehler, 1907, = *Notasterias armata* Koehler, 1911, according to A.M. Clark (1962a).

penicillaris Lamarck, 1816, see *Asterinopsis* (Asterinidae, part 2).

pentacanthus Delle Chiaje, 1827. A variety of *Astropecten irregularis* (Pennant, 1777) according to Döderlein (1917).

pentacanthus: Simroth, 1889:231, non *A. pentacanthus* Delle Chiaje, 1827, = *Astropecten hermatophilus* Sladen, 1883 according to Downey *in* Clark & Downey (1992).

pentacyphus Retzius, 1805. A synonym of *Pentaceros* [i.e. *Oreaster reticulatus* (Linnaeus, 1758) according to Sladen (1889).

pentagonula Lamarck, 1816, see *Anthenea* (Oreasteridae, part 2)

peregrina Bell an MS name according to Fisher (1928a:107), specimen (from Saya da Malha Bank, western Indian Ocean) probably = *Sclerasterias stenactis* (H.L. Clark, 1925), which itself is probably a forma of *S. mazophora* (Wood-Mason & Alcock, 1891) according to Macan (1938). Type labelled *Sclerasterias euplecta stenactis)* in collection of N.H.M., London.

perforata O.F. Müller, 1776, see *Henricia* (Echinasteridae, part 3)

perrieri E.A. Smith, 1876, see *Anasterias*

pertusa O.F. Müller , 1776, see *Henricia* (Echinasteridae, part 3)

philippii Bell, 1881:511. A synonym of *Leptasterias polaris acervata* (Stimpson, 1862) according to Fisher (1926a).

phrygiana Perelius, 1768, see *Hippasteria* (Goniasteridae, part 2)

placenta Pennant, 1777, see *Anseropoda* (Asterinidae, part 2)

planci Linnaeus, 1758, see *Acanthaster* (Acanthasteridae, part 2)

platyacantha Philippi, 1837, see *Astropecten* (Astropectinidae, part 2)

pleyadella Lamarck, 1816, see *Goniodiscaster* (Oreasteridae, part 2)

polaris Sabine, 1824:223. To *Ctenodiscus* according to Müller & Troschel (1842) and a synonym of *Ctenodiscus corniculatus* [i.e. *C. crispatus* (Retzius, 1805)] according to Sladen (1889).

polaris (Müller & Troschel, 1842) Lütken, 1857. Could be regarded as a secondary homonym of *Asterias polaris* Sabine, 1824, but not contemporaneus according to Fisher (1930:61). Possibly also threatened by *Asterias katherinae* Gray, 1840 [q.v.]. See *Leptasterias*.

polyplax: Perrier, 1875, see *Allostichaster*

polythela Verrill, 1909:68. A synonym of *Leptasterias polaris acervata* (Stimpson, 1862) according to Djakonov (1950a).

problema: Lütken, 1872:73. A synonym of *Asterias* [i.e. *Stephanasterias*] *albula* (Stimpson, 1853) according to Sladen (1889)

procyon Valenciennes *in* Cuvier, 1938, pl. 1 fig. 2. A NOMEN NUDUM, undescribed, probably = *Tosia australis* Gray, 1840 or *T. nobilis* (Müller & Troschel, 1843), according to A.M.C. (1953:404).

pulchella Blainville 1834:238. A synonym of *Asterina gibbosa* (Pennant, 1777) according to Perrier (1875)

pulvillus O.F. Müller, 1776, see *Porania* (Poraniidae, pt 2)

rarispina Perrier, 1875. A forma of *Marthasterias glacialis* (Linnaeus, 1758) according to Fisher (1940).

rathbuni (Verrill, 1909) (with subspecies *crassispinis* Djakonov, 1950 and formae *alveolata* Djakonov, 1950 and *nortonensis* (Verrill, 1909) also synonym *Allasterias rathbuni* var. *anomala* (Verrill, 1909, 1914))

rathbuni rathbuni (Verrill, 1909)

Verrill 1909:65,66 (vars *anomala* and *nortonensis*); 1914a:189 (as *Allasterias*). Djakonov 1950a:112 (with forma *alveolata*); 1950c:125; Baranova 1957:179.

Range: Bering Sea and Aleutian Is, 0-170 m.

A synonym of *A. amurensis* Lütken, 1871 according to Fisher (1930) but a valid species according to Djakonov (1950a).

rathbuni crassispinis Djakonov 1950

Djakonov 1950a:127.

Range: Okhotsk Sea.

reticulata Linnaeus, 1758, see *Oreaster* (Oreasteridae, pt 2)

reticulatum Philippi (MS) *in* Quijada, 1911. A NOMEN NUDUM according to Madsen (1956)

*richardi Perrier *in* Milne Edwards 1882:20; Perrier 1894:109 (as *Hydrasterias*); Fisher 1928:108 (probably *Sclerasterias*). Possibly conspecific with *Sclerasterias neglecta* (Perrier, 1891) according to Downey (MS) *in* Clark & Downey (1992) but holotype immature and Perrier's description does not conform so *richardi* proposed for suppression by the ICZN by A.M.C. *in* Clark & Downey (1992:511)

robusta Djakonov, 1950, a forma of *A. amurensis* Lütken, 1871

rodolphi Perrier, 1875, see *Astrostole*

rollestoni Bell, 1881

Bell 1881:514; Fisher 1930:23 (as subspecies of *A. amurensis* Lütken,

1871); Djakonov 1950a:123 (as separate species); 1950c:99; 1958:3323; Baranova & Wu 1962:111,116; Chang & Liao 1964:69; Baranova 1971:269.

Range: Japan Sea, Yellow Sea, 5-96 m.

rosacea Lamarck, 1816, see *Anseropoda* (Asterinidae, part 2)

rosacea: Delle Chiaje 1925:354, presumably non *A. rosacea* Lamarck, 1816, = *A. membranacea* Retzius, 1783 [i.e. *Anseropoda placenta* (Pennant, 1777)] according to Ludwig (1897).

rosacea: Verany, 1846:5, non *A. rosacea* Lamarck, 1816, = *Echinaster sepositus* (Retzius, 1783) according to Ludwig (1897).

rosea O.F. Müller 1776, see *Stichastrella*

roseum Philippi (MS) *in* Quijada, 1911. A NOMEN NUDUM according to Madsen (1956).

rubens Linnaeus, 1758 (with synonyms *A. minuta* Linnaeus, 1761, *A. clathrata* Pennant, 1777, *A. violacea* O.F. Müller, 1776 (as variety), *A. holsatica* Retzius, 1805, *A. spinosa* Say, 1825, *Asteracanthion distichum* Brandt, 1851, *A. pallidus* A. Agassiz, 1863 (nomen nudum) and *A. vulgaris* Verrill, 1866; variety *migratum* Sladen, 1878 now referred to synonymy of *A. amurensis* Lütken, 1871) Linnaeus 1758:661; Downey *in* Clark & Downey 1992:422. Nardo 1834:716 (as *Stellonia*). Müller & Troschel 1840a:102 (as *Asteracanthion*). Forbes 1841:83 (as *Uraster*.

Range: Scandinavia to S Portugal, Barents Sea, Greenland and Labrador to S Carolina (rarely Florida), LW to 900 m.

rubens: Olivi, 1792:65, non *A. rubens* Linnaeus, 1758, = *Echinaster sepositus* (Retzius, 1783) according to Ludwig (1897)

rubens: Johnston 1836:144, non *A. rubens* Linnaeus, 1758, = *Luidia ciliaris* (Philippi, 1837) according to Sladen (1889)

rubens: Murdoch, 1885:159, non *A. rubens* Linnaeus, 1758, = *A. amurensis* Lütken, 1871 according to Fisher (1930)

rugispina Stimpson, 1862:267. A synonym of *Anasterias antarctica* Lütken, 1857 according to Fisher (1940).

*rumphii Parelius, 1768:427 [1770:351]. Synonymous with (and antedates) *Solaster endeca* (Linnaeus, 1771) (Solasteridae, pt 3) according to von Martens (1902) and Engel (1959); proposed for suppression by the ICZN by A.M.C *in* Clark & Downey (1992).

rupicola Verrill, 1876:71, see *Anasterias*

rustica Gray, 1840:179. A synonym of *Meyenaster gelatinosus* Meyen, 1834) according to Madsen (1956).

saanichensis de Loriol, 1897:23. A synonym of *Leptasterias epichlora* (Brandt, 1835) according to Verrill (1914a), itself probably a synonym of *Evasterias troscheli* forma *alveolata* (Verrill, 1914) according to Fisher (1930) [in which case *saanichensis* would have priority].

sagena Retzius, 1805:21. A synonym of *Echinaster sepositus* (Retzius, 1783) according to Tortonese & Madsen (1979) after selection of lectotype.

sanguinolenta O.F. Müller, 1776, see *Henricia*

satsumana Döderlein, 1902, see *Sclerasterias*

savaresi Delle Chiaje, 1825:357; 1841:125. A synonym of *A*. [i.e. *Coscinasterias*] *tenuispina* Lamarck, 1816 according to Gray (1866)

savignyi Audouin, 1826, see *Luidia* (Luidiidae, part 1)

scabra: Farquharson, 1894, see *Astrostole*

scalprifera Sladen, 1889, see *Smilasterias*

schmideliana Retzius, 1805, see *Culcita* (Oreasteridae, part 2)

sebae de Blainville, 1830 [or 1834]. A synonym of *Pentaceros* [i.e. *Oreaster*] *reticulatus* (Linnaeus, 1758)

senegalensis Lamarck, 1816, see *Luidia* (Luidiidae, part 1)

sentus Say, 1825, see *Echinaster (Othilia)*

seposita Retzius, 1783, see *Echinaster (Echinaster)*

seposita: Retzius, 1805 = *Henricia* sp., probably *H. sanguinolenta* (O.F. Müller, 1776) according to Madsen (1987)

sertulifera Xantus, 1860, see *Astrometis*

similispinis H.L. Clark, 1908, see *Leptasterias*

sinusoida Perrier, 1875. A synonym of *Uniophora granifera* Lamarck, 1816 according to Shepherd (1967)

solaris Schreber, 1793:2. A synonym of *Acanthaster planci* (Linnaeus, 1758) (Acanthasteridae, part 2) according to Madsen (1955).

solaris: Carpenter, 1856:360, non *A. solaris* Schreber, 1793, probably = *Heliaster cumingi* (Gray, 1840) according to H.L. Clark (1909).

spectabilis: Sladen, 1889:566. A synonym of *Cosmasterias lurida* (Philippi, 1858) according to Leipoldt (1895).

spinosa Pennant, 1777:62. A synonym of *A*. [i.e. *Marthasterias*] *glacialis* Linnaeus, 1758 according to Sladen (1889).

spinosa Retzius, 1805:18. A junior homonoym of *A. spinosa* Pennant, 1777 and replaced by *A*. [i.e. *Echinaster (Othilia)*] *echinophora* Lamarck, 1816 (Echinasteridae, part 3) according to Fisher (1919).

spinosa Say, 1825:142. Also a junior homonym; a synonym of *A. rubens* Linnaeus, 1758 acccording to Downey *in* Clark & Downey (1992).

spinulosus Philippi, 1837, see *Astropecten* (Astropectinidae, part 1)

spirabilis Bell, 1881, see *Anasterias*

spitzbergensis Danielssen & Koren, 1881:177; 1884b:5. A synonym of *Leptasterias groenlandica* (Steenstrup, 1857) according to Fisher (1930)

spongiosa Fabricius, 1780, see *Henricia* (Echinasteridae, part 3)

steineni Studer, 1885, see *Neosmilaster*

stellionura: Perrier, 1875:46. A synonym of *Urasterias lincki* (Müller & Troschel, 1842) according to Fisher (1930)

stichantha Sladen, 1889, see *Distolasterias*

stimpsoni Verrill, 1866:349. At least partly = *A. vulgaris* Verrill, 1866 [i.e. *A. rubens* Linnaeus, 1758] according to Verrill (1895), Fisher (1930),

striata Lamarck, 1816, see *Valvaster* (Asteropseidae, part 2)

studeri Bell, 1881. A synonym of *Diplasterias meridionalis* (Perrier, 1875) according to Fisher (1940).

subinermis Philippi, 1837), see *Tethyaster* (Astropectinidae, part 2).

subulata Lamarck, 1816:568. A synonym of *Chaetaster longipes* (Retzius, 1805) according to Perrier (1875).

sulcifera: Perrier, 1875:58. A synonym of *Cosmasterias lurida* (Philippi, 1858) according to Fisher (1940).

tanneri Verrill, 1880, see *Sclerasterias*

tenella Retzius, 1783. A CRINOID (*Hathrometra tenella* according to A.H. Clark (1918) Siboga Exped.)

tenera Stimpson, 1862, see *Leptasterias*

tenuispina Lamarck, 1816, see *Coscinasterias*

tenuissima Risso, 1826:269. A printer's error for *tenuispina* [presumably *Coscinasterias tenuispina* (Lamarck, 1816)] according to Ludwig (1897) [though it is evidently the species later named *Luidia ciliaris* (Philippi, 1837)].

terwieli Goldschmidt, 1924:499 (subgenus *Smilasterias*). A synonym of *Luidia magellanica* Leipoldt, 1895 according to Madsen (1956).

tessellata (vars C and D) Lamarck, 1816, see *Goniaster* (Goniasteridae, part 2)

tomidata Sladen, 1889:576. A synonym of *Cosmasterias lurida* (Philippi, 1858) according to Madsen (1956).

torquata Sladen, 1889:570. A synonym of *Aphelasterias japonica* (Bell, 1881) according to Hayashi (1938).

triremis Sladen, 1889, see *Smilasterias*

*trochiscus Retzius, 1805:10. To *Asterina* (Asterinidae, see part 2) but of doubtful validity according to Sladen (1889).

troscheli Stimpson, 1862, see *Evasterias*

undulata O.F. Müller, 1784. A [superfluous] nom. nov. for *A. glacialis* Linnaeus, 1758 [i.e. *Marthasterias*] according to Madsen (1987).

umbilicata Konrad, 1814:4. A synonym of *Asterina gibbosa* (Pennant, 1777) according to Ludwig (1897).

vancouveri Perrier, 1875. A subspecies of *Leptasterias hexactis*

varia: Perrier, 1875, see *Anasterias*

variolata Retzius, 1805, see *Nardoa* (Ophidiasteridae, see part 2)

variolata: Risso, 1826:269, non *A. variolata* Retzius, 1805, = *Hacelia attenuata* (Gray, 1840) according to Ludwig (1897).

vernicina Lamarck, 1816, see *Petricia* (Asteropseidae, part 2)

verrilli Bell, 1881:513. A synonym of *Anasterias varia* (Philippi, 1870) according to Madsen (1956).

verrucosa Risso, 1826:271. A synonym of *Chaetaster longipes* (Retzius, 1805) according to Ludwig (1897).

verruculata Retzius, 1805:12. A synonym of *Asterina gibbosa* (Pennant, 1777) according to Sladen (1889).

versicolor Sladen, 1899

Sladen 1899:573; Fisher 1930:206; Hayashi 1936:19; Djakonov 1950a: 124; 1950c:101; Baranova & Wu 1962:111,116; Chiang & Liao 1964:70; A.M. Clark 1982:492; Liao & A.M. Clark 1995['96]:142.Hayashi 1973:106 (as subspecies of *A. amurensis* Lütken, 1871).

Range: S Japan, S China, 0-91 m.

vesiculosa Sladen, 1889, see *Diplasterias*

vestita Say, 1825, see *Tethyaster* (Astropectinidae, part 1)

victoriana Verrill, 1909:68; 1914:124. A synonym of *Evasterias troscheli* (Stimpson, 1862) according to Djakonov (1950a).

violacea: O.F. Müller, 1788:7. A synonym of *A. rubens* Linnaeus, 1758 according to Fisher (1930).

volsellatus Sladen, 1889, see *Coronaster* (Labidiasteridae)

volsellatus: Nutting, 1895:168, non *A. volsellatus* Sladen, 1889, = *Coronaster briareus* (Verrill, 1882) (Labidiasteridae)

vulgaris Packard, 1863:405. A NOMEN NUDUM according to Fisher (1930), = *A. rubens* Linnaeus, 1758.

vulgaris Verrill, 1866:347. A synonym of {A. rubens} Linnaeus, 1758 according to Downey *in* Clark & Downey (1992).

* wilkinsoni Gray, 1840:179. Northern Africa. Type lost; species unknown according to Sladen (1889).

ASTERODERMA Perrier, 1888:763, for Asteroderma papillosum undescribed and 1891a:96, a provisional name.

A synonym of *Anasterias* Perrier, 1875 according to Fisher (1940).

papillosum Perrier, 1891a:96 (1888:765 a nomen nudum), probably a synonym of Anasterias minuta Perrier, 1875 according to Fisher (1940).

ASTROMETIS Fisher, 1923

Fisher 1923:254; 1928a:118.

Type species: *Asterias sertulifera* Xantus, 1860.

*californica (Verrill, 1914)

Verrill 1914a:174 (as *Orthasterias*).

Fisher 1928:126 (as *Astrometis*).

Range: Type probably not off San Francisco but more likely San Diego or San Pedro, California according to Fisher.

Probably another variant of *A. sertulifera* (Xantus, 1860) according to Maluf (1988).

sertulifera (Xantus, 1860) (with synonyms *Orthasterias dawsoni* and *gonolena* Verrill, 1914a, probably also *Orthasterias californica* Verrill, 1914a)

Xantus 1860:568 (as *Asterias*).

Baker *in* Fisher 1912:89 (as *Coscinasterias*

Verrill 1914a:100 (as ? *Marthasterias*).

Fisher 1923:254; 1928:119; Caso 1961:97; Maluf 1988:46,126 (as *Astrometis*).

Range: S California to Peru, 0-156 m.

ASTROSTOLE Fisher, 1923 (nom. nov. for *Margaraster* Hutton, 1872, non Gray, 1866)

Fisher 1923:255; 1928a:130.

Type species: *Margaraster scaber* Hutton, 1872.

insularis H.L. Clark, 1938

H.L. Clark 1938:191.

Range: Lord Howe I, SE Australia.

multispina A.M. Clark, 1950

A.M. Clark 1950:808.

Range: Norfolk Island, N of New Zealand.

*paschae (H.L. Clark, 1920)

H.L. Clark 1920:105 (as *Stylasterias*).

Fisher 1928a:130; Marsh 1974:95 (as *Astrostole*).

Range: Easter I, shore.

'Not improbably' a synonym of *A. platei* (Meissner, 1896) according to Fisher (1928a).

platei (Meissner, 1896) (with probable synonym *Polyasterias fernandensis* Meissner, 1896)

Meissner 1896:103,104 (as *Polyasterias*).

Fisher 1928a:130 (as *Astrostole*).

Range: Juan Fernandez, W from Chile.

rodolphi (Perrier, 1875)

Perrier 1875:41; Farquhar 1897:192; Benham 1911:150 (as *Asterias*).

Fisher 1928:130 (as *Astrostole*).

Range: Kermadec Is, N of New Zealand.

scabra (Hutton, 1872)

Hutton 1872:5 (as *Margaraster?*).

Farquhar 1894:202 (as *Asterias (Strolasterias)* [sic] *scabra*).

Fisher 1923:255; Fell 1962:49; McKnight 1967:302; Dartnall 1969:54 (as *Astrostole*).

Range: Central and southern New Zealand, Tasmania, shore.

AUSTRALIASTER Fisher, 1923

Fisher 1923:253; 1928a:131.

Type species: *Coscinasterias dubia* (H.L. Clark, 1909).

dubia (H.L. Clark, 1909)

H.L. Clark 1909:532 (as *Coscinasterias*).

Fisher 1919a:489 (as *Distolasterias*).

Fisher 1923:253; Rowe & Pawson 1977:342 (as *Australiaster*).

Range: N.S.W., Australia, 40-90 m.

AUTASTERIAS Koehler, 1912

Koehler 1912a:152; 1912b:26.

Type species: *Asterias pedicellaris* Koehler, 1911.

A synonym of *Notasterias* Koehler, 1911, according to Fisher (1930).

bongraini Koehler, 1912, see *Notasterias*

pedicellaris: Koehler, 1911, see *Notasterias*

BATHYASTERIAS Fisher, 1930:231, as subgenus of *Diplasterias* Perrier, 1888 (type species *Asterias vesiculosa* Sladen, 1889) but listed in synonymy of *Diplasterias* by Fisher (1940).

vesiculosa (Sladen, 1889:568) (as *Asterias*) Fisher, 1930:231 (as *Diplasterias (Bathyasterias)*) supposedly from Arafura Sea, N of Australia. See under *Diplasterias*.

CAIMANASTER A.M. Clark, 1962

A.M. Clark 1962a:82.

Type species: *Caimanaster acutus* A.M. Clark 1962.

acutus A.M. Clark, 1962

A.M. Clark 1962a:82.

Range: Off Enderby Land, Antarctica, 193 m.

CALASTERIAS Hayashi, 1975

Hayashi 1975:199.

Type species: *Calasterias toyamensis* Hayashi, 1975.

toyamensis Hayashi, 1975

Hayashi 1975:199; Imaoka et al. 1991:100.

Range: Toyama Bay, Japan Sea, 100 m.

CALLIASTERIAS Fewkes, 1889:33.

Type species: *Asterias exquisita* de Loriol, 1888.

A synonym of *Pisaster* Müller & Troschel, 1840, according to Fisher (1926, 1930).

exquisita: Fewkes, 1889:33. A synonym of *Pisaster giganteus* (Stimpson, 1857) according to Fisher (1926, 1930).

CALVASTERIAS Perrier, 1875 (with synonym *Stichorella* Koehler, 1920)

Perrier 1875:84; Fisher 1930:225.

Type species: *Calvasterias asterinoides* Perrier, 1875.

*antipodum Bell, 1882

Bell 1882a:121.

Range: ? Given as 'Erebus' and 'Terror' (i.e. Ross' Antarctic

Expedition) so from anywhere between New Zealand and McMurdo Sound, Ross Sea.

Status doubtful.

asterinoides Perrier, 1875

Perrier 1875:84; Sladen 1889:589; Fisher 1922b:597.

Range: Torres Strait !? Falkland Is according to Sladen (1889).

laevigata (Hutton, 1879)

Hutton 1879:343 (as *Asterias rupicola* var. *laevigatus*).

Farquhar 1898a:180; Benham 1909:302 (as *Stichaster suteri* var. *laevigatus*).

Fisher 1923:606; Mortensen 1925a:311; Fell 1953:96; McKnight 1967:303 (as *Calvasterias*).

Range: Auckland, Campbell and Antipodes Is.

stolidota Sladen, 1889: A synonym of *Anasterias varia* (Philippi, 1870) according to Madsen (1956).

suteri (de Loriol, 1894)

de Loriol 1894:477; Farquhar 1898:197; Benham 1909a:302 (as *Stichaster*).

Koehler 1920:87 (as *Stichorella*).

Fisher 1922:597; Mortensen 1925a:310; Fell 1960:66; Pawson 1965:253; McKnight 1967:304.

Range: Southern New Zealand (S from Wellington) and offshore islands, Macquarie I, intertidal.

CARLASTERIAS da Costa, 1952:7. Nom. nov. for *Mortensenia* da Costa, 1941, preoccupied. Type species: *Mortensenia lusitanica* da Costa, 1941.

Holotype lost but probably a species of *Sclerasterias* Perrier, 1891, according to A.M. Clark *in* Clark & Downey (1992:415, 449).

lusitanica: da Costa, 1952:7, see above.

COELASTERIAS Verrill, [1867] 1871

Verrill 1871:247.

Type species: *Coelasterias australis* Verrill, 1871.

A synonym of *Stichaster* Müller & Troschel, 1840, according to Fisher (1922b, 1930).

australis Verrill, 1871, see *Stichaster*

COMASTERIAS Perrier, 1891a:K159. Lapsus for *Cosmasterias*.

*****COSCINASTERIAS** Verrill, 1870 (with synonyms *Polyasterias* Perrier, 1893 and *Lytaster* Perrier 1894, also subgenus *Stolasterias* Sladen, 1889*)

*The rank of *Stolasterias* in relation to *Coscinasterias* has been the subject of some dissention. Rowe *in* Rowe & Gates (1995), like Verrill (1907) and H.L. Clark (1922, 1933 and 1942), maintains that it is generically distinct, whereas Mortensen (1933) and Tortonese (1982) refuse to grant it even subgeneric rank. My own view is that the differences are less than generic, so the compromise position of treating it as a subgenus, following Fisher (1928), is followed here [A.M.C. **new obs.**]

COSCINASTERIAS (COSCINASTERIAS) Verrill, 1870
> Verrill 1870:248; 1914a:45; Fisher 1928a:128; Tortonese 1965:186; 1982:4; A.M. Clark *in* Clark & Downey 1992:423.
> Type species: *Coscinasterias muricata* Verrill, 1870.

aster (Gray, 1840) Verrill, 1914a:46. Types lost.

brucei (Koehler, 1908) 1911, see *Diplasterias*

calamaria: Fisher, 1928a, H.L. Clark 1946 and other records from Australasia, non *C. calamaria* Gray, 1840, = *C. muricata* Verrill, 1870

candicans: Koehler, 1917, see *Notasterias*

dubia H.L. Clark, 1909, see *Australiaster*

*echinata (Gray, 1840) (with possible synonym *Asteracanthion gemmifer* Perrier, 1869, or *Astrostole platei* (Meissner, 1896), if not itself a synonym of *Meyenaster gelatinosus* (Meyen, 1834) judging from locality)
> Gray 1840:179 (as *Asterias*).
> Fisher 1928a:128 (?*Coscinasterias*).
> Range: Valparaiso, Chile.
> Type not found in the Natural History Museum collection (1999).

euplecta Fisher, 1906, see *Sclerasterias*

gemmifera (Perrier, 1869) Fisher, 1928a:128, probably a synonym of *Asterias echinata* Gray, 1840 but see above.

gemmifera: H.L. Clark 1916:74, non *C. gemmifera* (Perrier, 1869), = *Coscinasterias calamaria* according to H.L. Clark (1946) i.e. *C. muricata)* (Verrill, 1870).

jehennesi (Perrier, 1869) Verrill, 1914:46 possibly a synonym of oscinasterias calamaria according to Perrier (1875).

*linearis (Perrier, 1881) Verrill 1915:21. See A.M. Clark in Clark & Downey 1992:425, affinities doubtful.

muricata (Verrill, 1870) (including Australian records of *C. calamaria*)
> Verrill 1870:249; Rowe 1989:290 (revived from the synonymy of *Coscinasterias calamaria*); A.M. Clark *in* Clark & Downey 1992:424; Rowe *in* Rowe & Gates 1995:28.
> Range: New Zealand and southern Australia.

victoriae Koehler, 1911:32. A synonym of *Diplasterias brucei* (Koehler, 1908) according to Fisher (1940).

COSCINASTERIAS (STOLASTERIAS) (Sladen, 1889, as subgenus of *Asterias*) (with synonyms *Polyasterias* Perrier, 1893 and *Lytaster* Perrier, 1894) Sladen 1889:583 (*Asterias (Stolasterias)*).
> H.L. Clark 1933:30 (as genus).
> Fisher 1928a:129 (as subgenus).
> Type species: *Asterias tenuispina* Lamarck, 1816.

acutispina (Stimpson, 1862) (with synonym *Asterias calamaria* var. *japonica* Döderlein, 1902)

Stimpson 1862:262 (as *Asterias*).

Verrill 1914:46; Fisher 1928a:129 (as *Coscinasterias (Stolasterias)*

Hayashi 1943:197; Chang & Liao 1964:73; Hayashi 1973:102; A.M. Clark 1982:487,491; Liao & A.M. Clark 1995 [1996]:144 (as *Coscinasterias*)

Range: Southern Japan, Korea, China, W Hawaiian Is, shore.

[The generic position of this species in relation to *Astrostole* Fisher, 1923 should be re-examined since larger specimens of *acutispina* have some diplacanthid proximal adambulacrals and no enlarged tooth on the crossed pedicellariae. A.M.C. **new obs.**]

calamaria (Gray, 1840) (with possible synonym *Asterias jehennesi* Perrier, 1875)

Gray 1840:179; de Loriol 1885:4; Bell 1905:251 (as *Asterias*

Dujardin & Hupé 1862:339 (as *Asteracanthion*).

H.L. Clark 1923a:306; Mortensen 1933a:227; A.M. Clark *in* Clark & Courtman-Stock 1876:92; A.M. Clark *in* Clark & Downey 1992:426 (as *Coscinasterias*).

Range: Mauritius, South Africa (at least Natal), shore.

calamaria: Fisher 1928a, H.L. Clark 1946, non (Gray, 1840), = *C. (Coscinasterias) muricata* (Verrill, 1870)

tenuispina (Lamarck, 1816) (with synonyms *Asterias savaresi* Delle Chiaje, 1827, *Asterias atlantica* Verrill, 1868 and *Lytaster inaequalis* Perrier, 1894))

Lamarck 1816:561; Perrier 1875:42 [306]; Ludwig 1897:344 (as *Asterias*).

Müller & Troschel 1842:16 (as *Asteracanthion*).

Sladen 1889:565 (as *Asterias (Stolasterias)*.

Perrier 1894:108 (as *Polyasterias*).

Verrill 1914:45; Koehler 1924:103; Madsen 1950:219; Tortonese 1965:186; 1982:3; A.M. Clark *in* Clark & Downey 1992:427 (as *Coscinasterias*).

Verrill 1907:324; H.L. Clark 1933:30 (as *Stolasterias*).

Fisher 1928a:129 (as *Coscinasterias (Stolasterias)*).

Range: Mediterranean, North Carolina, Brazil, Bermuda, Azores, St Helena, SW France (Biscay) S to Guinea, 1-165 m.

COSMASTERIAS Sladen, 1889

Sladen 1889:562, 576 (as subgenus of *Asterias*).

Perrier 1893:848; 1896:32 (as genus).

Type species: *Asteracanthion sulcifer* Perrier, 1869, a synonym of *A. luridum* Philippi, 1858.

*alba (Bell, 1891:92 (as *Asterias*)) Fisher 1940:250, possibly a *Cosmasterias*). [Type?]

capensis: Fisher 1926:198; Mortensen 1933a:276, see *Allostichaster*

capensis: Fisher, 1930:228, non *Asterias* (i.e. *Allostichaster capensis* Perrier, 1875, = *Cosmasterias felipes* (Sladen, 1889)
according to Mortensen (1933a).

dyscrita H.L. Clark, 1916
H.L. Clark 1916:71; Fell 1958:20; 1960:66; McKnight 1967:303; Baker & H.E.S. Clark 1970:5.
Range: Off Victoria, Australia, New Zealand and Chatham Is, 240-550 m.

felipes (Sladen, 1889)
Sladen 1889:433; H.L. Clark 1923a:304 (as *Stichaster*).
Perrier 1896:27 (as *Quadraster*).
Fisher 1923:596 (footnote); Mortensen 1933a:274; A.M. Clark & Courtman-Stock 1976:93; A.M.C. *in* Clark & Downey 1992:429; Stampanato & Jangoux 1997:31 (as *Cosmasterias*).
Range: West and South coasts of South Africa, 79-353 m, also the vicinity of Amsterdam & St Paul Islands, southern Indian Ocean, c.1100 (?880-1680) m.

fernandensis: Verrill, 1914a:360. Probably a juvenile *Astrostole platei* according to Fisher (1930).

germaini (Philippi, 1858) Fisher 1940:263. Probably a forma of *C. lurida* according to Fisher but listed as one of many synonyms by A.M.C. *in* Clark & Downey (1992).

lurida (Philippi, 1858) (with synonyms *Asteracanthion germaini*
Philippi, 1858, *A. sulcifer* Perrier, 1869, *A. clavatum, fulvum, spectabile* and *mite* Philippi, 1870, *Asterias sulcifer* Perrier, 1875, *A. alba* and *obtusispinosa* Bell, 1881a and *A. (Cosmasterias) tomidata* Sladen, 1889)
Philippi 1858:265 (as *Asteracanthion*).
Ludwig 1903:40; Koehler 1912b:22; Fisher 1940:263; Madsen 1956a:42; Bernasconi 1973a:318; Hernandez & Tablado 1985:7; A.M. Clark *in* Clark & Downey 1992:429 (as *Cosmasterias*).
de Loriol 1904:39 (as *Asterias (Cosmasterias)*).
Range: S Chile, Falkland-Magellan area, South Georgia and S Argentina N to c.41°S., 0-650 m.

obtusispinosa: Fisher 1940:250. A synonym of *C. lurida* (Philippi, 1858 according to A.M.C. *in* Clark & Downey (1992).

*polygramma (Sladen, 1889)
Sladen 1889:434 (as (*Stichaster*).
Verrill 1914a:360; Fisher 1930:228 (as *Cosmasterias*).
Synonymized with *C. lurida* (Philippi, 1858) by Leipoldt (1895) but a distinct species according to Verrill (1914a).

272

Range: West of Patagonia, 448 m.

radiata Koehler, 1923, see *Diplasterias*

sulcifera: Leipoldt, 1895:553. A synonym of *Cosmasterias lurida* (Philippi, 1858) according to Fisher (1940).

tomidata: Leipoldt, 1895:552. A synonym of *Cosmasterias lurida* according to Madsen (1956).

CRYPTASTERIAS Verrill, 1914

Verrill 1914b:15; 1914a:362.

Type species: *Diplasterias turqueti* Koehler, 1906.

brachiata Koehler, 1923

Koehler 1923:25.

Range: Falkland Is.

turqueti (Koehler, 1906) (with probably synonym *Diplasterias induta* Koehler, 1906)

Koehler 1906:19 (as *Diplasterias*).

Verrill 1914a:362; Koehler 1920:57; Fisher 1940:257 (as *Cryptasterias*).

Range: South Orkney Is, 0-36 m.

CTENASTERIAS Verrill, 1914

Verrill 1914a:148.

Type species: *Asterias spitzbergensis* Danielssen & Koren, 1884

A synonym of *Leptasterias groenlandica* (Steenstrup, 1857) according to Djakonov (1938).

A synonym of *Leptasterias* Verrill, 1866 according to Fisher (1930).

cribraria (Stimpson, 1862) Verrill, 1914:148. A synonym of *Leptasterias groenlandica* (Steenstrup, 1857) according to Djakonov (1950).

georgianus: Koehler, 1923:40, see *Neosmilaster*

DIPLASTERIAS Perrier, 1891 (Diplasterias Perrier, 1888:765 a nomen nudum) (synonyms *Podasterias* Perrier, 1894, *Koehleraster* Fisher, 1922) and *Bathyasterias* Fisher, 1930)

Perrier 1891a:77; Fisher 1908a:89; 1908b:358; 1930:229; 1940:248; Bernasconi 1964b:273; 1970:270; A.M. Clark *in* Clark & Downey 1992:430.

Type species: *Diplasterias lutkeni* Perrier, 1891, a synonym of *Asterias brandti* Bell, 1881, replacement designation by Fisher (1930, awaiting validation by the ICZN, see A.M.C., 1992:431)

brandti (Bell, 1881) (with synonyms *Asterias neglecta* Bell, 1881 (forma), *Asterias belli* Studer, 1884), *Asterias glomerata* Sladen, 1889, also *Diplasterias loveni* and *lutkeni* Perrier, 1891)

Bell 1881a:91 (as *Asterias*).

Meissner 1904:7; Koehler 1908:572; Barattini 1938:23; Fisher 1940:249 (with forma neglecta Bell); Bernasconi 1966b:172; 1973a:319; A.M.C. *in* Clark & Downey 1992:431 (as *Diplasterias*). [Non *D. bradtii* [sic]

Tommasi 1970:20, nec *D. brandtii*: Carrera-Rodriguez & Tommasi 1977; see note in Clark & Downey]

Koehler 1917:26; 1920:40,51; 1923:27 (as *Podasterias*).

Range: Uruguay to the Falkland-Magellan area and South Georgia; also South Shetlands, Antarctic Peninsula and Bellingshausen Sea but with some doubt - see Clark & Downey (1992); 0-450 m.

brucei (Koehler, 1908) (with synonyms *Coscinasterias victoriae* Koehler, 1911 and *Podasterias fochi* Koehler, 1920)

Koehler 1908a:569 (as *Stolasterias*).

Koehler 1911a:30 (as *Coscinasterias*).

Koehler 1920:42 (as *Podasterias*).

Fisher 1930:231; 1940:253; Bernasconi 1956:18; A.M. Clark 1962a:83; H.E.S. Clark 1963:75; Bernasconi 1967b:10 (as *Diplasterias*).

Range: South Orkney Is, probably circum-polar antarctic, 18-730 m.

eunota Fisher 1940:255. A form of *Diplasterias octoradiata* (Studer, 1885).

fochi: Fisher 1930:231. A synonym of *D. brucei* (Koehler, 1908) according to A.M. Clark (1962).

georgiana: Perrier, 1891, see *Neosmilaster*

germaini (Philippi, 1858) Meissner, 1904:7. A synonym of *Cosmasterias lurida* (Philippi, 1858) according to Madsen (1956).

induta Koehler, 1908a:575. Probably young *Cryptasterias turqueti* Koehler, 1906 according to Fisher (1940).

kerguelensis (Koehler, 1917)

Koehler 1917:24 (as Podasterias).

Fisher 1930:231 (as (Diplasterias).

Range: Kerguelen, Southern Ocean.

loveni Perrier, 1891a:K81. A synonym of *Diplasterias brandti* (Bell, 1881) according to Fisher (1940).

luetkeni Perrier, 1891a:K81. Also a synonym of *Diplasterias brandti* according to Fisher (1940).

meridionalis (Perrier, 1875) (with synonyms *Asterias mollis* Studer 1884 and *A. studeri* Bell, 1881, also probably *Asterias vesiculosa* Sladen, 1889, see below under *D. vesiculosa*)

Perrier 1875:76; E.A. Smith 1879:272 (as *Asterias*).

Perrier 1891a:7; Fisher 1940:251; Bernasconi 1956:21; A.M. Clark 1962a:84; Guille 1974:38 (as *Diplasterias*).

Koehler 1917:20; Döderlein 1928:294 (as *Podasterias*).

Range: Kerguelen, Marion I, South Georgia and Shag Rocks, 0-234 m.

octoradiata (Studer, 1885) (with forma eunota Fisher 1940)

Studer 1885:147 (as *Pedicellaster*).

Koehler 1914b:64 (as *Anasterias* but given as n.sp.!).

Fisher 1922b:593; 1930:234 (as *Koehleraster*).

Fisher 1940:254 (as *Diplasterias*) 255 (forma eunota).
Range: South Georgia, 15-55 m.
papillosa Koehler, 1906:21, see *Adelasterias*
radiata (Koehler, 1923)
Koehler 1923:36 (as *Cosmasterias*).
Fisher 1940:257 (as *Diplasterias*).
Range: Shag Rocks, Falkland Plateau [?], 160-177 m.
steineni: Perrier, 1891, see *Neosmilaster*
sulcifera: Perrier, 1891c:77, = *Cosmasterias lurida* (Philippi, 1858)
according to Madsen (1956)
turqueti Koehler, 1906, see *Cryptasterias*
*vesiculosa (Sladen, 1889) Fisher 1930:231 (subgenus *Bathyasterias*), 232.
'Challenger' station almost certainly wrong and not Arafura Sea, N. of
Australia but probably 151, near Kerguelen and vesiculosa a synonym
of *Diplasterias meridionalis* (Perrier, 1875) (also with six arms. **New
observation**
DISTOLASTERIAS Perrier, 1896
Perrier 1896:34; Fisher 1928a:102.
Type species: *Asterias (Stolasterias) stichantha* Sladen, 1889.
chelifera Verrill, 1914a:185. Reduced to a subspecies of
Lethasterias nanimensis (Verrill, 1914) by Fisher (1928).
dubia: Fisher, 1919, see *Australiaster*
elegans Djakonov, 1931
Djakonov 1931:67; 950a:116; Chang & Liao 1964:71; Baranova
1971:257.
Range: Japan Sea, N China, 4-68 m.
euplecta: (Fisher, 1919), see *Sclerasterias*
eustyla: Fisher, 1919, see *Sclerasterias*
hypacantha Fisher 1917c:92. Reduced to a forma of *Sclerasterias euplecta*
(Fisher, 1906) by Fisher (1928).
mazophora: Fisher, 1919, see *Sclerasterias*
mollis: Fisher, 1919, see *Sclerasterias*
nanimensis: Fisher, 1923, see *Lethasterias*
nipon (Döderlein, 1902) (with synonym *Distolasterias tricolor*
Djakonov, 1927)
Döderlein 1902:234; Uchida 1928:798 (as *Asterias*).
Fisher 1928a:103; Djakonov 1950a:116; Chang & Liao 1964:71; Rho
1971:71; Baranova 1971:256; A.M. Clark 1982:496.
Range: Japan Sea, N China, 30-150 m.
robusta (Ludwig, 1905)
Ludwig 1905:228 [non 1908:350 cited by Fisher (1928)] (as *Stolasterias*).
Verrill 1914a:179 (as *Stylasterias*).

Fisher 1928a:109 (as *Distolasterias*).
Range: Galapagos Is, 704 m.

stichantha (Sladen, 1889)
Sladen 1889:586 (as *Asterias (Stolasterias)*).
Perrier 1896:34 (as *Distolasterias*).
Range: Japan, 630 m.

tricolor Djakonov, 1927:315. A synonym of *Distolasterias nipon* (Döderlein, 1902) according to Fisher (1928a).

ENDOGENASTERIAS Djakonov, 1938:844. As subgenus of *Leptasterias* Verrill, 1866, for *Leptasterias groenlandica* (Steenstrup, (1857) and *L. siberica* Djakonov, 1930, but neither designated as type species, so not strictly valid. See under *Leptasterias*.

EOLEPTASTERIAS Djakonov, 1938:814. As subgenus of *Leptasterias* Verrill, 1866, for 7 species 'which stand very near *L. ochotensis* (Brandt, 1851)' which may stand as type validation. See under *Leptasterias*.

EREMASTERIAS Fisher, 1930, for type species *Pisaster antarcticus* Koehler, 1917. A synonym of *Anasterias* Perrier, 1875, according to A.M. Clark (1962).

antarctica: Fisher, 1930:323. A synonym of *Anasterias perrieri* (E.A. Smith, 1876) according to A.M. Clark (1962).

*robusta Mortensen, 1941:4, from Tristan da Cunha. Left in limbo by A.M. Clark (1962); position uncertain.

EUSTOLASTERIAS Fisher, 1923
Fisher 1923:255.
Type species: *Coscinasterias (Distolasterias) euplecta* Fisher, 1906.
A synonym of *Sclerasterias* Perrier, 1891 according to Fisher (1928a).

euplecta: Fisher, 1923, see *Sclerasterias*
mollis: Fisher, 1923, see *Sclerasterias*
stenactis: H.L. Clark 1926:23, see *Sclerasterias*

EVASTERIAS Verrill, 1914
Verrill 1914a:51,151; Fisher 1930:139.
Type species: *Asterias troscheli* Stimpson, 1862).

acanthostoma (Verrill, 1909). Reduced to a forma of *E. troscheli* by Fisher (1930) and Djakonov (1950).

alveolata Verrill, 1914, with synonym *Asterias saanichensis* de Loriol, 1897. A forma of *E. troscheli*.

derjugini (Djakonov, 1938)
Djakonov 1938:842 (as *Leptasterias*).
Djakonov 1950a:131; 1958b:322 (as *Evasterias*).
Range: Peter the Great Bay, Japan Sea, 25 m.

echinosoma Fisher, 1926
Fisher 1926:2; 1930:152; Hayashi 1943:229; Djakonov 1950a:131;

1950c:90; Baranova 1957:128; Djakonov 1958a:263 (with forma tro-scheliformis nov.)

Range: Japan, Okhotsk and Bering Seas, 4-112 m.

retata Djakonov, 1950a. A forma of *E. retifera* below

retifera Djakonov, 1938 (with forms retata, retifera and tabulata Dja-konov)

Djakonov 1938:444; 1950a:132; 1950c:79; Baranova 1957:177.

Range: Japan, Okhotsk and Bering Seas, 0-68 m.

tabulata Djakonov, 1938. A forma of *E. retifera*.

troscheli (Stimpson, 1862) (with synonyms *Asterias saanichensis* de Loriol, 1897, *Leptasterias macouni* Verrill, 1914, *L.*? *inaequalis* Verrill, 1914 and probably also *Asterias epichlora* Brandt, 1835, and forms troscheli (Stimpson, 1862), acanthostoma and alveolata (Verrill, 1909))

Stimpson 1862:267 (as *Asterias*).

Verrill 1914a:151; Fisher 1926:1; 1930:139; Djakonov 1950a:130; 1950c:73; Baranova 1957:178; Shin 1995a:120 (as *Evasterias*).

Range: Kamtchatka and Korea to California, 0-70 m.

GASTRASTER Perrier, 1894

Perrier 1894:103; Fisher 1930:206; Mortensen 1927:137.

Type species: *Pedicellaster margaritaceus* Perrier *in* Milne-Edwards, 1882.

A synonym of *Neomorphaster* Sladen, 1889 according to Downey *in* Clark & Downey (1992).

margaritaceus (Perrier *in* Milne-Edwards, 1882), see *Neomorphaster* (Neomorphasteridae)

*studeri de Loriol, 1904:34.

Range: Port San Antonio, Argentina.

Left in limbo by the synonymy of *Gastraster* with *Neomorphaster*; posi-tion uncertain.

GRANASTER Perrier, 1894 (with synonym *Hemiasterias* Verrill, 1914)

Perrier 1894:129; Fisher 1940:264.

Type species: *Stichaster nutrix* Studer, 1885.

biseriatus Koehler, 1906:11. A synonym of *Granaster nutrix* (Studer, 1885) according to Fisher (1940).

nutrix (Studer, 1885) (with synonym *G. biseriatus* Koehler, 1906)

Studer 1885:1554 (as *Stichaster*).

Perrier 1894:129; Koehler 1912:29; Fisher 1940:264; Bernasconi 1956:12; 1967:13 (as *Granaster*).

Range: South Georgia to Antarctic Peninsula, 0-250 m.

HEMIASTERIAS Verrill, 1914

Verrill 1914a:362.

Type species: *Granaster biseriatus* Koehler, 1906.

A synonym of *Granaster* Perrier, 1894 according to Fisher (1940).

biseriata: Verrill, 1914:362. A synonym of *Granaster nutrix* (Studer, 1885) according to Fisher (1940).

HEXASTERIAS Fisher, 1930. A subgenus of *Leptasterias* Verrill, 1866

ICASTERIAS Fisher, 1923

Fisher 1923:601; 1930:210.

Type species: *Asterias panopla* Stuxberg, 1879.

panopla (Stuxberg, 1879)

Stuxberg 1879:32; Danielssen & Koren 1884:17; Döderlein 1900:204 (as Asterias}).

Fisher 1930:210; Mortensen 1932:20; Djakonov 1933:64; 1950a:114; Smirnov 1990:444 (as *Icasterias*).

Range: Arctic Atlantic: Greenland, Spitzbergen, Barents and Kara Seas, 18-560 m.

KALYPTASTERIAS Koehler, 1923

Koehler 1923:43.

Type species: *Kalyptasterias conferta* Koehler, 1923.

A synonym of *Anasterias* Perrier, 1875 according to Fisher (1940).

conferta Koehler, 1923:43. A synonym of *Anasterias spirabilis* (Bell, 1881) according to A.M. Clark (1962a).

KENRICKASTER A.M. Clark, 1962

A.M. Clark 1962a:80-81.

pedicellaris A.M. Clark, 1962

A.M. Clark 1962a:81.

Range: [?Enderby or MacRobertson Land, Antarctica.]

KOEHLERASTER Fisher, 1922

Fisher 1922b:596; 1930:234.

Type species: *Anasterias octoradiata* Koehler, 1914.

A synonym of *Diplasterias* Perrier, 1891 according to Fisher (1940).

octoradiatus: Fisher, 1922:593, see *Diplasterias*

LEPTASTERIAS Verrill, 1866 (with synonym *Ctenasterias* Verrill, 1914; subgenera *Hexasterias* and *Nesasterias* Fisher, 1930, also *Eoleptasterias* Djakonov, 1938 - though *Endogenasterias* Djakonov not strictly valid, no type designation)

Verrill 1866:350; 1914a:116; Fisher 1930:23; Djakonov 1938:749-764; 1950a:133; Official List of ICZN 1940 [?1942]: ; A.M. Clark *in* Clark & Downey 1992:432.

Note: As with *Pteraster* (see part 3), all the species are treated below under the generic heading rather than the subgeneric, pending re-evaluation, especially of the incredible multiplicity of taxa described by Djakonov (1929 and 1938a).

LEPTASTERIAS (LEPTASTERIAS) Verrill 1866 [To end with other subgenera?]

Verrill 1866:350; Fisher 1930:3,23; Djakonov 1938a:763 (emended).

Type species: *Asteracanthion muelleri* M. Sars, 1846.

acervata (Stimpson, 1862). A subspecies of *L. polaris* (Müller & Troschel, 1842)

aequalis (Stimpson, 1862:273) Fisher, 1930:120. A synonym of *L. hexactis* (Stimpson, 1862) according to Chia (1966).

alaskensis (Verrill, 1909) (subgenus *Hexasterias*; with subspp. *asiatica* and *multispina* Fisher, 1930, and forms *shumaginensis* and *pribilofensis* Fisher)

alaskensis alaskensis (Verrill, 1909)

Verrill 1909b:549 (as var. of *Asterias epichlora* Brandt).

Verrill 1914a:136 (as *Leptasterias epichlora alaskensis*).

Fisher 1930:124 (emended, non *epichlora* (Brandt), as species of *Leptasterias* with forms *alaskensis, shumaginensis* and *pribilovensis*); Baranova 1957:184.

Range: Alaska, Pribilov and Shumagin Is, 0-10 m.

alaskensis asiatica Fisher, 1930

Fisher 1930:131; Djakonov 1938:901; 1950a:151; Baranova 1957:184; Hayashi 1973:110.

Range: Kurile Is, Bering Sea, Tatar Strait, 0-28 m.

alaskensis multispina Fisher, 1930

Fisher 1930:133.

Range: Alaska to northern Vancouver I.

aleutica Fisher, 1930

Fisher 1930:101; Baranova, 1957:185.

Range: Aleutian Is.

aphelonota Fisher, 1930. A forma of *L. polaris acervata* (Stimpson, 1862)

***arctica** (Murdoch, 1885) (conspecific with *Asterias nautarum* Bell, 1883) a long-forgotten name which needs to be supressed; see under *Asterias nautarum*; possible synonym *L. islandica* (Levinsen, 1887); forms *beringensis* Fisher, 1930, *arctica* and *glomerata* Djakonov, 1938)

Murdoch 1885:159 (as *Asterias*).

Verrill 1914a:120; Fisher 1930:24, 227 (forma *beringensis*); Djakonov 1930:46; 1938:764, 773 (forma *arctica*, 775 (f. *glomerata*); 1950a:139; Baranova 1957:179 (as *Leptasterias*).

Range: N Alaska, Bering Sea, 0-80 m.

asiatica Fisher, 1930, as subspecies of L. alaskensis

asteira Fisher, 1930

Fisher 1930:103; Baranova 1957:186.

Range: Aleutian Is, Bering Sea, 0-26 m.

austera (Verrill, 1895)

Verrill 1895:209 (as *Asterias* perhaps *Leptasterias*).

H.L. Clark 1904:555 (as *Asterias*).

Fisher 1930:207; Gosner 1971:577; A.M.C. *in* Clark & Downey 1992:435 (as *Leptasterias*).

Range: Off Cape Cod, possibly S to Rhode Island, 33-64 m.

bartletti A.H. Clark, 1940 (Leptasterias *Hexasterias*)

'Not readily distinguishable from *L. polaris* (Müller & Troschel, 1842)' according to Grainger (1966:50).

beringensis Fisher, 1930, as forma of *L. arctica*

camtschatica (Brandt, 1835) (subgenus *Hexasterias*; with synonym *Asterias multiclava* Verrill, 1914a) subsp. *dispar* Verrill, 1914 and forma *nitida* Fisher, 1930)

Brandt 1835:270; Ludwig 1900:485 (as *Asterias*).

Brandt 1851:32 (as *Asteracanthium*).

Fisher 1930:91, 94 (*dispar* as forma), 95 (forma *nitida*); Djakonov 1938:889 (*dispar* as subsp.); 1950:151; Baranova 1957:185.

Range: Kamchatka, Bering Sea, Aleutians and Kurile Is, 0-27 m.

camtschatica dispar Verrill, 1914

Verrill 1914a:142. See above.

***canuti** Heding, 1936

Heding 1936:13.

Range: SE Greenland

carinata Heding, 1935. A forma of *L. hyperborea* (Danielssen & Koren).

clavispina Heding, 1936

Heding 1936:14

Both overlooked by A.M.C. *in* Clark & Downey (1992).

coei (*coei*) Verrill, 1914 (subgenus *Hexasterias*); with subspp. *truculenta* Fisher, 1930 and *shantarica* Djakonov, 1938.)

coei coei Verrill, 1914

Verrill 1914a:123; Fisher 1930:84.

Range: Alaska, 18-186 m.

coei shantarica Djakonov, 1938

Djakonov 1938:881; 1950a:149.

Range: NW Ochotsk Sea.

coei truculenta Fisher, 1930

Fisher 1930:186; Baranova 1957:184.

Range: Aleutian Is, Barents Sea, 79-102 m.

compta (Stimpson, 1862)

Stimpson 1862:270 (as *Asterias*).

Verrill 1866:340 (as *Asterias* (*Asteracanthion*)).

Verrill 1879:14; 1885b:540 (as (?*Leptasterias*).

Sladen 1889:583 (as *Asterias* (*Leptasterias*)).

Fisher 1930:208 (as *Leptasterias tenera* forma *compta*).

A.M. Clark *in* Clark & Downey 1992:436 (as *Leptasterias*).

Range: South of Halifax, Nova Scotia to New Jersey, 32-275 m.

compacta and concinna Verrill 1914:130, 132 as varieties of *L. aequalis* (Stimpson, 1862), a synonym of *L. hexactis* (Stimpson, 1862).

confinis Djakonov, 1950. A forma of *L. polaris katherinae* (Gray)

crassa Heding, 1935. A variety of *L. floccosa* (Levinsen, 1903)

cribraria (Stimpson, 1862) Djakonov 1930:41. A synonym of *L. groenlandica* (Steenstrup, 1857) according to Fisher (1930) and Djakonov (1950a).

danica (Levinsen, 1887)

Levinsen 1887:393 (as *Asterias Mülleri* var. *danica*).

Heding 1935:49 (neotype; raised to specific rank); A.M. Clark *in* Clark & Downey 1992:436 (as *Leptasterias*).

Range: Southern Kattegat, between Denmark and Sweden.

degerboelli Heding, 1935

Heding 1935:56; 1936:12.

Range: SE Greenland.

derbeki Djakonov, 1938 (subgenus *Eoleptasterias*, with subspecies *tatarica*)

derbeki derbeki Djakonov, 1938

Djakonov 1938a:829; 1950a:144.

Range: Okhotsk Sea. 31-500 m.

derbeki tatarica Djakonov, 1938

Djakonov 1938a:8291 1950a:144.

Range: Tatar Strait and Okhotsk Sea, 66-73 m.

derjugini Djakonov, 1938 (subgenus *Eoleptasterias*)

Djakonov 1938a:842.

Range: Japan Sea, 25 m.

dispar Verrill, 1914. A subspecies of *L. camtschatica* Brandt, 1835.

epichlora (Brandt, 1835) (to *Leptasterias* according to Verrill, 1914a:132).

Probably a synonym of *Evasterias troscheli* forma *alveolata* Verrill, 1914, according to Fisher (1930).

epichlora: Verrill 1914, non *Asterias epichlora* Brandt, 1835, partly = *L. hexactis* formae *regularis* and *aspera* and partly *L. alaskensis* according to Fisher (1930).

fascicularis (Perrier, 1881) Verrill, 1915, see *Tarsaster* (Pedicellasteridae)

fisheri Djakonov, 1929 (with subspecies *meridionalis* Djakonov, 1938)

fisheri fisheri Djakonov, 1929

Djakonov 1929a:283; Fisher 1930:42; Djakonov 1938a:801;1950a:142.

Range: Okhotsk Sea and Tatar Strait, 40-160 m.

fisheri meridionalis Djakonov, 1938

Djakonov 1938a:806.

Range: Japan Sea, 10-20 m.

floccosa (Levinsen, 1887)

Levinsen 1887:392 (as *Asterias Mulleri* var. *floccosa*).

Heding 1935:54 (as *L. floccosa*, with var. *crassa* nov.).

Range: Baffin Bay, NE Greenland.

floccosoides Heding, 1935. A forma of *L. hyperborea* (Danielssen & Koren)

glomerata Djakonov, 1938. A forma of *L. arctica* (Murdoch, 1885).

granulata Djakonov, 1938 (subgenus *Eoleptasterias*)

Djakonov 1938a:836; 1950a:145.

Range: NW Okhotsk Sea, 53-157 m.

groenlandica (Steenstrup, 1857) (with synonyms *A. inermis* Bell, 1881, *A. spitzbergensis* Danielssen & Koren, 1881 and *L. obtecta* Verrill, 1914; also *Asterias cribraria* Stimpson, 1862 but a distinct forma according to Fisher (1930))

Steenstrup 1857:228; Duncan & Sladen 1881:27 (as *Asteracanthion*). Stimpson 1863:142; Döderlein 1900:202; Ludwig 1900:482 (as *Asterias*).

Verrill 1879:151; 1895:210; Fisher 1930:45; Djakonov 1930:38; Mortensen 1932:21; Djakonov 1938a:844 and f. *cribraria*; 1950a:146; Baranova 1957:181; Grainger 1964:44; A.M. Clark *in* Clark & Downey 1992:433.

Range: Circum-North Polar, possibly S in Atlantic to Nova Scotia - see A.M.C. (1992), 5-276 m.

hartii Rathbun, 1879, see *Allostichaster*

hexactis (Stimpson, 1862) (subgenus *Hexasterias*) (with synonyms *Asterias aequalis* Stimpson, 1862, along with *L. compacta* and *L. concinna* Verrill, 1914 and *aequalis* forma *nana* Verrill, 1914; subspp. *vancouveri* (Perrier, 1875) and *occidentalis* Djakonov, 1938)

hexactis hexactis (Stimpson, 1862)

Stimpson 1862:82 (as *Asterias*.

Fisher 1930:114; Lambert 1981:114; Maluf 1988:46,126 (as *Leptasterias*).

Range: South of Vancouver Island to Catalina Island, S California, intertidal - 3 m.

hexactis occidentalis Djakonov, 1938

Djakonov 1938a:897; 1950a:150.

Range: Okhotsk Sea.

hexactis vancouveri (Perrier, 1875)

Perrier 1875:328 (as *Asterias*).

Verrill 1914a:125; Fisher 1930:115 (as *Leptasterias*).

Range: Vicinity of Vancouver Island.

hirsuta Djakonov, 1938

Djakonov 1938a:785; 1950a:140.

Range: Okhotsk Sea, 17-106 m.

hispidella Verrill, 1895

Verrill 1895:210; Fisher 1930:207; A.M. Clark *in* Clark & Downey 1992:437.

Range: Off Nova Scotia, 91 m.

hylodes Fisher, 1930 (with subsp. *reticulata* Djakonov, 1938)

hylodes hylodes Fisher, 1930

Fisher 1930:35; Djakonov 1938a:780.

Range: S Bering Sea, Aleutian Is, S Alaska, 45-125 m.

hylodes reticulata Djakonov, 1938

Djakonov 1938a:781; 1950a:140; Baranova 1957:180.

Range: W side of Bering Sea.

hyperborea (Danielssen & Koren, 1882) (with synonym *Asterias normani* Danielssen & Koren, 1883 with formae *carinata*, *floccosoides* and *intermedia* all of Heding, 1935)

Danielssen & Koren 1882:269; 1884b:10; Ludwig 1900:484 (as *Asterias*).

Fisher 1930:207; Heding 1935:44 (formae).

Range: Spitzbergen, Bear Island, Faeroe Is, formae from S Greenland and Iceland.

A subspecies of *L. muelleri* (M. Sars, 1846) according to Djakonov (1930, 1933) but a valid species according to Heding (1935).

inaequalis Verrill, 1914a:117. A synonym of *Evasterias troscheli* according to Fisher (1930).

intermedia Heding, 1935. A forma of *L. hyperborea* (Danielssen & Koren)

intermedia Djakonov, 1938b:867 (a homonym of the above). A forma of *L. polaris acervata* (Stimpson, 1862)

insolens Djakonov, 1938

Djakonov 1938a:807; 1950a:142.

Range: Tatar Strait, Okhotsk Sea, 100-128 m.

*islandica Levinsen, 1877 (with variety *gracilis* Heding, 1935)

Levinsen 1877:393 (as *Asterias Mulleri* var. *islandica*).

Heding 1935:47 (as species, with var. *gracilis*); A.M.C. *in* Clark & Downey 1992:433 (in key), 434 (remarks).

Range: Iceland and the Faeroe Islands.

A synonym of *L. arctica* (Murdoch, 1885) according to Djakonov (1950a), apparently overlooked by A.M.C. (1992).

kussakini Baranova, 1962

Baranova 1962:348.

Range: Kurile Is.

leptalea Verrill, 1914

Verrill 1914a:119; Fisher 1930:44 (type lost).

Range: Alaska.

leptodoma Fisher, 1930 (subgenus *Hexasterias*)

Fisher 1930:105; Djakonov 1938a:899; 1950a:150; Baranova 1957:184.

Range: Aleutian Is, Commander Is, mid-low water.

littoralis (Stimpson, 1853)

Stimpson 1853:14 (as *Asteracanthion*).

Verrill 1866:349; Ganong 1888:39; H.L. Clark 1905:4 (as *Asterias*).

Verrill 1879:14; 1895:211; Fisher 1930:207; Gosner 1971:577; A.M.C. *in* Clark & Downey 1992:438 (as *Leptasterias*).

Range: Bay of Fundy, Gulf of St Lawrence S to Maine, intertidal - 42 m.

macouni Verrill, 1914a:124. A synonym of *Evasterias troscheli* (Stimpson, 1862) according to Fisher (1923).

meridionalis Djakonov, 1938a. Subspecies of *L. fisheri* Djakonov, 1929.

mexicana (Lütken, 1860 [1859 in part 1 references])

Lütken 1860:94 (as *Asteracanthion*).

Verrill 1867:344 (as *Asterias*).

Verrill 1915:24 (as ? *Leptasterias*, or young *Asterias*).

A.M.C. *in* Clark & Downey 1992:434 (under *Leptasterias*).

Range: 'Vera Cruz, (atlantic coast of) Mexico'.

Verrill (1915) doubted the generic affinity of the type material and A.M.C. (1992) thought the locality incompatible.

muelleri (M. Sars, 1846) (with subsp. *celtica* A.M.C. in Clark & Downey, 1992 – conspecific with *Asterias hispida* Pennant, 1777, suppressed; var. *nobilis* Heding, 1935)

muelleri muelleri (M. Sars, 1846)

M. Sars 1846:56; 1861:88 (as *Asteracanthion*; *A. muelleri* M. Sars, 1844 a nomen nudum, undescribed).

Norman 1865:127; Danielssen & Koren 1884b:21 (as *Asterias*).

Verrill 1866:350; Mortensen 1927:141; Brun 1970:238, Opinion ICZN 984 1972:115 (on Official List of Specific Names); Fisher 1930:208; Heding 1935:41 (and var. *nobilis*); Djakonov 1950a:138; A.M.C.

In Clark & Downey 1992:439 (as *Leptasterias*).

Sladen 1889:582 (as *Asterias (Leptasterias)*).

Range: Southern Norway to the Øresund, E Denmark, and Dogger Bank (mid-North Sea), ?Shetlands, E of Faeroe Is., North Norway and Murman coast, intertidal to ?140 m.

muelleri celtica A.M. Clark *in* Clark & Downey 1992 (nom. nov. for *Asterias hispida* Pennant, 1777, suppressed)

Bell 1881b:510 (*Asterias muelleri*).

Mortensen 1927:141 (*L. muelleri*).

A.M.C. *in* Clark & Downey 1992:440.

Range: Northern Ireland, Scotland and NE England, intertidal.

Brun's premature rejection through the ICZN of the name *Asterias hispida* Pennant, 1777 necessitated a new name for British specimens believed by A.M.C. to be subspecifically distinct from *L. muelleri* (M. Sars) of Scandinavia.

obtecta Verrill, 1914a:144. A synonym of *L. groenlandica* (Steenstrup, 1857), according to Fisher (1930).

ochotensis (Brandt, 1851) (could be constructed as type species of subgenus *Eoleptasterias* Djakonov, 1938)
> Brandt 1851:28; Perrier 1869: (as *Asteracanthion*).
> Fisher 1930:57; Djakonov 1938a:815; 1950a:143.
> Range: Okhotsk Sea, 0-30 m.

orientalis Djakonov, 1929 (with subsp. *japonica* Djakonov, 1938)
orientalis orientalis Djakonov, 1929
> Djakonov 1929b:277; Fisher 1930:40; Djakonov 1938a:791; 1950a:141; 1958b:324.
> Range: Okhotsk Sea to Sea of Japan, 16-161 m.

orientalis japonica Djakonov, 1938
> Djakonov 1938a:796.
> Range: Japan Sea, 37-128 m.

*polaris (Müller & Troschel, 1842) (type species of subgenus *Hexasterias* Fisher, 1930; synonyms: *Asterias borealis* Perrier, 1875 and possibly *L. bartletti* A.H. Clark, 1940; subspp.: *acervata* (Stimpson, 1862); *katherinae* (Gray, 1840*) and *ushakovi* Djakonov, 1938 and formae *aphelonota* Fisher 1930, *confinis* Djakonov, 1938, *intermedia* Djakonov, 1938 and *polythela* (Verrill, 1909))
> *Fisher and Djakonov's treatment of *Asterias katherinae* Gray, 1840 as a subspecies of *L. polaris* Müller & Troschel, 1842, renders the latter a junior synonym. Since *polaris* is such a widely-used name there is a good case for suppression of *katherinae* by the ICZN.

polaris polaris (Müller & Troschel, 1842)
> Müller & Troschel 1842:16; Duncan & Sladen 1881:23 (as *Asteracanthion*).
> Lütken 1857:28; Stimpson 1862:271; Verrill 1866:356; 1895:208; Mortensen 1932a:22 (as *Asterias*).
> Fisher 1923:599; Djakonov 1950a:148; Tortonese 1955:681; Grainger 1966:48; Gosner 1979:262 (as *Leptasterias*).
> Fisher 1930:60; A.M.C. *in* Clark & Downey 1992:441 (as *Leptasterias (Hexasterias)*).
> Range: Greenland*, Circum-N polar, S to George's Bank off Cape Cod in W Atlantic, and to Okhotsk Sea and Washington State in Pacific, 0-360 m.

polaris acervata (Stimpson, 1862) (with forms: *aphelonota* Fisher, 1930, *intermedia* Djakonov, 1938a and *polythela* (Verrill, 1909))
> Stimpson 1862:271 (as *Asterias*).

Fisher 1930:66 (and forms); Djakonov 1938a:867 (f. *intermedia*); Baranova 1957:182 (forms),

Range: Okhotsk and Bering Seas, S Alaska, 3-150 m.

*polaris katherinae (Gray, 1840) (with synonym *Pisaster grayi* Verrill, 1914 and forma *confinis* Djakonov, 1938)

Gray 1840:179 (as *Asterias katherinae*).

Fisher 1930:77 (as subsp.); Djakonov 1938a:876 (with forma *confinis*).

See remarks under *polaris* above.

polaris ushakovi Djakonov, 1938

Djakonov 1938a:870.

Range: W Okhotsk Sea.

polymorpha Djakonov, 1938 (subgenus *Hexasterias*)

Djakonov 1938a:876; 1950a:149.

Range: Okhotsk Sea, 66-100 (?183) m.

polythela: Fisher, 1930:255. A forma of *L. polaris acervata* (Stimpson).

pribilofensis Fisher, 1930. A forma of *L. alaskensis* (Verrill, 1909).

pusilla Fisher, 1930

Fisher 1930:118.

Range: Monterey Bay, California.

schmidti Djakonov, 1938 (subgenus *Hexasterias*)

Djakonov 1938a:885; 1950a:148.

Okhotsk Sea, 30-68 m.

shantarica Djakonov, 1938a. A subspecies of *L. coei* (Verrill, 1909).

shumaginensis Fisher, 1930. A forma of *L. alaskensis* Verrill, 1909.

siberica Djakonov, 1930

Djakonov 1930:44; 1933:70; 1938a:855; 1952b:294.

Range: Bering Sea (c.67°N, 175°W), 16 m.

similispinis (H.L. Clark, 1908) (subgenus *Eoleptasterias*)

H.L. Clark 1908:288 (as *Asterias*).

Fisher 1930:59 (as subsp. of *L. ochotensis*).

Djakonov 1938a:821; 1938b:446; 1950a:143 (as species).

Range:NE Hokkaido, Japan, Tatar Strait, 0-10 m.

squamulata Djakonov, 1938 (subgenus *Eoleptasterias*)

Djakonov 1938a:839; 1950a:145; Baranova 1957:181.

Range: Commander Is, Bering Sea, littoral.

stolacantha Fisher, 1930 (subgenus *Nesasterias*)

Fisher 1930:135; Baranova 1957:186.

Range: Unalaska, Aleutian Is, Bering Sea, 64-108 m.

subarctica Djakonov, 1938

Djakonov 1938a:798; 1950a:141.

Range: SW Okhotsk Sea, 30-50 m.

tatarica Djakonov, 1938a. A subspecies of *L. derbeki* Djakonov, 1938.

tenera (Stimpson, 1862) (with synonym *Asteracanthion flaccida* A. Agassiz, 1863) Stimpson 1862:269; H.L. Clark 1904:554(pt); Coe 1912:63 (as *Asterias*).

Verrill 1866:349 (as *Asterias (Leptasterias)*).

Verrill 1873b:504; Ganong 1888:40; Gray et al. 1968:160; Gosner 1971:583; Franz et al. 1981:404; A.M.C. *in* Clark & Downey 1992:441 (as *Leptasterias*).

Range: Massachusetts Bay*, Nova Scotia and the Gulf of Maine to Cape Hatteras, 18-180 m.

truculenta Fisher, 1930. A subspecies of *L. coei* Verrill, 1914.

ushakovi Djakonov, 1938. A subspecies of *L. polaris* (Müller & Troschel, 1842)

vancouveri (Perrier, 1875): Verrill, 1914. A subspecies of *L. hexactis* (Stimpson, 1862) according to Fisher (1930).

vinogradovi Djakonov, 1938 (subgenus *Eoleptasterias*)

Djakonov 1938b:826; 1950a:144.

Range: SE Kamtchatka, 0-10 m.

LEPTASTERIAS (ENDOGENASTERIAS) Djakonov, 1938

Djakonov, 1938b:844 (for *L. groenlandica* (Steenstrup, 1857) and *L. siberica* Djakonov, 1930 but neither designated as type species so *Endogenasterias* not strictly valid.

LEPTASTERIAS (EOLEPTASTERIAS) Djakonov, 1938.

Djakonov 1938b:814.

Type species: *Asteracanthion ochotensis* Brandt, 1851.

Other species included: *L. derbeki, L. derjugini* and *L. granulata*, all of Djakonov, 1938, *L. similispinis* (H.L. Clark, 1908), *L. squamulata* and *L. vinogradovi* Djakonov, 1938.

LEPTASTERIAS (HEXASTERIAS) Fisher, 1930

Fisher, 1930:59; Djakonov 1938b:856.

Type species: *Asteracanthion polaris* Müller & Troschel, 1842.

Other species included: *alaskensis* (Verrill, 1909),

camtschatica (Brandt, 1835), *coei* Verrill, 1914, *hexactis* (Stimpson, 1862), *leptodoma* Fisher, 1930, *polymorpha* and *schmidti* Djakonov, 1938.

LEPTASTERIAS (NESASTERIAS) Fisher, 1930

Fisher 1930:135.

Type and only species: *Leptasterias stolacantha* Fisher, 1930

LETHASTERIAS Fisher, 1923

Fisher 1923:258; 1928:131.

Type species: *Asterias nanimensis* Verrill, 1914.

acutispina Hayashi, 1973

Hayashi 1973a:12; 1973b:103.

Range: Sagami Bay, SE Japan, 160-230 m.

australis Fisher, 1940

Fisher 1940:223; A.M. Clark 1962a:

Range: Between Falkland Is and Argentina, 155 m.

fusca Djakonov, 1931

Djakonov 1931:79; 1950a:118; 1958:320.

Range: N Japan Sea, Tatar Strait, 16-40 m.

nanimensis (Verrill, 1914) (with subspecies *beringiana* Djakonov, 1931 and *chelifera* (Verrill, 1914))

nanimensis nanimensis (Verrill, 1914)

Verrill 1914a:105 (as *Asterias*).

Fisher 1923:258 (as *Distolasterias* and *Lethasterias*).

Fisher 1928a:132 (as *Lethasterias*).

Range: Nanaimo, Vancouver I, Washington, 46-73 m.

nanimensis beringiana Djakonov, 1931

Djakonov 1931:74.

Range: Bering Sea.

nanimensis chelifera (Verrill, 1914)

Verrill 1914a: .

Fisher 1928a:132; Baranova 1957:177; Djakonov 1958b:319; Hayashi 1973b:103; Imaoka et al. 1991:104 (as subsp. of *Lethasterias nanimensis*).

Range: Okhotsk and Japan Seas, 0-247 m.

LYSASTERIAS Fisher, 1908 (with synonym *Paedasterias* Verrill, 1914)

Fisher 1908a:88; 1922b:594; 1930:235; 1940:239; A.M. Clark 1962a:89; H.E.S. Clark 1963b:72.

Type species: *Anasterias perrieri* Studer, 1885 (not Perrier, 1891)

adeliae (Koehler, 1920)

Koehler 1920:26 (as *Anasterias*).

Fisher 1930:236; 1940:244; H.E.S. Clark 1963b:72 (as *Lysasterias*).

Range: Adelie Land, Palmer Archipelago, Antarctic Peninsula, 22-235 m.

belgicae (Ludwig, 1903)

Ludwig 1903:51 (as *Anasterias*).

Fisher 1930:236; Bernasconi 1956:16 (affinity with *L. joffrei*);

Jangoux & Massin 1986:90 (syntypes)(as *Lysasterias*).

Range: Bellingshausen Sea, South Shetland Is.

chirophora (Ludwig, 1903)

Ludwig 1903:43 (as *Anasterias*).

Verrill 1914a:371 (as *Paedasterias*).

Fisher 1930:236; Jangoux & Massin 1986:90 (type material) (as *Lysasterias*).

Range: Bellingshausen Sea, 450-56- m.

cupulifera (Koehler, 1908) Fisher, 1930:236. A synonym of *L. perrieri* (Studer, 1885) according to Fisher (1940).

digitata A.M. Clark, 1962
A.M. Clark 1962a:91.
Range: Enderby and MacRobertson Lands, Antarctica, 163-300 m.

***hemiora** Fisher, 1940
Fisher 1940:245.
Validity in relation to *L. joffrei* (Koehler, 1920) questioned by Bernasconi (1956:16).
Range: Palmer Archipelago, Antarctic Peninsula, 70 m.

heteractis Fisher, 1940
Fisher 1940:247.
Range: Clarence I, Antarctic, 342 m.

joffrei (Koehler, 1920) (with possible synonym *L. hemiora* Fisher, 1940)
Koehler 1920:30 (as *Paedasterias*).
Fisher 1930:236; 1940:245; Bernasconi 1956:15 (cf. *hemiora*);
A.M. Clark 1962a:93; H.E.S. Clark 1963:74 (as *Lysasterias*).
Range: Adelie Land, Ross Sea, South Shetland Is, 163-810 m.

lactea (Ludwig, 1903) Ludwig 1903:50 (as *Anasterias*).
Fisher 1930:236; Jangoux & Massin 1986:91 (holotype) (as *Lysasterias*).
Range: Bellingshausen Sea.

perrieri (Studer, 1885) (with synonyms *Anasterias cupulifera* Koehler, 1908, *A. lysasteria* Verrill, 1914, *A. tenera* Koehler, 1906 and *A. victoriae* Koehler, 1920; possibly also *Sporasterias* pedicellaris} Koehler, 1923)
Studer 1885:153 (as *Anasterias*).
Fisher 1930:236; 1940:241; Bernasconi 1967b:11 (as *Lysasterias*).
Range: Probably Circum-S Polar, including South Georgia, ? - 650 m.

studeri: Fisher, 1930, see *Anasterias*

tenera (Koehler, 1906) Fisher 1930:236. A synonym of *L. perrieri* (Studer, 1885) according to Fisher (1940)

victoriae (Koehler, 1920) Fisher 1930:237. A synonym of *L. perrieri* (Studer, 1885) according to Fisher (1940).

LYTASTER Perrier, 1894
Perrier 1894:98.
Type species: *Lytaster inaequalis* Perrier, 1894.
A synonym of *Coscinasterias* Verrill, 1870, subgenus *Stolasterias* Sladen, 1889, according to Fisher (1928).

inaequalis Perrier, 1894:98. A synonym of *Coscinasterias tenuispina* (Lamarck, 1815) according to Fisher (1926a, 1928a).

MARGARASTER Gray, 1866
Gray 1866:2.

A synonym of *Uniophora* Gray, 1840. Fisher (1923:597) noted 'at first sight it appears to be a synonym of *Uniophora* but is described as having monacanthid ambulacrals'. However, apart from *Margaraster graniferus* – author given as the pre-linnaean Linck, not Lamarck, 1816 – Gray listed, apparently as a synonym, *Asterias janthina* Brandt, later synonymised with *Pisaster ochraceus* by Fisher (1930), which may have been the source of the error.

graniferus: Gray, 1866:2 = *Uniophora granifera* (Lamarck, 1816).

MARGARASTER: Hutton, 1872, a homonym of *Margaraster* Gray, 1866, = *Astrostole* Fisher, 1923.

scaber Hutton, 1872, see *Astrostole*

MARTHASTERIAS Jullien, 1878

Jullien 1878:141; Koehler 1921:21; 1924:95; Fisher 1928a:129; Tortonese 1937a:104; A.M.C. *in* Clark & Downey 1992:443.

Type species: *Marthasterias foliacea* Jullien, 1878, a synonym of *Asterias glacialis* Linnaeus, 1758.

africana (Müller & Troschel, 1842) H.L. Clark 1923a:306, reduced to a variety of *glacialis* by Mortensen (1933) but a forma by A.M.C. (1974).

foliacea Jullien, 1878:141. A synonym of *Asterias* (i.e. *Marthasterias*) *glacialis* Linnaeus, 1758, according to Sladen (1889).

glacialis (Linnaeus, 1758) (with synonyms *Asterias undulata* O.F. Müller, 1784, *A. spinosa* Pennant, 1777, *A. angulosa* O.F. Müller, 1788, *Stellonia webbiana* D'Orbigny, 1839, *A. madeirensis* Stimpson, 1862 and *Marthasterias foliacea* Jullien, 1878; formae *africana* (Müller & Troschel, 1842) and *rarispina* (Perrier, 1875))

Linnaeus 1758:661; Lamarck 1816:561; Viguier (1878) 1879:100; Ludwig 1897:364 (full references to date and description) (as *Asterias*). Nardo 1834:716 (as *Stellonia*).

Müller & Troschel, 1840a:101; Perrier 1869:28 (as *Asteracanthion*).

Forbes 1841:78 (as *Uraster*).

Perrier 1894:109 (as *Stolasterias*).

Fisher 1906a:1105 (as *Coscinasterias (Marthasterias)*). Verrill 1914a:47,100; Koehler 1921:22; 1924:96; Mortensen 1927:143; Nataf & Cherbonnier 1975:825; A.M. Clark & Courtman-Stock 1976:94 (with forms *africana and rarispina*); Jangoux & De Ridder 1987:89 (syntypes of *africana*); A.M.C. *in* Clark & Downey 1992:443 (as *Marthasterias*).

Range: Scandinavia from Finmark to West Sweden, N and W of British Isles, S to Mediterranean, Canary, Cape Verde and Azores Islands, Annobon in Gulf of Guinea, W and S of Cape Province, South Africa (forms *africana* and *rarispina*) (see A.M.C., 1992).

rarispina (Perrier, 1875) H.L. Clark, 1923:305. A forma of *M. glacialis* (Linnaeus, 1758).

MEYENASTER Verrill, 1913

Verrill 1913:485; Fisher 1928a:130.

Type species: *Asterias gelatinosa* Meyen, 1834.

gelatinosus (Meyen, 1834)

Meyen 1834:222; H.L. Clark 1910:337 (as *Asterias*).

Müller & Troschel 1842:15 (as *Asteracanthion*).

Verrill 1913:485; Fisher 1928a:130; Madsen 1956:36 (as *Meyenaster*).

Range: Chile.

MORTENSENIA da Costa. 1941:11. Type species: *M. lusitanica* da Costa, 1941. A NOMEN NUDUM according to Da Costa (1952), preoccupied by *Mortensenia* Döderlein, 1905 – Echinoidea and renamed *Carlasterias*.

C. lusitanica probably a *Sclerasterias* According to A.M. C. *in* Clark & Downey (1992:415, 449).

NANASTER Perrier, 1894

Perrier 1894:129, 131, 133; 1896:27.

Type species: *Asteracanthion albulus* Stimpson, 1853.

A synonym of *Stephanasterias* Verrill, 1871 according to Fisher (1930).

albulus Perrier, 1894, see *Stephanasterias*

NEOSMILASTER Fisher, 1930

Fisher 1930:237.

Type species: *Asterias georgiana* Studer, 1885.

georgianus (Studer, 1885)

Studer 1885:15 (as *Asterias*).

Perrier 1891a:7 (as *Diplasterias*).

Koehler 1917:26 (as *Podasterias*).

Koehler 1923:40 (as *Ctenasterias*).

Fisher 1930:237; 1940:258; Bernasconi 1956:22; 1967b:12 (as *Neosmilaster*).

Range: South Georgia, South Shetlands, Palmer Archipelago, 0-335 m.

steineni (Studer, 1885)

Studer 1885:152 (as *Asterias*).

Perrier 1891a:84 (as *Diplasterias*).

Koehler 1917:26; 1920:41; 1923:30 (as *Podasterias*).

Fisher 1920:237; 1940:259 (as *Neosmilaster*).

Range: Falkland-Magellan area, South Georgia, 99-160 m.

NESASTERIAS Fisher, 1930. A subgenus of *Leptasterias* Verrill, 1866.

NOTASTERIAS Koehler, 1911 (with synonym *Autasterias* Koehler, 1911)

Koehler 1911a:35; Fisher 1930:243; A.M. Clark 1962a:77; H.E.S. Clark 1963:67.

Type species: *Notasterias armata* Koehler, 1911.

armata Koehler,1911

> Koehler 1911a:39; Fisher 1940:225; A.M. Clark 1962a:79; H.E.S. Clark 1963:67.
>
> Range: McMurdo Sound, Ross Sea and Bellingshausen Sea, probably circum-S polar, 30-647 m.

bongraini (Koehler, 1912)

> Koehler 1912a:152; 1912b:26 (as *Autasterias*).
>
> Fisher 1940:227; Bernasconi 1956:14; A.M. Clark 1962a:79 (as *Notasterias*).
>
> Range: South Shetland Is, Antarctic Peninsula, Kemp and Adelie Lands, 110-830 m.

candicans (Ludwig, 1903)

> Ludwig 1903:41 (as *(Stolasterias)*).
>
> Koehler 1917:30 (as *Coscinasterias*).
>
> Fisher 1940:75 (listed as *Sclerasterias*).
>
> Jangoux & Massin 1986:90 (syntype; as *Notasterias*).
>
> Range: Bellingshausen Sea, Antarctic, 450-560 m.

haswelli Koehler, 1920

> Koehler 1920:70; Fisher 1930:244; A.M. Clark 1962a:78.
>
> Range: Off Queen Mary and Adelie Lands, Ross Sea, 110-647 m.

pedicellaris (Koehler, 1907)

> Koehler 1907a:145 (as *Asterias*).
>
> Koehler 1911a:38 (as *Autasterias*).
>
> Fisher 1940:75 (listed as *Notasterias*).
>
> Range: East Weddell Sea, Southern Ocean, 1410-2580 m.

*stolophora Fisher, 1940

> Fisher 1940:226; A.M. Clark 1962a:77; H.E.S. Clark 1963:70; McKnight 1976:29.
>
> Range: Ross Sea, 247-274 m.
>
> Possibly a synonym of *N. armata* Koehler, 1911 according to A.M. Clark (1962a).

ORTHASTERIAS Verrill, 1914

> Verrill 1914a:168; Fisher 1923:257 (emended).
>
> Type species: *O. columbiana* Verrill, 1914, a synonym of *Asterias koehleri* de Loriol, 1897.

biordinata Verrill, 1914a:173. A forma of *O. koehleri* (de Loriol, 1897) according to Fisher (1928).

californica Verrill, 1914a, see *Astrometis*

columbiana Verrill, 1914a:48. A synonym of *Orthasterias koehleri* (de Loriol, 1897) according to Fisher (1923).

contorta (Perrier, 1881) Verrill, 1914:48, see *Sclerasterias*

dawsoni Verrill, 1914a:175. A synonym of *Astrometis sertulifera* (Xantus, 1860), according to Fisher (1928).

eustyla (Sladen, 1889) Verrill, 1914a, see *Sclerasterias*

forreri (de Loriol, 1888) Verrill, 1914, see *Stylasterias*

gonolena Verrill, 1914a:184. A synonym of *Astrometis sertulifera* (Xantus, 1860) according to Fisher (1928).

koehleri (de Loriol, 1897) (with formae *biordinata*, *leptostyla* and *montereyensis* Fisher, 1938)

de Loriol 1897:21 (as *Asterias*).

Verrill 1914a:175; Fisher 1928:139 (with formae).

Range: Alaska to California, 0-130 m.

leptolena Verrill, 1914a:182. A synonym of *Stylasterias forreri* (de Loriol, 1887).

leptostyla Fisher, 1928:139. A forma of *O. koehleri* (de Loriol).

merriami Verrill, 1914a:177. A synonym of *Leptasterias coei* Verrill, 1914, according to Fisher (1930).

montereyensis Fisher 1928. A forma of *O. koehleri* (de Loriol).

subangulosa Verrill, 1914a:168, 370, see *Sclerasterias*

tanneri: Verrill, 1914a, see *Sclerasterias*

PAEDASTERIAS Verrill, 1914

Verrill 1914a:355.

Type species: *Anasterias chirophora* Ludwig, 1903.

A synonym of *Lysasterias* Fisher, 1908, according to Fisher (1922).

chirophora: Verrill, 1914, see *Lysasterias*

joffrei Koehler, 1920, see *Lysasterias*

PARASTERIAS Verrill, 1914

Verrill 1914a:53, 187.

Type species: *Parasterias albertensis* Verrill, 1914.

A synonym of *Asterias* Linnaeus, 1758, according to Fisher (1930).

albertensis Verrill, 1914a:167. A synonym of *Asterias amurensis* Lütken, 1871, according to Fisher (1930).

PARASTICHASTER Koehler, 1920

Koehler 1920:89.

Type species: *Parastichaster mawsoni* Koehler, 1920.

A synonym of *Anasterias* Perrier, 1875, according to Fisher (1940).

directus Koehler, 1920, see *Anasterias*

mawsoni Koehler, 1920, see *Anasterias*

sphaerulatus Koehler, 1920, see *Anasterias*

PERISSASTERIAS H.L. Clark, 1923

H.L. Clark 1923:307, Fisher 1930:238; A.M. Clark & Courtman-Stock 1976:95; A.M.C. *in* Clark & Downey 1992:445.

Type species: *P. polyacantha* H.L. Clark, 1923.

heptactis H.L. Clark, 1926 H.L. Clark 1926a:26; A.M. Clark & Courtman-Stock 1976:95.

Range: W of Lambert's Bay, Cape Province, South Africa, 385 m.

monacantha McKnight, 1973
McKnight 1973:231.
Range: South from New Zealand (c.48°S, 168°E), 668 m.
***obtusispina** H.L. Clark, 1926
H.L. Clark 1926a:28; A.M. Clark 1974:440; A.M.C. & Courtman-Stock
1976:95; A.M.C. *in* Clark & Downey 1992:446.
Range: W of Cape Province, South Africa, 392 m.
Possibly a synonym of *P. polyacantha* H.L. Clark, 1923, according to
A.M. Clark (1974:440).
polyacantha H.L. Clark, 1923
H.L. Clark 1923:307; 1926a:29; Mortensen 1933a:278; A.M. Clark
1974:439; A.M.C. & Courtman-Stock 1976:95; Tablado & Maytia
1988:1; A.M.C. *in* Clark & Downey 1992:447.
Range: W and SW coasts of Cape Province, South Africa, Uruguay and
N Argentina, 96-760 m.
PISASTER Müller & Troschel, 1840 (with synonym *Calliasterias* Fewkes,
1889)
Müller & Troschel 1840:367; Fisher 1926:556; 1930:162.
Type species: *Asterias ochracea* Brandt, 1835.
antarcticus Koehler, 1917:30. A synonym of *Anasterias perrieri* (E.A.
Smith, 1876), according to A.M. Clark (1962a).
australis Verrill 1914a:88, as a variety of *Pisaster luetkeni* (Stimpson,1862.
A synonym of *P. giganteus* (Stimpson, 1857) according to Fisher (1926a,
1930).
brevispinus (Stimpson, 1857) (with synonyms *Asterias paucispina*
Stimpson, 1862 – reduced to forma – and *A. (Pisaster) papulosa* Verrill,
1909)
Stimpson 1857:528 (as *Asterias*).
Verrill 1909a:63; 1914a:77; Fisher 1926a:564; 1930:180;
Maluf:1988:47, 127 (as *Pisaster*).
Range: Sitka, Alaska to San Diego, Southern California, intertidal – 102
m.
capitatus (Stimpson, 1862) Verrill, 1909a. A subspecies of *P. giganteus*
(Stimpson, 1857), according to Fisher (1926a, 1930).
fissispinus (Stimpson, 1862) Verrill, 1909a:63; 1914a:76. A synonym of *P.
ochraceus* (Brandt, 1835) according to Fisher (1926a, 1930).
giganteus (Stimpson, 1857) (with synonyms *Asterias luetkeni* (Stimpson,
1862 and var. *australis* Verrill, 1914a, also *A. exquisita* de Loriol, 1888;
subspecies *capitatus* (Stimpson, 1862))
giganteus giganteus (Stimpson, 1857)
Stimpson 1857:528 (as *Asterias*).
Verrill 1909b:545; Fisher 1926a:561; 1930:172; Maluf:1988:47, 127
(as *Pisaster*).

294

Range: Oregon to San Cristobal, Mexico, intertidal - 374 m.

giganteus capitatus (Stimpson, 1862)

Stimpson 1862:264 (as species of *Asterias*).

Verrill 1909a:63 (as *Pisaster*).

Fisher 1930:177; Maluf 1988:47, 127 (as subspecies of *P. giganteus*).

Range: California, intertidal – 156 m.

grayi Verrill, 1914a:97. A synonym of *Leptasterias polaris katherinae* Gray, 1840, according to Fisher (1930).

luetkeni (Stimpson, 1862) Verrill, 1909a:63, 1914a:83. A synonym of *Pisaster giganteus* (Stimpson, 1857) according to Fisher (1930).

ochraceus (Brandt, 1835) (with synonyms *Asteracanthion margaritifer* Müller & Troschel, 1842 and *Asterias fissispina* Stimpson, 1862; subspecies *segnis* Fisher, 1926 and formae *confertus* (Stimpson, 1862) and *nodiferus* Verrill)

ochraceus ochraceus (Brandt, 1835)

Brandt 1835:69; A. Agassiz 1877:96 (as *Asterias*).

Fisher 1908:89; 1926a:557; 1930:164; Maluf 1988:47, 127 (as *Pisaster*),

Range: Alaska - Point Concepcion, California, intertidal – 97 m.

ochraceus segnis Fisher, 1926

Fisher 1926a:560; 1930:171.

Range: Southern and N part of Lower California.

papulosus: Verrill, 1914:91. A synonym of *P. brevispinus* (Stimpson, 1857) according to Fisher (1926a, 1930).

paucispinus (Stimpson, 1862) Verrill, 1914. A forma of *P. brevispinus* (Stimpson, 1857) according to Fisher (1930).

PODASTERIAS Perrier, 1891a:160. Apparently an MS name of Perrier's for the *Asterias scalprifera* group, only mentioned as a synonym of *Cosmasterias* Sladen, 1889, see Fisher (1940:249).

PODASTERIAS Perrier, 1894 (Non Podasterias Perrier, 1891)

Perrier 1894:107; 1896:35.

Type species: *Diplasterias luetkeni* Perrier, 1891, a synonym of *D. brandti* (Bell, 1881).

A synonym of *Diplasterias* Perrier, 1891, according to Fisher (1940).

brandti: Koehler 1917:26, see *Diplasterias*.

brucei (Koehler, 1908) Koehler 1920:42, see *Diplasterias*

fochi Koehler, 1920:35. A synonym of *Diplasterias brucei* (Koehler, 1908) according to A.M. Clark (1962a).

georgiana (Studer, 1885) Koehler, 1917, see *Neosmilaster*

glomerata (Sladen, 1889) Koehler, 1923:29, as a variety of *P. brandti*, = *Diplasterias brandti* forma *neglecta* (Bell, 1881), according to Fisher (1940).

kerguelensis Koehler, 1917, see *Diplasterias*

luetkeni (Perrier, 1891) 1894:107. A synonym of *Diplasterias brandti* (Bell, 1881) according to Fisher (1940).

meridionalis: Koehler, 1917, see *Diplasterias*

steineni: Koehler, 1917, see *Neosmilaster*

POLYASTERIAS Perrier, 1894

Perrier 1894:108.

Type species: *Asterias tenuispina* Lamarck, 1816.

A synonym of *Coscinasterias* (*Stolasterias*) Sladen, 1889, according to Fisher (1928).

tenuispina: Perrier, 1894, see Coscinasterias (*Stolasterias*)

PSALIDASTER Fisher, 1940

Fisher 1940:229.

Type species: *Psalidaster mordax* Fisher, 1940.

mordax Fisher, 1940 (with subspecies *rigidus* A.M. Clark, 1962)

mordax mordax Fisher, 1940

Fisher 1940:229; Bernasconi 1973a:320. [Non *mordax*: H.E.S. Clark 1963:71, = *P. mordax rigidus* according to McKnight (1976)].

Range: Falkland Is, Magellan area, Buenos Aires Province, Argentina, 75-600 m.

mordax rigidus A.M. Clark, 1962

A.M. Clark 1962a:79; McKnight 1976:30.

Range: MacRobertson Land and Ross Sea, Antarctica, 163 m.

PSEUDECHINASTER H.E.S. Clark, 1962

H.E.S. Clark 1962:41.

Type species: *Pseudechinaster rubens* H.E.S. Clark, 1962

rubens H.E.S. Clark, 1962

H.E.S. Clark 1962:41; McKnight 1967a:303.

Range: Cook Strait and Chatham Rise, New Zealand, 102-402 m.

QUADRASTER Perrier, 1896

Perrier 1896:27.

Type species: *Stichaster felipes* Sladen, 1889.

A synonym of *Cosmasterias* Sladen, 1889, according to Fisher (1923).

SALIASTERIAS Koehler, 1920

Koehler 1920:52; Fisher 1930:238.

Type species: *Saliasterias brachiata* Koehler, 1920.

brachiata Koehler, 1920

Koehler 1920:54; Fisher 1940:75 (listed); A.M. Clark 1962a:88; McKnight 1976:30 (*Soliasterias*).

Range: off Adelie Land, Enderby and MacRobertson Lands, Ross Sea, 27-647 m.

SCLERASTERIAS Perrier, 1891 (with synonyms *Eustolasterias* Fisher, 1923, *Triplasterias* Engel & Schroevers, 1960 and probably *Mortensenia* and *Carlasterias* da Costa, 1941 and 1952)

296

Perrier 1891c:1227; 1896:35; Fisher 1928a:105 (emended); A.M. Clark & Courtman-Stock 1976:96; Jangoux 1986a:91; Downey *in* Clark & Downey 1992:448.

Type species: *Sclerasterias guernei* Perrier, 1891.

alexandri (Ludwig, 1905) (with var. *crassa* H.L. Clark, 1940 and synonym *Hydrasterias diomedeae* Ludwig, 1905)

Ludwig 1905:221 (as *Stolasterias*).

Fisher 1928a:107; H.L. Clark 1940:335 (var. *crassa*); Maluf 1988:47, 127 (as *Sclerasterias*).

Range: Panama, Lower California to Malpelo Ridge, off Colombia, 46-384 m.

candicans: Fisher, 1940:75 (listed). See *Notasterias*.

contorta (Perrier, 1881) (with synonyms *Asterias angulosa* Perrier, 1881 and *Orthasterias subangulosa* Verrill, 1914)

Perrier 1881:1 (as *Asterias*).

Verrill 1914a:48 (as *Orthasterias (Stylasterias)*).

Fisher 1928a:107; Downey *in* Clark & Downey 1992:449 (as *Sclerasterias*).

Range: Florida, Gulf of Mexico, Caribbean, Venezuela, Brazil, 344(?20)-424 m.

crassa H.L. Clark, 1940:335. A variety of *S. alexandri*.

*euplecta (Fisher, 1906) (with forms *hypacantha* Fisher and *stenactis* (H.L. Clark, 1923))

Fisher 1906a:1105 (as *Coscinasterias*).

Fisher 1919a:487 (as *Distolasterias*).

Fisher 1923:255 (as *Eustolasterias*).

Fisher 1928a:107 (as *Sclerasterias*).

Range: Hawaiian Is, Indo-W. Pacific, 250-270 m.

Probably a form of *Sclerasterias mazophora* (Wood-Mason & Alcock, 1891) according to Macan (1938).

eustyla (Sladen, 1889)

Sladen 1889:587 (as *Asterias (Stolasterias)*).

Koehler 1907a:145 (as *Stolasterias*).

Verrill 1914a:168 (as *Orthasterias*).

Fisher 1928a:107; Mortensen 1933a:278; A.M. Clark & Courtman-Stock 1976:96; Downey *in* Clark & Downey 1992:450 (as *Sclerasterias*).

Range: Tristan da Cunha and W coast of South Africa, 183-275 m.

guernei Perrier, 1891

Perrier 1891c:1227; 1891b:264; Koehler 1895b:441; 1896:41; Perrier 1896:35; Fisher 1928a:107; Downey *in* Clark & Downey 1992:450.

Range: Bay of Biscay, 160-490 m.

heteropaes Fisher, 1924

Fisher 1924a:7; 1928a:107; Ziesenhenne 1937:220; Maluf 1988:47,127.

Range: Southern and Lower California, 18-457 m.

hypacantha: Fisher 1928a:108. Probably a forma of *S. euplecta* (Fisher, 1906) according to Fisher (1928a) but both probably forms of *S. mazophora* (Wood-Mason & Alcock, 1891) according to Macan (1938).

mazophora (Wood-Mason & Alcock, 1891) (probably with synonym *S. nitida* Koehler, 1910 and forms *euplecta* (Fisher, 1906) and *hypacantha* (Fisher, 1917))

Wood-Mason & Alcock 1891:436; Alcock 1893:115 (as *Asterias*).

Fisher 1919a:489 (as *Distolasterias*).

Fisher 1928a:107; Macan 1938:416 (as *Sclerasterias*).

Range: Bay of Bengal, Gulf of Aden, also probably the Philippines and Hawaiian Is, 220-550 m.

mollis (Hutton, 1872)

Hutton 1872b:4 (as *Asterias*).

Fisher 1919a:489 (as *Distolasterias*).

Fisher 1923:255 (as *Eustolasterias*).

Fisher 1924a:6; Mortensen 1925:318; Fell 1958:19; McKnight 1967a:302; H.E.S. Clark 1970b:23; Baker & Clark 1970:6 (as *Sclerasterias*).

Range: New Zealand, Chatham Is, 22-697 m.

*****neglecta** (Perrier, 1891) (with synonym *Asterias edmundi* Ludwig 1897)

Perrier 1891b:226; von Marenzeller 1895:136; Perrier 1896:37 (as *Stolasterias*).

Verrill 1914a:48, 370 (as *Stylasterias*).

Fisher 1928a:107; Downey *in* Clark & Downey 1992:451 (as *Sclerasterias*).

Range: Bay of Biscay, Mediterranean, 160-887 m.

Possibly conspecific with *Asterias richardi* Perrier *in* Milne-Edwards 1882, which may need suppression by the ICZN according to A.M. Clark *in* Clark & Downey (1992:448,451,511).

*****nitida** Koehler, 1910:176. Probably a juvenile *S. mazophora* (Wood-Mason & Alcock, 1891) according to Fisher (1928a).

*****parvulus** (Perrier, 1891b:258) Fisher 1928a:59. A young *Sclerasterias* according to Fisher (1928). If so, from locality SE of Newfoundland could be *S. tanneri* (Verrill, 1895), so needing invocation of the ICZN (A.M.C. **new obs.**, not mentioned by Downey *in* Clark & Downey (1992)).

peregrina (Bell, date?) Fisher 1928a:107, indexed in italics so probably an MS name omitted from Bell (1909) and a NOMEN NUDUM. Specimen in the Natural History Museum, London labelled *S. euplecta stenactis* from locality – western Indian Ocean – probably a form of *S. mazophora* (Wood-Mason & Alcock, 1891).

*****richardi** (Perrier, 1882)

Perrier *in* Milne-Edwards 1882:20; Marenzeller 1893b:8 (as *Asterias*).

Perrier 1894:109 (as *Hydrasterias*).

Fisher 1928a:108 (as *Sclerasterias.*

Range: Off Corsica, Mediterranean.

See comments under *S. neglecta*, also *Asterias richardi* and *Hydrasterias richardi* (Pedicellasteridae). Needs suppression.

satsumana (Döderlein, 1902)

Döderlein 1902:334 (as *Asterias*).

Hayashi 1943:193; 1973b:100; Imaoka et al. 1991:98 (as *Sclerasterias*).

Range: Southern Japan, 750-100 m.

stenactis (H.L. Clark, 1926)

H.L. Clark 1926:23 (as *Eustolasterias*).

Fisher 1928a:107; Mortensen 1933a:277; A.M. Clark 1952:199; A.M.C. & Courtman-Stock 1976:96 (as *Sclerasterias*).

Range: South and SE of South Africa, 146-410 m.

Reduced to a forma of *S. euplecta* (Fisher, 1906) by Fisher (1928a) but probably a forma of *S. mazophora* (Wood-Mason & Alcock, 1891) according to Macan (1938). [No comment on that by A.M.C.!]

subangulosa: Fisher 1928. A synonym of *Sclerasterias contorta* (Perrier, 1881) according to Fisher (1928a) and Downey (1973).

tanneri (Verrill, 1880) (with synonym *Triplasterias mercatoris* Engel & Schrovers, 1960)

Verrill 1880b:401; 1895:209; Gray, Downey & Cerame-Vivas 1968 (as *Asterias*).

Verrill 1914a:48,168 (as *Orthasterias*).

Fisher 1928a:107,108; Franz et al. 1981:397; Jangoux & Massin 1986:91; Downey *in* Clark & Downey 1992:452.

Range: Newfoundland to Cape Hatteras, N Carolina, 62-700 m.

SMILASTERIAS Sladen, 1889

Sladen 1889:562,578 (as *Asterias (Smilasterias)*).

Perrier 1893:848; 1896:32 (as genus); Fisher 1930:239; 1940:260; A.M. Clark 1962a:85; O'Laughlin & O'Hara 1990:307 (reviewed); Downey *in* Clark & Downey 1992:452.

Type species: *Asterias (Smilasterias) scalprifera* Sladen, 1889.

clarkailsa O'Laughlin & O'Hara, 1990

O'Laughlin & O'Hara 1990:316.

Range: Macquarie Island, Southern Ocean, 69-135 m.

irregularis H.L. Clark, 1928

H.L. Clark 1928:402; Shepherd 1968:752; O'Laughlin & O'Hara 1990:317.

Range: Spencer Gulf, S Australia to N.S.W., 1-30 m.

multipora O'Laughlin & O'Hara, 1990

O'Laughlin & O'Hara 1990:311.

Range: Victoria and Tasmania, 0-3 m.

scalprifera (Sladen)

Sladen 1889:578 (as *Asterias (Smilasterias)*.

Fisher 1930:239 (as *Smilasterias*); 1940:261; A.M. Clark 1962a:85; Guille 1974:38; O'Laughlin & O'Hara 1990:318.

Range: Falklands Is, Marion I, Kerguelen and Heard Islands, 40-267 m.

tasmaniae O'Laughlin & O'Hara, 1990

O'Laughlin & O'Hara 1990:315.

Range: SE Tasmania, 0-8 m.

terweili (Goldschmidt, 1924), as *Asterias (Smilasterias) terweili, a synonym of uidia magellanica* Leipoldt, 1895, Luidiiae, see part 1, according to Madsen (1956).

triremis (Sladen, 1889)

Sladen 1889:578 (as *Asterias (Smilasterias)*).

Fisher 1930:239 (as *Smilasterias*); 1940:262; A.M. Clark 1962a:85; Bernasconi 1971:285; Downey *in* Clark & Downey 1992:453.

Range: S of Kerguelen, Palmer Archipelago, Antarctic Peninsula and off Argentina (36°S)(Downey), 93-335 and 2707 m.

SOLIASTERIAS McKnight, 1976:30, lapsus for *Saliasterias* Koehler, 1920.

SPORASTERIAS Perrier, 1893

Perrier 1893:848; 1894:107.

Type species: *Asterias spirabilis* Bell, 1881.

A synonym of *Anasterias* Perrier, 1875 according to Fisher (1940).

antarctica: Ludwig, 1903:39, see *Anasterias*

*borbonica (Perrier, 1875)

Perrier 1875:61 (as *Asterias*).

Fisher 1930:216 (as ? *Sporasterias*).

Range: Reunion I, western Indian Ocean.

Either tropical locality or generic affiliation must be incorrect. [A.M.C.]

cocosana Ludwig, 1905, see *Tarsaster* (Pedicellasteridae)

directa: Fisher, 1930:241, see *Anasterias*

galapagensis Ludwig, 1905, see *Tarsaster* (Pedicellasteridae)

mariana Ludwig, 1905, see *Ampheraster* (Pedicellasteridae)

mawsoni: Fisher 1930:241, see *Anasterias*

pedicellaris Koehler, 1923, see *Anasterias*

perrieri: Verrill, 1914:236, see *Anasterias*

rugispina: Perrier, 1896: A synonym of *Anasterias antarctica* (Lütken, 1857) according to Fisher (1940)

rupicola (Verrill, 1881) Ludwig 1903:40 (as var. of *S. antarctica*; Döderlein 1928:294 (as species), see *Anasterias*

sphaerulata: Fisher 1930:241, see *Anasterias*

spirabilis Perrier, 1894, see *Anasterias*

STELLONIA Nardo, 1834

Nardo 1834:716.

Type species: *Asterias rubens* Linnaeus, 1758, designated by Fisher (1913) to render this conglomerate of species a synonym of *Asterias* Linnaeus.

angulosa: L. Agassiz, 1836b:192. A synonym of *Asterias* (i.e. *Marthasterias*) *glacialis* Linnaeus, 1758, according to Sladen (1889).

echinites: L. Agassiz, 1836b:192. A synonym of *Acanthaster planci* (Linneaus, 1758), according to Verrill (1914a).

endeca: L. Agassiz, 1836b:192, see *Solaster* (Solasteridae, part 3).

glacialis: Nardo, 1834:716, see *Marthasterias*

helianthus: L. Agassiz, 1836b:192, see *Heliaster* (Heliasteridae)

hispida (Pennant, 1777) Forbes, 1839:123. Suppressed and replaced by *Leptasterias muelleri* (M. Sars, 1846), see A.M.C. *in* Clark & Downey (1992).

papposa: L. Agassiz, 1836b:192, see *Crossaster* (Solasteridae, see part 3)

rubens: Nardo, 1834:716, see *Asterias*

seposita: Nardo, 1834:716. Validated as species of *Echinaster* (Echinasteridae, see part 3) by Tortonese & Madsen (1979).

spinosa: Nardo, 1834:716. Invalid and a synonym of *Echinaster (Othilia) echinophorus* (Lamarck, 1816)

tenuispina: d'Orbigny, 1839:148, see *Coscinasterias*

webbiana: d'Orbigny, 1839:148. A synonym of *Asterias* (i.e. *Marthasterias glacialis* (Linnaeus, 1758) according to Sladen (1889).

*STENASTERIAS Verrill, 1914

Verrill, 1914a:145; Fisher 1930:138.

Type species: *Asterias macropora* Verrill, 1909.

'Name has status only by technicality' according to Fisher (1930).

*macropora (Verrill, 1909)

Verrill 1909:65 (as *Asterias*).

Verrill 1914a:145; Fisher 1930:138 (as *Stenasterias*).

Range: Sitka, Alaska and Queen Charlotte Is.

Type material juvenile (R 15 mm.) and lost.

STEPHANASTERIAS Verrill, 1871 (with synonym *Nanaster* Perrier, 1894)Verrill 1871c:5; Perrier 1875:84 [348];Verrill 1914a:146; Fisher 1930:156; Downey *in* Clark & Downey 1992:453.

Type species: *Asteracanthion albulus* Stimpson, 1853.

albula (Stimpson, 1853) (with synonyms *Asteracanthion problema* Steenstrup, 1855 and *Asterias gracilis* Perrier, 1881)

Stimpson 1853:14 (as *Asteracanthion*).

Stimpson 1864:142 (as *Asterias*).

Verrill 1866:351; Duncan & Sladen 1881:29; Danielssen & Koren 1884b:31; Ludwig 1900:479 (as *Stichaster*).

Verrill 1871:5; 1914a:147; Fisher 1930:156; Djakonov 1950a:121; John & A.M. Clark 1954:151; Grainger 1966:43; Downey *in* Clark & Downey 1992:454 (as *Stephanasterias*).

Range: Circumboreal, S in W Atlantic to Caribbean, in Pacific to Japan Sea, 33-2300 m.

gracilis (Perrier, 1881) Verrill 1899:223; Gray, Downey & Cerame-Vivas 1968:160. A synonym of *S. albula* (Stimpson, 1853) according to Downey *in* Clark & Downey (1992).

hebes Verrill, 1915:26. A synonym of *Allostichaster capensis* (Perrier, 1875) according to A.M.C. *in* Clark & Downey (1992).

STICHASTER Müller & Troschel, 1840 (with synonyms *Tonia* Gray, 1840 and *Coelasterias* Verrill, 1871)

Müller & Troschel 1840a:102; 1840b:323.

Type species: *Stichaster striatus* Müller & Troschel, 1840 (non *Asterias striatus* Lamarck, 1816).

albulus (Stimpson, 1853) Verrill, 1866, see *Stephanasterias*

arcticus Danielssen & Koren, 1884

Danielssen & Koren 1884b:27; Ludwig 1900:478.

Range: Lofoten Is, Norway.

aurantiacus: Verrill 1870:293. Invalidated as junior homonym, = *Stichaster striatus* Müller & Troschel, 1840, according to Fisher (1930).

australis (Verrill, 1870)

Verrill 1870:247 (as *Coelasterias*).

Sladen 1889:431; Fisher 1922b:598; Mortensen 1925:313; Fell 1953:93; 1962:48; McKnight 1967:303.

Range: New Zealand, Snares and Chatham Is.

felipes Sladen, 1889, see *Cosmasterias*

insignis Farquhar, 1894, see *Allostichaster*

littoralis Farquhar, 1894:206. A synonym of *Calvasterias suteri* (de Loriol, 1894) according to Farquhar (1897).

nutrix Studer, 1885, see *Granaster*

*polygrammus Sladen, 1889. A synonym of *Cosmasterias lurida* (Philippi, 1858) according to Leipoldt (1895) but a valid species according to Verrill (1914a:360), though neither Fisher (1940) nor Madsen (1956) mention.

polyplax Sladen, 1889, see *Allostichaster*

roseus: M. Sars, 1861, see *Stichastrella*

striatus Müller & Troschel, 1840 (non *Asteracanthion striatum*: Müller & Troschel, 1842, = *Valvaster striatus* (Lamarck, 1816) - see part 2) (includes *Asterias aurantiacus*) Meyen, 1834, a junior homonym of *A. aurantiacus* Tiedemann, 1816, a synonym of *Astropecten aranciacus* Linnaeus, 1758)

Müller & Troschel, 1840a:102; 1840b:323; Fisher 1930:241; Madsen

1956:37; Bernasconi 1961c:319.

Range: Peru to Chile.

suteri de Loriol, 1894, see *Calvasterias*

talismani Perrier, 1885. A synonym of *Neomorphaster margaritaceus* (Perrier, 1882) according to Downey *in* Clark & Downey (1992)

STICHASTER: Perrier, 1896, emended, for type species *Asterias rosea* O.F. Müller, 1776, = *Stichastrella* Verrill, 1914.

STICHASTRELLA Verrill, 1914

Verrill 1914a:40; Koehler 1921:28; 1924:117; Mortensen 1927:136; Fisher 1930:209; A.M. Clark *in* Clark & Downey 1992:455.

Type species: *Asterias rosea* O.F. Müller, 1776.

ambigua (Farran, 1913)

Farran 1913:18 (as var. *ambiguus* of *Stichaster roseus*)

Mortensen 1927:136; Gage et al. 1983:286; Harvey et al. 1988:167 (as *Stichastrella rosea* var. *ambigua*).

Fisher 1930:210 (as *S. rosea* forma *ambigua*).

A.M. Clark *in* Clark & Downey 1992:456 (as species).

Range: W slope of Porcupine Bank, W of Ireland, Porcupine Seabight, Rockall Trough to SW of Faeroe Is, 245-1630 m.

rosea (O.F. Müller, 1766) (with synonym *Chaetaster hermanni* Müller & Troschel, 1842)

O.F. Müller 1776:234; 1788: 35 (as *Asterias*).

Thompson 1840:245 (as *Linkia*).

Forbes 1841:106 (as *Cribella*).

Müller & Troschel 1842:17(pt); Düben & Koren 1846b:241; Perrier 1869:37 (as *Asteracanthion*).

M. Sars 1861:86; Norman 1865:125; Perrier 1875:83; Danielssen & Koren 1884b:30; Bell 1893a:85; Ludwig 1900:477; Süssbach & Breckner 1911:228; Farran 1913:18; Grieg 1921 [1932]:24 (as *Stichaster*).

Verrill 1914a:40; Koehler 1920:82; 1924:118; Mortensen 1927:136; Cherbonnier 1951: Gage et al. 1983:286, pt); Jangoux 1986b:128; Harvey et al. 1988:167; A.M. Clark *in* Clark & Downey 1992:456 (as *Stichastrella*).

Range: Lofoten Is, southern Norway, NE England, Scotland, Rockall Bank, S to Bay of Biscay, 4-c.200 [366] m.

STICHORELLA Koehler, 1920

Koehler 1920:89.

Type species: *Stichaster suteri* de Loriol, 1894.

A synonym of *Calvasterias* Perrier, 1875 according to Fisher (1922b,1930).

suteri: Koehler, 1920, see *Calvasterias*

STOLASTERIAS Sladen, 1889. Treated as a subgenus of *Coscinasterias* Verrill, 1870.

alexandri Ludwig, 1905, see *Sclerasterias*

brucei Koehler, 1908, see *Diplasterias*

candicans Ludwig, 1903, see *Notasterias*

***edmondi** (Benham, 1911:151) (as *Asterias (Stolasterias)*. Kermadec Is. ? Identity.

eustyla (Sladen, 1889), see *Sclerasterias*

glacialis: Perrier, 1894, see *Marthasterias*

neglecta Perrier, 1891, see *Sclerasterias*

robusta Ludwig, 1905, see *Distolasterias*

tenuispina (Sladen, 1889), see *Coscinasterias (Stolasterias)*

STOLASTERIAS: Perrier, 1894, 1896, non Verrill, 1870, = *Marthasterias* Jullien, 1878.

STYLASTERIAS Verrill, 1914

Verrill 1914a:48, 65, 179 (as subgenus of *Orthasterias*), 50 (as genus); Fisher 1928a:96 (as genus, restricted).

Type species: *Asterias forreri* de Loriol, 1887.

contorta (Perrier, 1881) Verrill, 1914a, see *Sclerasterias*

forreri (de Loriol, 1887) (with synnyms *Asterias forcipulata* Verrill, 1909 and *Orthasterias leptolena* Verrill, 1914)

Verrill 1887:401 (as *Asterias*).

Verrill 1914a:179 (as *Orthasterias Stylasterias*).

Fisher 1928a:96; Jangoux 1985:21 (lectotype selected); Maluf 1988:48, 127 (as *Stylasterias*).

Range: Southern Alaska to San Diego, California, 29-532 m.

neglecta (Perrier, 1891) Verrill, 1914a, see *Sclerasterias*

paschae H. L. Clark, 1920, see *Astrostole*

reticulatus (H.L. Clark, 1916)

H.L. Clark 1916:69 (as *Pedicellaster*).

A.M. Clark 1962a:99; H.E.S. Clark 1970:24 (as *Stylasterias*).

Range: Tasmania and NW of New Zealand, 128-421 m.

robusta (Ludwig, 1905) Verrill, 1914a, see *Distolasterias*

subangulosa Verrill, 1914a (*Orthasterias (Stylasterias)*. A synonym of *Sclerasterias contorta* (Perrier, 1881) according to Fisher (1928a).

TARANUIASTER McKnight 1973

McKnight 1973b:233.

Type species: *Taranuiaster novaezealandiae* McKnight, 1973.

novaezealandiae McKnight, 1973

McKnight 1973b:233.

Range: East of South Island, New Zealand, 198 m.

TARSASTROCLES Fisher, 1923

Fisher 1923:605.

Type species: *Hydrasterias verrilli* Fisher, 1906.

verrilli (Fisher, 1906)

Fisher 1906a:1106 (as *Hydrasterias*).

Fisher 1923:605; 1928a:68 (as *Tarsastrocles*).

Range: Hawaiian Is, 520-530 m.

TONIA Gray, 1840 (December)

Gray 1840:180.

Type species: *Tonia atlantica* Gray, 1840.

A synonym of *Stichaster* Müller & Troschel, April, 1840, according to Sladen (1889).

atlantica Gray, 1840:180. A synonym of *Stichaster striatus* Müller & Troschel, 1840 according to Sladen (1889).

aurantiaca: Perrier, 1894:129; 1896:27. Invalid and conspecific with *Stichaster striatus* Müller & Troschel, 1840 according to Fisher (1930) indirectly.

TRIPLASTERIAS Engel & Schroevers, 1960

Engel & Schroevers 1960:6.

Type species: *Triplasterias mercatoris* Engel & Schroevers, 1960.

A synonym of *Sclerasterias* Perrier, 1891 according to Jangoux & Massin (1986).

mercatoris Engel & Schroevers, 1960:7. A synonym of *Sclerasterias tanneri* (Verrill, 1880) according to Jangoux & Massin (1986).

UNIOPHORA Gray, 1840 (with synonym *Margaraster* Gray, 1866)

Gray 1840:288; Fisher 1930:242; H.L. Clark 1938:416; Shepherd 1967:3.

Type species: *Uniophora globifera* Gray, 1840, a synonym of *Asterias granifera* Lamarck, 1816.

dyscrita H.L. Clark, 1923

H.L. Clark 1923b:244; Shepherd 1967b:7.

Range: Western Australia, 'shallow' and 120-200 m.

fungifera (Perrier, 1875) Fisher 1926:198. A synonym of *U. granifera* (Lamarck, 1816) according to Shepherd (1967b).

globifera Gray, 1840:288. A synonym of *U. granifera* (Lamarck, 1816) according to Shepherd (1967b).

granifera (Lamarck, 1816) (with synonyms *U. globifera* Gray, 1840, *Asterias fungifera* and *sinusoida* Perrier, 1875, *U. obesa, multispina* (form) and *uniserialis* H.L. Clark, 1928) (non *U. granifera*: Farquhar, 1878)

Lamarck 1816:560 (as *Asterias*).

Gray 1840:288; H.L. Clark 1946:159; Shepherd 1967b:4; 1968:753 (as *Uniophora*).

Müller & Troschel 1842:20 (as *Asteracanthion*).

Gray 1866:2 (as *Margaraster*).

Range: N.S.W., South Australia and Tasmania, 5-40 m.

granifera: Farquhar, 1878, non *U. granifera* (Lamarck, 1816), probably = *Stichaster australis* (Verrill, 1870), according to Benham (1909b).

gymnonota H.L. Clark, 1928. A synonym of *U. nuda* (Perrier, 1875) according to Shepherd (1967b).

multispina H.L. Clark, 1928:407. A form of *U. granifera* according to Shepherd (1967b).

nuda (Perrier, 1875) (with synonym *U. gymnonota* H.L. Clark, 1928)
Perrier 1875:71 [335] (as *Asterias*).
Fisher 1926:198; Shepherd 1967b:6; 1968:753 (as *Uniophora*).
Range: South Australia, 8-20 (40) m.
'Intergrades with *U. granifera*' according to Shepherd (1968).

obesa H.L. Clark, 1928:409. A synonyn of *U. granifera* (Lamarck, 1816) according to Shepherd (1967b).

sinusoida (Perrier, 1875) Fisher 1923:597. A synonym of *U. granifera* (Lamarck, 1816) according to Shepherd (1967b).

uniserialis H.L. Clark, 1928:413. A synonym of *U. granifera* (Lamarck, 1816) according to Shepherd (1967b).

URASTER L. Agassiz, 1836b:191 (part). A substitute name for *Stellonia* Nardo, 1834. A synonym of *Asterias* Linnaeus, 1758 according to Fisher (1930).

glacialis: Forbes, 1841:78, see *Marthasterias*

hispida (Pennant, 1777) Forbes, 1841:95. Suppressed by the ICZN, 1972, = *Leptasterias muelleri celtica* A.M. Clark {*in* Clark & Downey (1992).

rubens: Forbes, 1841:83, see *Asterias*

violaceus Forbes, 1841:91. A synonym of *Asterias rubens* Linnaeus, 1758, according to Fisher (1930).

URASTERIAS Verrill, 1909
Verrill 1909:67; 1914a:51,187; Fisher 1930:211.
Type species: *Asteracanthion lincki* Müller & Troschel, 1842.

enopla (Verrill, 1895) Fisher 1930:211. A synonym of *O. lincki* Müller & Troschel, 1842 according to Downey *in* Clark & Downey (1992).

lincki (Müller & Troschel, 1842) (with synonyms *Asteracanthion stellionura* Perrier, 1869, *Asterias gunneri* Danielssen & Koren 1884 and *Asterias enopla* Verrill, 1895) *vars. siberica* and *robusta* Kalischewsky, 1907)
Müller & Troschel 1842:18 (as *Asteracanthion*).
Döderlein 1900:200; Ludwig 1900:486 (as *Asterias*).
Kalischewevsky 1907:51 (vars *robusta* and *siberica* of *Asterias lincki*).
Verrill 1909:67; 1914a:52; Fisher 1930:211; Mortensen 1932:20; Djakonov 1933:63; 1950a:114; 1952b:295; Grainger 1964:43; 1966: 44; Smirnov 1990:442; Downey *in* Clark & Downey 1992:458 (as *Urasterias*).
Range: Circumboreal arctic, S in Atlantic to Nova Scotia, 5-2000 m.

Family HELIASTERIDAE Viguier

Heliasteridae Viguier 1878:111; Fisher 1923:248; 1928a:3; Madsen 1956:35.
Genus-group name: **Heliaster**

HELIASTER Gray, 1840
  Gray 1840:179 (as subgenus of *Asterias* Linnaeus, 1758).
  Verrill 1870[1867]:289; Rathbun 1887:440; H.L. Clark 1907:23; A.H. Clark 1946:10; Caso 1943 [1961a]:112 (as genus).
  Type species: *Asterias helianthus* Lamarck, 1816, designated by H.L. Clark (1907).
canopus Perrier, 1875
  Perrier 1875:88(352); H.L. Clark 1907:45.
  Range: Juan Fernandez Is, W from Chile, probably from shore.
cumingi (Gray, 1840)
  Gray 1840:180 (as *Asterias (Heliaster) cumingii*.
  Dujardin & Hupé, 1862:343; A.H. Clark 1939c:11 (as *Heliaster*).
  Range: Galapagos Is, shore.
cumingi: Verrill, 1870 [1867]:291,333, non *H. cumingi* (Gray, 1840), = *H. polybrachius* H.L. Clark, 1907, according to Caso (1943).
helianthus (Lamarck, 1816)
  Lamarck 1816:558 (as *Asterias*).
  L. Agassiz 1836b:192 (as *Stellonia*).
  Gray 1840:179 (as *Asterias (Heliaster)*.
  Müller & Troschel 1842:18 (as *Asteracanthion*).
  Dujardin & Hupé, 1862:344; Verrill 1870 [1867]:289; H.L. Clark 1910:338; Madsen 1956:34; Caso 1943 [1961a]:113; Maluf 1988:45, 125 (as *Heliaster*).
  Range: Colombia to Valparaiso, Chile, ?W Mexico, shore.
kubingii Rathbun, 1887. Lapsus for
kubiniji Xantus, 1860
  Xantus 1860:568; H.L. Clark 1907:48; Ziesenhenne 1937:220; Caso 1943 [1961a]:110, 115, 118 (var. *nigra]; 1944:224 (var. nigra*); Dungan et al. 1982:989; Maluf 1988:45,125.
  Range: Lower California S to Pacific Nicaragua, 0-9 m.
kubiniji: Verrill, 1869:387 (from Peru) = ?
microbrachius Xantus, 1860
  Xantus 1860:568; H.L. Clark 1926b:4; Caso 1943 [1961a]:123; Maluf 1988:45.
  Range: Lower California to Panama, shore.
microbrachius var. polybrachius: Caso, 1943 [1961a]:121; 1980:207 (NW Mexico), ? non *H. polybrachius* H.L. Clark, 1907, = ?

*morrisoni A.H. Clark, 1949

A.H. Clark 1946:9.

Range: Pearl Is, Panama.

Probably conspecific with *H. microbrachius* Xantus, 1860
according to Madsen (1956).

multiradiatus (Gray, 1840:180, as *Asterias (Heliaster)*); Dujardin & Hupé,
1862:343; H.L. Clark 1907:46. A NOMEN NUDUM, junior homonym
of *Asterias multiradiatus* Linnaeus, 1758 (a crinoid) according to A.H.
Clark 1920, = *H. solaris* nom. nov.

*polybrachius H.L. Clark, 1907

H.L. Clark 1907:41,54; 1910:338; Tortonese 1955:681; Madsen 1956:9.

Range: Acapulco, Mexico to N Chile.

Probably conspecific with *H. cumingi* (Gray, 1840) according to Madsen
(1956).

solaris A.H. Clark, 1920

A.H. Clark 1920:183 (nom. nov. for *Asterias (Heliaster)*
multiradiata Gray, 1840, non *A. multiradiata* Linnaeus, 1758); 1946:10;
Maluf 1988:45,126.

Range: Galapagos Is.

*Family PYCNOPODIIDAE Fisher

Pycnopodiinae Fisher 1928a:57 (in key), 148.

When the other subfamilies of Asteriidae sensu Fisher 1928 were raised to
family rank, primarily by Downey *in* Clark & Downey 1992: 400-401, the
Pycnopodiinae, being unrepresented in the Atlantic, was left in limbo. For
the sake of consistency, it is here treated as of family rank but this needs
to be reassessed.

Genus-group names: **Lysastrosoma, Pycnopodia**

LYSASTROSOMA Fisher, 1922b

Fisher 1922b:590; 1928a:148.

Type species: *Lysastrosoma anthosticta* Fisher, 1922

anthosticta Fisher, 1922 (with subspecies *crassispina* Djakonov, 1938 and
forma desmiora H.L. Clark, 1925[?6])

anthosticta anthosticta Fisher, 1922

Fisher 1922b:591; 1928a:149; Djakonov 1950a:112; 1958:317;
Baranova 1971:256; Shin 1995: .

Range: S coast of Hokkaido, Japan, Korea to Kurile Is, 0-50 m.

anthosticta crassispina Djakonov, 1938

Djakonov 1938b:499.

Range: Japan Sea.

desmiora H.L. Clark, 1925:5; Fisher 1928a:151. Reduced to forma of
L. anthosticta by Djakonov (1950a).

PYCNOPODIA Stimpson, 1861
Stimpson 1861:261; Fisher 1928a:153.
Type species: *Asterias helianthoides* Brandt, 1835.

helianthoides (Brandt, 1835)
Brandt 1835:271 (as *Asterias*).
Stimpson 1961:261; A. Agassiz 1877:100; Verrill 1914a:198;
Fisher 1928a:154; Baranova 1957:176 (as *Pycnopodia*).
Range: Alaska, Aleutian Is to Southern California, 0-455 m

Order BRISINGIDA Fisher 1928

Fisher 1928: 3; Spencer and Wright 1966: U77 (as Brisingina)
Tortonese 1958: 1 (as Euclasteroidea)
Downey 1973: 98 (as Euclasterida)
Downey 1986: 1; Blake 1987: 517; Downey in Clark and Downey 1992:
462; Mah 1998 (as Brisingida)

Family ODINELLIDAE Mah

Genus-group name: **Odinella**

ODINELLA Fisher, 1940
Fisher 1940: 204; A.M. Clark, 1962: 68
Type species: *Odinella nutrix* Fisher, 1940.

nutrix Fisher, 1940
Fisher 1940: 207; A.M. Clark 1962: 68; Mein 1992: 251
Range: Southern Ocean: South Georgia, the South Shetland Islands, End-
erby to Adelie Lands and from southern Argentina, 120-640 m.

Family BRISINGASTERIDAE Mah

Genus-group name: **Brisingaster**

BRISINGASTER de Loriol, 1883
deLoriol 1883: 55; Fisher 1917f: 419; 1919: 502; 1928: 5; 1940: 205
Rowe 1989: 274 (as *Novodinia helenae*).
Type species: *Brisingaster robillardi* de Loriol, 1883

robillardi de Loriol, 1883

de Loriol 1883: 55; Fisher 1919: 502

Range: off Mauritius, New Caledonia, Norfolk Sea Mount, 100-1200 m.

Family NOVODINIIDAE Mah

Genus-group names: **Novodinia,** Odinia

NOVODINIA Dartnall, Pawson, Pope and Smith, 1969.

Dartnell, Pawson, Pope and Smith 1969: 211; Downey 1986: 21; Downey in Clark and Downey 1992: 471. (nom. nov. for *Odinia* Perrier 1885, preoccupied)

Type species: *Odinia semicoronata* Perrier, 1885.

americana (Verrill, 1880)

Verrill 1880: 139 (as *Brisinga*)

Verrill 1894: 279; 1895: 211 (as *Odinia*)

Downey 1986: 21; Downey in Clark and Downey 1992: 472 (as *Novodinia*)

Range: Nova Scotia to Colombia, 320-732 m.

antillensis (A.H. Clark, 1934)

A.H. Clark 1934: 1; Downey 1973: 99 (as *Odinia*)

Downey 1986: 23; Downey in Clark and Downey 1992: 472 (as *Novodinia*)

Range: West Indies, Gulf of Mexico. 366-622 m.

australis (H.L. Clark, 1916)

H.L. Clark 1916: 75; H.L. Clark 1946: 153 (as *Odinia*)

Range: SE Ocean. NSW Victoria. Known only from Shoalhaven Heads NSW to off Cape Everand, Victoria, 360-549 m

austini (Koehler, 1909) with forma *japonica* Hayashi, 1943

Koehler 1909c: 124; H.L. Clark 1946: 153. (as *Odinia*)

Hayashi 1943: 140 (as forma *japonica*)

Range: Indian Ocean/Philippines: Arabian Sea-vicinity of Sri Lanka, Japan: Misaki and Suruga Bay, 400-802 m.

clarki (Koehler, 1909)

Koehler 1909c: 120 (as *Odinia*)

Range: Indian Ocean/Philippines: Arabian Sea -Lakshadweep islands? 918m

helenae Rowe, 1989. A junior synonym of *Brisingaster robillardi* according to Mah

homonyma Downey, 1986

Perrier 1885c: 9 (as *Brisinga*, non *Brisinga elegans* Verrill, 1884)

Perrier 1894: 71 (as *Odinia*)

Downey 1986: 25; Downey in Clark and Downey 1992: 474 (as *Novodinia*)

Range: South of Canary Islands 1056-1435 m.

magister Fisher, 1917

Fisher 1917f: 93; 1919: 507 (as *Odinia*)

Range: Philippines, off S. Panay Juraojurao Island.

novaezelandiae (H.E.S. Clark, 1962)

H.E.S. Clark 1962: 6; McKnight 1967: 304 (as *Odinia*)

Range: Sub-Antarctic, off Chatham Islands. 625-690 m.

pacifica Fisher, 1906. With forma *sagamiana* Hayashi, 1943

Fisher 1906: 1108 (as *Odinia*)

Hayashi 1943: 136, (as forma *sagamiana*)

Range: Hawaiian Islands: South coast of Molokai Island, vicinity of Kuai Island to Misaki Japan,600-1056 m .

pandina (Sladen, 1889)

Sladen 1889: 598; Bell 1893: 105; Mortensen 1927: 123; H.L. Clark 1941: 64 (as *Odinia*)

Downey 1986: 27; Harvey et. al. 1988: 166 (as *Novodinia*)

Range: Faeroe Channel; North Carolina, off Cuba and Hbridean Slope; 278-990 m

penichra Fisher, 1916

Fisher 1916b: 31; 1919: 505; Jangoux 1981: 459 (as *Odinia*)

Range: Philippines between Burias and Luzon, 187-210 m.

radiata Aziz and Jangoux, 1985

Aziz and Jangoux 1985a: 289

Range: Off Mindoro, Philippines, 215 m

robusta Perrier, 1894. A junior synonym of *N. semicoronata* by Downey (1986)

semicoronata (Perrier,1885)

Perrier 1885d: 442; 1885c: 9 (as *Brisinga*)

Perrier 1885d: 11; 1885c (as *B. robusta*)

Perrier 1885a: 885; 1894: 75 (as *Odinia*)

Perrier 1894: 78; Koehler 1895b: 440, (as *O. robusta*)

Downey 1986: 29; Downey in Clark and Downey 1992: 475 (as *Novodinia*)

Range: Bay of Biscay and off Senegal, 1056-2160 m

ODINIA Perrier, 1885

Perrier 1885c: 9; Sladen 1889: 598; Bell 1892 (1893): 105; Perrier 1894: 71; Verrill 1894: 279; 1895: 211; Koehler 1895b: 420; Fisher 1906: 1108; 1916: 31; 1917: 419; A.H. Clark 1916b: 57; Fisher 1918: 105; 1919: 505; A.H. Clark 1934: 1; Hayashi 1943: 136.

Odinia Perrier, 1885 was replaced by *Novodinia* after being shown to be a junior homonym of *Odinia* Robineau-Desvoidy, 1830 (Insecta: Diptera.

americana (Verrill,1880), see *Novodinia*
antillensis A.H. Clark,1934, see *Novodinia*
australis H.L. Clark,1916 see *Novodinia*
austini Koehler, 1909, see *Novodinia*
clarki Koehler, 1909, see *Novodinia*
elegans Perrier, 1885, syn. with *Novodinia homonyma*
magister Fisher, 1917, see *Novodinia*
novaezelandiae H.E.S. Clark, 1962, see *Novodinia*
pacifica Fisher, 1906, see *Novodinia*
pandina Sladen, 1889, see *Novodinia*
penichra Fisher, 1916, see *Novodinia*
robusta Aziz and Jangoux, 1985, see *Novodinia*
emicoronata (Perrier, 1885), see *Novodinia*

Family BRISINGIDAE Sars

Sars 1875: 1; Verrill 1880: 139; Perrier 1885c: 3; Sladen 1889: 603; Perrier 1891c: 198; Verrill 1895: 199; Ludwig 1897a: 418; 1900: 488; Fisher 1906: 1108; Koehler 1907: 3; 1906: 141; 1908: 578; Fisher 1916b: 31; 1917: 419; 1928: 7; Hayashi 1943: 136; Madsen 1951b: 84; Tortonese 1958: 1; A.M. Clark 1962: 68; Bernasconi 1964: 248; Spencer and Wright 1966: U76; H.E.S. Clark 1970: 27 (arm fragments); McKnight 1975: 17; A.M. Clark and Courtman-Stock 1976: 97; Korovchinsky N.M. 1976a: 1205; 1976b: 165; Galkin and Korovchinsky 1984: 164; Korovchinsky and Galkin 1984: 1187; Downey 1986: 4; Downey in Clark and Downey 1992: 462
Genus-group names: **Astrostephane**, **Brisinga**, **Brisingenes**, Craterobrisinga, **Midgardia**, **Stegnobrisinga**.

ASTROSTEPHANE Fisher, 1917
     Fisher 1917f: 421; Fisher 1919: 525; Baker and Clark 1970: 7
     Type species: *Brisinga moluccana* Fisher, 1916
acanthogenys (Fisher, 1916)
     Fisher 1916b: 33 (as *Brisinga*)
     Fisher 1917f: 421; 1919: 528 (as *Astrostephane*)
     Range: Mouth of Lingayan Gulf, Luzon. 344m.
moluccana (Fisher, 1916) with synonym *Brisingenes delli* Fell, 1958
     Fisher 1916b: 32 (as *Brisinga*)
     Fell 1958: 18 (as *Brisingenes delli*)

Fisher 1917f: 421; Baker and Clark 1970: 7 (as *Astrostephane*)
Range: Celebes (Burton Strait) and Molucca Islands, Indonesia.
New Zealand: Mayor Island, Bay of Plenty. Betw. Alderman & Major
Is, 290-1118m.

BRISINGA Asbjornsen, 1856

Asbjornsen 1856: 95; G.O. Sars: 1; Perrier 1882a: 61; 1885a: 441; Sladen 1889: 604; Wood-Mason and Alcock 1891: 12; Alcock 1893b: 170; Perrier 1894: 61: Verrill 1894: 280; 1895: 211; Fisher 1916bb: 31; 1917c: 427; 1918: 103; 1919: 516; Döderlein 1927: 292; Fisher 1928: 7; McKnight 1967: 304; McKnight 1973: 238; A.M. Clark and Courtman-Stock 1976: 97; Downey 1986: 4; Downey in Clark and Downey 1992.

Fisher 1916b: 33; 1917: 427 [as *Brisinga (Craterobrisinga)*]

alberti Fisher, 1907

Fisher 1906: 1111 (as *Brisinga*)

Fisher 1919: 513; 1928: 8 (as *Craterobrisinga*)

Range: Hawaiian Islands: Niihau Islands, Kauai Island, 638-1000 m

acanthogenys syn. *Astrostephane acanthogenys*

americana syn. with *Novodinia americana*

analoga (Fisher, 1919) (new comb.)

Fisher 1919: 516 (as *Craterobrisinga*)

Range: Philippines: Palawan Passage, 750 m^2

andamanica Wood-Mason and Alcock, 1891

Wood-Mason and Alcock 1891: 430; Alcock 1893a: 118

Range: Andaman Sea, 810 m.

armillata syn. with *Brisingella armillata*

bengalensis Wood-Mason and Alcock, 1891

Wood-Mason and Alcock 1891: 439; Alcock 1893a: 119;

Range: Bay of Bengal, 1122 m.

chathamica McKnight, 1973

McKnight 1973: 238

Range: SE New Zealand, 309-512 m

coronata (G.O. Sars, 1871), see *Hymenodiscus*

costata Verrill, 1884a

Verrill 1884a: 382; 1894: 280; 1895: 211; Downey 1986: 5; Downey in Clark and Downey 1992: 465.

Range: Western Atlantic, George's Bank to Venezuela, 1514-2377 m.

cricophora Sladen, 1889 (new comb.)

Sladen 1889: 606; H.L. Clark 1923a: 309; 1926a: 30; A.M. Clark and Courtman-Stock 1976: 97; Downey 1986: 7; Downey in Clark and Downey 1992: 466.

Fisher 1917fc: 426; 1919a: 513 (as *Craterobrisinga*)

Range: Florida, the Bahamas, the Virgin Islands, Sargasso Seas, western South Africa, 360-1340 m.

distincta Sladen, 1889
> Sladen 1899:
> Range: Sub-Antarctic; South of Australia, 5200 m

edwardsi Perrier, 1882 see *Colpaster*

elegans see *Novodinia homonyma*

endecacnemos Asbjornsen, 1856
> Asbjornsen 1856: 95; Sars 1861: 34; Bell 1892: 520; 1893: 104; Perrier 1894: 62; Koehler 1909: 123; Farran 1913: 27; Mortensen 1927: 125; Cherbonnier and Sibuet 1972: 1353, Gage et al. 1983: 284; Downey 1986: 9; Harvey et al. 1988:166; Downey *in* Clark and Downey 1992: 466;
> Range: Norway to Portugal, 183-2000 m.

eucoryne Fisher, 1916 (new comb.)
> Fisher 1916b: 33 [as *Brisinga (Craterobrisinga)*]
> Fisher 1917f: 426 (as *Craterobrisinga*)
> Range: Philippines: Palawan Passage, 750 m

evermanni Fisher, 1906 (new comb.)
> Fisher 1906: 1113 (as *Brisinga*)
> Fisher 1917f: 426 (as *Craterobrisinga*)
> Range: Hawaian Islands: Kauai Island, 800-1000 m.

exilis Fisher, 1905, see *Hymenodiscus*

fragilis Fisher, 1906 see *Hymenodiscus*

gracilis Koehler, 1909 see *Stegnobrisinga*

gunni Alcock, 1893
> Alcock 1893: 120
> Range: Indian Ocean, off Konkan Coast, 1118 m.

hirsuta Perrier, 1894
> Perrier 1894: 66; Downey 1986: 11; Downey in Clark and Downey 1992: 467.
> Range: Iberian Peninsula, Gulf of Guinea, 640-2525 m.

insularum Wood-Mason and Alcock, 1891
> Wood-Mason and Alcock 1891: 439; Alcock 1893: 117
> Range: Laccadive Sea, 2086 m.

mediterranea Perrier, 1885: 442. See *Hymenodiscus*

membranacea Sladen, 1889: See *Hymenodiscus*

mimica Fisher, 1916: see *Brisingenes*

moluccana Fisher, 1916: see *Astrostephane*

multicostata Verrill, 1894: see *Brisingenes*

panamensis Ludwig, 1905 see *Astrolirus*

panopla Fisher, 1906
> Fisher 1906: 1109; Koehler 1909: 117 (as *Brisinga*)
> Fisher 1919: 513; 1928: 8 (as *Craterobrisinga*)
> Range: Hawaii to Indian Ocean (Arabian Sea). Nihau Island, South coast

of Molokai Island, vicinity of Kauai Island, Arabian Sea vicinity of Sri Lanka, 600-1136 m

parallela Koehler, 1909
  Koehler 1909c: 118 (as *Brisinga*)
  Fisher 1919: 513 (as *Craterobrisinga*)
  Range: Indian Ocean/Philippines: Indian Ocean/Arabian Sea off Sri Lanka, 1136 m

placoderma Fisher, 1916. see *Stegnobrisinga*

robusta Perrier, 1894 (a synonym of *Novodinia semicoronata*)

semicoronata Perrier, 1885 see *Novodinia*

synaptoma (Fisher, 1917)
  Fisher 1917f: 426; 1928: 8; Lambert 1978: 8
  Range: British Columbia to Mendocino Ridge, CA, 1353-3176 m.

tasmani H.E.S. Clark, 1970
  H.E.S. Clark 1970: 25
  Range: New Zealand, west of Cape Farewell, 908-915 m.

tenella syn. with *Brisingella tenella*

trachydisca Fisher, 1916
  Fisher 1916bb: 31; Fisher 1919: 510
  Range: Philippines-between Leyte and Mindanao, 1472 m.

variispina Ludwig, 1905
  Ludwig 1905a: 249 (as *Brisinga)*
  Fisher 1928: 8 (as *Craterobrisinga*)
  Range: Paumotu (Tuamotu) Islands,1614 m.

verticellata Sladen, 1889: 604 see *Hymenodiscus*

BRISINGENES Fisher 1917f
  Fisher 1917f: 427; 1919: 517; Downey 1986: 16; Downey in Clark and Downey 1992: 470
  Type Species: *Brisinga mimica* Fisher 1917f

anchista (Fisher, 1919)
  Fisher 1919: 521
  Range: Celebes, Burton Strait, North Island. 1118 m.

delli Fell, 1958: 18; 1963: 41. A synonym of *Astrostephane moluccana* according to Baker and Clark (1970).

mimica (Fisher, 1916)
  Fisher 1916bb: 32 (as *Brisinga*)
  Fisher 1917f: 427 (as *Brisingenes*)
  Range: Philippines, Burton strait, Celebes-1118 m.

multicostata (Verrill, 1894)
  Verrill 1894: 280 (as *Brisinga*)
  Fisher 1917f: 426; 1919: 513 (as *Craterobrisinga*)
  Downey 1986: 17; Downey *in* Clark and Downey 1992: 470 (as *Brisingenes*)

Range: Georges Bank to the Straits of Florida, 805-3186 m.

plurispinula Aziz & Jangoux, 1985

Aziz & Jangoux 1985a: 271

Range: Indo-Malaysian region Indonesia, vicinity of Java/Sumatra, 694-794 m

CRATEROBRISINGA Fisher (A synonym of *Brisinga* according to Downey 1986)

Fisher 1916bb: 33; 1917: 427 [as *Brisinga (Craterobrisinga)*]

Fisher 1919: 512; 1928: 7; McKnight 1973: 238 (as *Craterobrisinga*).

Type Species: *Brisinga panopla* Fisher, 1906

alberti Fisher, 1906 see *Brisinga*

analoga Fisher, 1919 see *Brisinga*

chathamica McKnight, 1973 see *Brisinga*

cricophora Sladen, 1889 see *Brisinga*

eucoryne Fisher, 1916 see *Brisinga*

evermanni Fisher, 1906 see *Brisinga*

multicostata (Verrill, 1894) see *Brisingenes*

panopla Fisher, 1906 see *Brisinga*

parallela Koehler, 1909 see *Brisinga*

synaptoma Fisher, 1906 see *Brisinga*

variispina Ludwig, 1905 see *Brisinga*

GYMNOBRISINGA see LABIDIASTERIDAE

MIDGARDIA Downey, 1972

Downey 1972: 422; 1973: 99; 1986: 19; Downey *in* Clark and Downey 1992: 471.

Type Species: *Midgardia xandaros* Downey, 1972

xandaros Downey, 1972

Downey 1972: 422; 1973: 99; 1986: 19; Downey *in* Clark and Downey 1992: 471

Range: Southern Gulf of Mexico, Yucatan Strait; and off Honduras, 366-467 m

STEGNOBRISINGA Fisher, 1916

Fisher 1916bb: 34 [as *Brisinga (Stegnobrisinga)*]

Fisher 1917f: 428; 1919: 530; H.L. Clark 1926: 31; A. M. Clark and Courtman-Stock 1976: 98; Downey 1986: 29; Downey *in* Clark and Downey 1992: 476. (as *Stegnobrisinga*)

Type Species: *Stegnobrisinga placoderma* Fisher, 1916

gracilis (Koehler, 1909)

Koehler 1909c: 115 (as *Brisinga*)

Fisher 1919: 535

Range: Indian Ocean/Philippines: Bay of Bengal east of the Andaman Islands, 1920 m

splendens H.L. Clark, 1926 [A possible synonym of *S. placoderma* according to Downey (1986)]

H.L. Clark 1926: 31; A.M. Clark and Courtman-Stock 1976: 98; Downey 1986: 31; Downey *in* Clark and Downey 1992: 476.

Range: Forida Strait to S. Caribbean off Venezuela and off Namibia (SW Africa); also

Natal, SE Africa, 860-4000 m.

placoderma Fisher, 1916

Fisher 1916bb: 34 [as *Brisinga (Stegnobrisinga)*]Fisher 1919: 531; Downey 1986: 31; Downey *in* Clark and Downey 1992: 476 as *Stegnobrisinga*)

Range: China Sea, off southern Luzon and ButonStrait, Celebes, 1050-1118 m.

Family FREYELLIDAE Downey

Freyellidae Downey, 1986: 33.; Downey in Clark and Downey 1992: 477.

Genus-group names: **Astrocles, Belgicella, Colpaster, Freyastera, Freyella, Freyellaster,** Freyellidea.

ASTROCLES Fisher, 1917 syn. with *Freyella* by Downey 1986: 43 but synonymy not supported by Mah (1998)

Fisher 1917f: 430

actinodetus Fisher, 1917

Fisher 1917f: 430; Fisher 1928: 29; Djakonov 1950 (1968): 104; Korovchinsky 1976b: 173

Range: Kamchatka opposite Avacha Bay to British Columbia and Oregon, 2870-4200 m.

djakonovi Gruzov, 1964

Gruzov 1964: 1394

Range: North Pacific/Arctic: Okhotsk Sea westward from the Isle Onekotan, 2975 m

japonicus Korovchinsky, 1976

Korovchinsky 1976b: 175

Range: NW Pacific, 3880-3900 m

BELGICELLA Ludwig, 1903

Ludwig 1903: 59

Type species: *Belgicella racowitzana* Ludwig, 1903

racowitzana Ludwig, 1903

Ludwig 1903: 59; Doederlein 1927: 293; Koehler 1907: 141; A.M. Clark 1962: 68; Jangoux and Massin 1986: 91

Range: Antarctica: off Ellsworth, Dronning Maud and Princess Elizabeth Lands, 2450-2800 m.

COLPASTER Sladen, 1889

Sladen 1889: 647; Downey 1986: 33; Downey in Clark and Downey 1992: 477

Type Species: *Colpaster scutigerula* Sladen 1889

edwardsi (Perrier, 1882)

Perrier 1882: 61 (as *Brisinga*)

Perrier 1885d: 441; 1894: 82; Sladen 1889: 616; Koehler 1907: 6; 1909c: 124; Mortensen 1927: 128; Galkin and Korovchinsky 1984: 165 (as *Freyella*)

Downey 1986: 33; Downey in Clark and Downey 1992: 478 (as *Colpaster*)

Range: Golfe de Gascogne, Canary Islands, 1700-2300 m.

scutigerula Sladen, 1889

Sladen 1889: 648; Downey 1986: 34; Downey *in* Clark and Downey: 479

Range: S. of Canary Islands: Gulf of Guinea; off Honduras, 933-2789 m

FREYASTERA Downey, 1986

Korovchinsky and Galkin 1984: 1205 (part) (as *Freyella*)

Downey 1986: 36; Downey in Clark and Downey 1992: 479

Type Species: *Freyella sexradiata* Perrier 1885

Because there are many species of non-Atlantic *Freyella* which potentially belong within *Freyastera* it is possible that the diversity within this genus is currently underepresented.

benthophila (Sladen, 1889).

Sladen 1889: 641; Wood-Mason and Alcock 1891: 430; Alcock 1893a: 121; Fisher 1919: 538; Madsen 1950: 184; Cherbonnier and Sibuet 1972: 1356; Sibuet 1975: 292; Korovchinsky and Galkin 1984: 1215 (as *Freyella*)

Fisher 1917f: 429 (as *Freyellidea*)

Downey 1986: 37; Downey *in* Clark and Downey 1992: 480 (as *Freyastera*)

Range: South Pacific, eastern Pacific off California up to Oregon, Washington Pacific Northwest, Bay of Bengal, north to mid-Atlantic (between Azores and Spain), Bay of Biscay, Gulf of Gascogne, 4250-5000 m.

mexicana (A.H. Clark, 1939)

A.H. Clark 1939a: 442; Korovchinsky and Galkin 1984: 1213 (as *Freyella*)

Downey 1986: 38; Downey *in* Clark and Downey 1992: 481

Range: Gulf of Mexico (type only)2683 m.

sexradiata (Perrier, 1885)

Perrier 1885c: 6 ;1894: 89; Koehler 1909c: 129; Grieg 1921: 30; Madsen 1951b: 84 (as *Freyella*)

318

Downey 1986: 40; Downey *in* Clark and Downey 1992: 482 (as *Freyastera*)

Range: Off east coast of North America, near Azores, Porcupine Abyssal Plain, SW of Ireland, to west of Gibraltar; 4020-5110 m.

tuberculata (Sladen, 1889)

Sladen 1889: 638; Alcock 1893a: 121(as *Freyella)*

Fisher 1917f: 429; H.L. Clark 1920: 113. (as *Freyellidea*)

Downey 1986: 41; Downey *in* Clark and Downey 1992: 482(as *Freyastera*)

Range: Eastern Atlantic (22 deg. N to 3 deg. S); also Bay of Bengal and E. Pacific, 3365-5618 m.

FREYELLA Perrier, 1885. With synonym *Freyellidea.*

Perrier 1885d: 443; 1885c: 5; Sladen 1889: 618; Alcock 1893a: 121; Verrill 1894: 283; Perrier 1894: 81; Verrill 1895: 212; Ludwig 1905a: 267; Fisher 1905: 319; Koehler 1907: 3; 1909c: 127; Fisher 1916bb: 35; 1917: 428; 1918: 104; 1919: 538; 1928: 24; Grieg 1921: 30; A.H. Clark 1939: 442; H.L. Clark 1941: 65; Hayashi 1943: 169; Madsen 1956: 29; Cherbonnier and Sibuet 1973: 1353; Sibuet: 1975: 281; Korovchinsky 1976: 1188; Galkin and Korovchinsky 1984: 161; Korovchinsky and Galkin 1984: 1205-1215; Downey 1986: 43; Downey in Clark and Downey 1992: 483 (as *Freyella*)

Fisher 1917f: 424; 1919; 538; H.L. Clark 1920: 108; Doederlein 1927: 293 (as *Freyellidea*)

Type Species: *Freyella spinosa* Perrier 1885

abyssicola A.H. Clark 1949. A synonym of *Freyella elegans* according to Downey (1986).

*****americana**. Mentioned in Sladen 1889 NOMEN NUDUM

Range: Continental zone off Nova Scotia, 350 m

aspera Verrill, 1884. A synonym of *Freyella elegans* according to Downey (1986).

attenuata Sladen, 1889

Sladen 1889: 645; Galkin and Korovchinsky 1984: 165; Korovchinsky and Galkin 1984: 1215

Range: Philippine waters, 4210 m

benthophila Sladen, 1889: 641. see *Freyastera*

bracteata Sladen, 1889: A synonym of *Freyella elegans* according to Downey (1986).

breviispina (H.L. Clark, 1920)

H.L. Clark 1920: 108 (as *Freyellidea*)

Galkin and Korovchinsky 1984: 165; Korovchinsky and Galkin 1984: 1213 (as *Freyella*)

Range: Indonesian waters 4430m.

*dimorpha Sladen, 1889

Sladen 1889: 623; Fisher 1928: 24; Galkin and Korovchinsky 1984: 165; Korovchinsky and Galkin 1984: 1214

Range: Torres Strait (Australia), 2560 m.

Downey (1986: 46) suggests that this may be a synonym of *F. elegans.*

drygalskii Döderlein, 1927

Döderlein 1927: 293, (as *Freyellidea*)

A.M. Clark 1962: 68; Galkin and Korovchinsky 1984: 165; Korovchinsky and Galkin 1984: 1213 (as *Freyella*)

Range: Antarctic Ocean: off Princess Elizabeth Land, 3425 m.

echinata (Sladen, 1889)

Sladen 1889: 623; McKnight 1975: 59; Galkin and Korovchinsky 1984: 165; Korovchinsky and Galkin 1984: 1213

Range: Vicinity of China Sea,1922-3930 m.

Fisher (1919: 540) states that this might be a species of *Freyellaster*

edwardsi (Perrier, 1882) see *Colpaster*

elegans (Verrill, 1884)

Verrill 1884: 382, (as *Brisinga elegans*) (non *B. elegans* Perrier 1885 = *Novodinia homonyma* Downey 1986, nom. nov.)

Perrier 1885c: 5; 1894: 85; Koehler 1907: 6; Mortensen 1927: 129; Sibuet 1975: 281; 1980: pl 1A; Gage et al. 1983: 285; Galkin and Korovchinsky 1984: 167; Korovchinsky and Galkin 1984: 1213 (in key) (as *Freyella spinosa*)

Sladen 1889: 629; Galkin and Korovchinsky 1984: 165; (as *Freyella bracteata*)

Verrill 1894: 283; 1895: 212; Galkin and Korovchinsky 1984: 165; Korovchinsky and Galkin 1984: 1214; Downey 1986: 43; Downey *in* Clark and Downey 1992: 484 (as *Freyella elelgans*)

Perrier 1894: 85; Sibuet 1975: 292; (as *Freyella spinosa* var. *abyssicola*)

Verrill 1895: 212; Galkin and Korovchinsky 1984: 165; Korovchinsky and Galkin 1984: 1214 (as *Freyella aspera*)

A.H. Clark 1949: 375 (as *Freyella abyssicola*)

Cherbonnier and Sibuet 1973: 1353; Galkin and Korovchinsky 1984: 166; Korovchinsky and Galkin 1984: 1213 (as *Freyella laubieri*)

Range: Greenland south to North Carolina in the west and the northern part of the West European Basin to Angola in the east, 1600-4500m.

Downey(1986) suggests that *F. mutabilis, F. dimorpha,* and *F. pennata* are all conspecific with *F. elegans.*

fecunda, see *Freyellaster*

flabellispina Korovchinsky and Galkin, 1984

Galkin and Korovchinsky 1984: 167; Korovchinsky and Galkin 1984: 1206

Range: SE Pacific. 4160 m.

formosa Korovchinsky, 1976

Korovchinsky 1976a: 1192; Galkin and Korovchinsky 1984: 165; Korovchinsky and Galkin 1984: 1212

Range: South Atlantic: Scott Sea, SE from the Falkland Islands, 3852-3858 m.

*fragilissima Sladen, 1889

Sladen 1889: 626; Fisher 1928: 24; A.M. Clark 1962: 68; Galkin and Korovchinsky 1984: 165; Korovchinsky and Galkin 1984: 1212;

Range: Southern Ocean: between Marion Island and the Crozets and north from Queen Mary land 515-3950 m.

Fisher (1928) notes that without knowledge of the gonad morphology placement of this species within the genus *Freyella* is tentative.

giardi Koehler, 1908

Koehler 1907: 145; 1908: 578; Fisher 1940: 75; A.M. Clark 1962: 68; Galkin and Korovchinsky 1984: 165; Korovchinsky and Galkin 1984: 1213

Range: Southern Ocean: Weddell Sea and Kron Prinsesse Martha Kyst, 4570-4790 m.

heroina Sladen, 1889

Sladen 1889: 543; Fisher 1928: 24; Galkin and Korovchinsky 1984: 165; Korovchinsky and Galkin 1984: 1212

Range: Arabian Sea (Indian Ocean) to North Pacific off Japanese coast, 3000-5305 m.

hexactis Baranova, 1957

Baranova 1957: 174; Galkin and Korovchinsky 1984: 166; Korovchinsky and Galkin 1984: 1215

Range: Bering Sea, south-easterly from Point Oliutarsky. 3939 m

indica Koehler, 1909

Koehler 1909: 126; Galkin and Korovchinsky 1984: 166; Korovchinsky and Galkin 1984: 1213

Range: Bay of Bengal. 2552-2592m. 3190-3244m? (in Koehler)

insignis Ludwig, 1905

Ludwig 1905a: 272; Fisher 1928: 27; Galkin and Korovchinsky 1984: 166; Korovchinsky and Galkin 1984: 1213

Range: Southern California: Santa Catalina Island, California to Gulf of Panama, 3182m-3436 m.

kurilokamchatica Korovchinsky, 1976b

Korovchinsky 1976: 165; Galkin and Korovchinsky 1984: 166; Korovchinsky and Galkin 1984: 1212

Range: Region of Kurilo-Kamchatcka, Japanese Idzu-Bonine trenches and in the northwestern basin of the Pacific Ocean, 4890-6860 m.

laubieri Cherbonnier and Sibuet, 1972. A synonym of *F. elegans* according to Downey (1986: 46).

loricata Korovchinsky and Galkin, 1984

Galkin and Korovchinsky 1984: 167; Korovchinsky and Galkin 1984: 1208

Range: NW Pacific: regions of Kurilo-Kamchatka and Idzu-Bonine fissures and NW region, 4995-5998 m.

macropedicellaria Korovchinsky and Galkin, 1984

Korovchinsky and Galkin 1984: 1209; Galkin and Korovchinsky 1984: 167.

Range: Indian Ocean Bay of Bengal, 5120 m.

mexicana Clark, 1939. See *Freyastera*

microplax (Fisher, 1917)

Fisher 1919: 538; 1928: 25; Galkin and Korovchinsky 1984: 166; Korovchinsky and Galkin 1984: 1214 (as *Freyella*)

Fisher 1917f: 430 (as *Freyellidea*)

Range: British Columbia, Central California, 1722m-3176 m.

microspina Verrill, 1894

Verrill 1894: 285; Galkin and Korovchinsky 1984: 166; Korovchinsky and Galkin 1984: 1214 (key); Downey 1986: 46; Downey *in* Clark and Downey 1992: 485 (as *Freyella elegans*)

H.L. Clark 1941: 65; Galkin and Korovchinsky 1984: 167; Korovchinsky and Galkin 1984: 1214 (key) (as *Freyella trispinosa*)

Range: Off Nantucket; southern coast of Cuba, Surinam (Africa), 1847-2734 m.

mortenseni Madsen, 1956

Madsen 1956: 29; Galkin and Korovchinsky 1984: 166; Korovchinsky and Galkin 1984: 1213

Range: Kermadec Trench (close to New Zealand): 5850-6160 m.

This is another species which might possibly be included in *Freyastera*.

mutabilia Korovchinsky, 1976

Korovchinsky 1976: 1188; Galkin and Korovchinsky 1984: 166; Korovchinsky and Galkin 1984: 1212; Downey 1986: 46

Range: Sub-Antarctic; Southern Ocean side of the Southern Sandwiche fissure, Southern Orknian fissure north Scott Sea, 664-6070 m.

Downey (1986: 46) states this is a possible junior synonym of *Freyella elegans*

octoradiata (H.L. Clark, 1920)

H.L. Clark 1920: 110 (as *Freyellidea*)

Galkin and Korovchinsky 1984: 166; Korovchinsky and Galkin 1984: 1213 (as *Freyella*)

Range: East Pacific, Southern California, Central America, 4085-4430 m.

oligobrachia (H.L. Clark, 1920)

H.L. Clark 1920: 112 (as *Freyellidea*)

Bernasconi 1967: 418; Galkin and Korovchinsky 1984: 166; Korovchinsky and Galkin 1984: 1215 (as *Freyella*)

Range: East Pacific to Chile, 4030-4241 m

Bernasconi (1967) states that this may be related to or synonymous with *F.benthophila*

oligobrachia polyspina Korovchinsky, 1976

Korovchinsky 1976b: 170; Galkin and Korovchinsky 1984: 166

Range: NW part of the Pacific Ocean to the east from Bonin Island to the south from Kuriles and to the South from W. Aleutians.

pacifica Ludwig, 1905

Ludwig 1905: 267; Galkin and Korovchinsky 1984: 166; Korovchinsky and Galkin 1984: 1213

Range: Paumotu Archipelago (aka Tuamotu Archipelago). 1485-3200 m.

pennata Sladen, 1889

Sladen 1889: 618; Hayashi 1943: 160; Galkin and Korovchinsky 1984: 166 ; Korovchinsky and Galkin 1984: 1213

Range: Japan: south of Kawatsu, 1200-3429 m.

Downey (1986: 46) suggests that this is a synonym of *Freyella elegans*.

polycnema syn. with *Freyellaster polycnema*

propinqua Ludwig, 1905

Ludwig 1905: 280; Galkin and Korovchinsky 1984: 166; Korovchinsky and Galkin 1984: 1214

Range: Cape San Francisco, Gulf of Panama 2877 m.

recta Koehler, 1907

Koehler 1907: 3; 1909; 127; Galkin and Korovchinsky 1984: 166; Korovchinsky and Galkin 1984: 1214; Downey 1986: 48; Downey *in* Clark and Downey 1992: 485

Range: Mid-Atlantic ridge between Azores and Virgin Islands, 3465m.

Downey 1986: 49 notes that this species may yet be another junior synonym of *Freyella elegans*.

remex Sladen, 1889

Sladen 1889: 635; Fisher 1928: 24; Galkin and Korovchinsky 1984: 167; Korovchinsky and Galkin 1984: 1212

Range: Coral Sea , SE of New Guinea. 4880 m

Fisher (1928) notes that without knowlede of the gonad morphology placement of this species within the genus *Freyella* is tentative.

scalaris A.H. Clark, 1916 see *Freyellaster*

sexradiata Perrier, 1885 see *Freyastera*

spatulifera Fisher, 1916 see *Freyellaster*

spinosa Perrier 1885. A synonym of *Freyella elegans* according to Downey (1986)

trispinosa H.L. Clark, 1941 a synonym of *Freyella microspina* according to Downey (1986)

tuberculata Sladen, 1889 see *Freyastera*

vitjazi Korovchinsky and Galkin, 1984

Galkin and Korovchinsky 1984: 167; Korovchinsky and Galkin 1984: 1211

Range: East Indian Ocean. 3680-3900 m.

This might be another species belonging to *Freyastera*.

FREYELLASTER Fisher, 1918

Fisher 1918: 104; 1919: 537; 1928: 21

Type Species: *Freyella fecunda* Fisher 1905

fecundus (Fisher, 1905) with forma *ochotensis* Hayashi 1943

Fisher 1905: 319 (as *Freyella*)

Fisher 1918: 104; 1919: 538; 1928: 21; Lambert 1978a: 10; Lambert 1978b: 62 (as *Freyellaster*)

Hayashi 1943: 153 (as forma *ochotensis*)

Range: Okhotsk Sea (Japan), British Columbia to Central California (Point Pinos, Monterey Bay), 880-2124 m.

intermedius Hayashi, 1943

Hayashi 1943: 156

Range: Japan: Omaesaki, 1836 m.

polycnema (Sladen, 1889)

Sladen 1889: 621(as *Freyella*)

Fisher 1918: 104; 1928: 21 (as *Freyellaster*)

Range: E of the Kermadec islands, 1200 m

scalaris (A.H. Clark, 1916)

A.H. Clark 1916a: 115; Fisher 1918: 104 (as *Freyella*)

Fisher 1919: 538 (as *Freyellaster*)

Range: Galapagos Islands, 1480 m.

spatulifer (Fisher, 1916)

Fisher 1916b: 34; 1917: 429 (as *Freyella*)

Fisher 1918: 104; 1919: 538 (as *Freyellaster*)

Range: Celebes,Mascassar Strait off Mamuji Island, 1802 m.

FREYELLIDEA Fisher 1917. A synonym of *Freyella* but some *F. benthophila* and *F. tuberculata* later placed into *Freyastera* Downey 1986.

Fisher 1917f: 429

benthophila Sladen, 1889 see *Freyastera*

breviispina H.L. Clark, 1920 see *Freyella*

drygalskii Döderlein, 1927 see *Freyella*

microplax Fisher, 1917 see *Freyella*

octoradiata H.L. Clark, 1920 see *Freyella*

oligobrachia H.L. Clark, 1920 see *Freyella*

tuberculata Sladen, 1889 see *Freyastera*

Family HYMENODISCIDIDAE Mah

Genus-group names: **Astrolirus,** Brisingella, **Hymenodiscus, Parabris-inga**.

ASTROLIRUS Fisher, 1917
panamensis (Ludwig, 1905)
  Ludwig 1905a: 259; Macan 1938: 419; (as *Brisinga*)
  Fisher 1917f: 428; 1928: 20 (as *Astrolirus*)
  Range: East Pacific-Cocos Islands, Malpelo Island, the Galapagos, Gulf of California , and Gulf of Panama to Oregon, Pacific Northwest, 1820-2418m.
BRISINGELLA Fisher,1917 (a synonym of *Hymenodiscus* according to Mah 1998)
  Fisher 1917f: 427
aotearoa McKnight, 1973 see *Hymenodiscus*
armillata (Sladen, 1889) see *Hymenodiscus*
beringiana Korovchinsky, 1976 see *Hymenodiscus*
coronata (G.O. Sars 1871) see *Hymenodiscus*
distincta Sladen,1889 see *Hymenodiscus*
exilis (Fisher, 1905) see *Hymenodiscus*
fragilis (Fisher, 1906) see *Hymenodiscus*
mediterranea (Perrier, 1881) see *Hymenodiscus*
membranacea (Sladen, 1889) see *Hymenodiscus*
monacantha H.L. Clark, 1920 see *Hymenodiscus*
ochotensis Djakonov, 1950 see *Hymenodiscus*
pannychia Fisher, 1928 see *Hymenodiscus*
pusilla Fisher, 1917 see *Hymenodiscus*
submembranacea (Döderlein, 1927) see *Hymenodiscus*
tenella (Ludwig, 1905) see *Hymenodiscus*
verticellata (Sladen, 1889)see *Hymenodiscus*
HYMENODISCUS Perrier, 1884 (with synonym *Brisingella*)
  Perrier 1882: 61; 1884: 189; Fisher 1918: 104; 1919: 502; Spencer and Wright 1962: U78 (as *Hymenodiscus*);
  Fisher 1917f: 419; 1918: 104; 1919: 515; H.L. Clark 1920: 107; Fisher 1928: 13; 1940: 206; Hayashi 1943: 152; Djakonov 1950; Dilwyn John and A.M. Clark 1954: 150; Spencer and Wright 1962: U78; McKnight 1973: 235; Downey 1986: 13; Downey in Clark and Downey 1992: 468 (as *Brisingella*)

Yamaoka 1987: 5 (as fossil *Brisingella* sp.)

Fisher 1918: 104

Type species: *Hymenodiscus agassizi* Perrier 1882

Many nominal species of *Hymenodiscus* are known only from type material, which in many cases is based on arm tips and other fragments. The widespread distributions of the better known taxa and the taxonomic implications of recognizing juvenile brisingidans dictate that all the species within this genus be seriously re-examined.

agassizi Perrier, 1882

Perrier 1882: 61; Fisher 1918: 104; Dilwyn-John and A.M. Clark 1954: 151

Range: Santa Cruz, Dominica. Lesser Antillies.

aotearoa McKnight, 1973 (new comb.)

McKnight 1973: 235 (as *Brisingella*)

Range: SE New Zealand waters, 113m-1518m

armillata (Sladen, 1889) (new comb.)

Sladen 1889: 608; Hayashi 1943: 152 (as *Brisingella*)

Range: off the coast of Japan, south of Kawatsu. 3750m

beringiana Korovchinsky, 1976 (new comb.)

Korovchinsky 1976b: 172 (as *Brisingella*)

Range: Bering Sea, 3661 m.

coronata (G.O. Sars, 1871) with synonym *Brisinga mediterranea* Perrier 1885 (new comb.)

G.O. Sars 1872: 5; Thomson 1873: 66(pt); G.O.Sars 1875: 4; Ludwig 1878: 216; Perrier 1882: 61; Danielssen and Koren 1884: 104; Carus 1885: 91; Perrier 1885d: 442; 1885: 4; Sladen 1889: 598; Bell 1889: 433; Sladen 1891: 698; Bell 1893: 105; Norman 1893; 347; Perrier 1894: 50; Koehler 1896a: 440; von Marenzeller 1895b: 137; Koehler 1896b: 38; Perrier 1896; 20; Ludwig 1897a: 308; 1897b: 418; 1900: 488 (as *Brisinga coronata*)

Marion 1883: 129 (as *Brisinga* sp.)

Perrier 1885d: 442; 1885b: 3; 1894: 70; von Marenzeller 1895b: 137; Boone 1933: 93(as *Brisinga mediterranea*)

Grieg 1927b: 127; Mortensen 1927: 127; John and A.M. Clark 1954: 151; Gage et al. 1983: 285; Downey 1986: 13 (as *Brisingella coronata*)

Range: Norway to Azores and Cape Verde Islands; Mediterranean 100-2600 m.

*distincta Sladen, 1889 (new comb.)

Sladen 1889: ;A.M.Clark 1962: 68; A.M. Clark 1962: 68

Range: Antarctic Ocean: south from Australia. 43°S, 134°E. 4755 m

Fisher (1928) tentatively referred this species and *B. membranacea* to *Brisingella* (now *Hymenodiscus*). However, only detached arms of this species are known and an unambiguous placement seems unlikely without collection and examination of less damaged material. It also seems unlikely that species based on such limited material will remain valid after more material is studied.

exilis (Fisher, 1905) (new comb.)

Fisher 1905: 318 (as *Brisinga*)

Fisher 1917f: 427; 1919: 524; 1928: 13;Alton 1966: 1708 (as *Brisingella*)

Range: Off Northern Oregon, central California, southern California: off San Diego to Santa Barbara Island, 592-2118 m.

fragilis (Fisher, 1906) (new comb.)

Fisher 1906: 1115 (as *Brisinga*)

Fisher 1917f: 423 (as *Brisingella*)

Range: Hawaii, 256-414 m.

*membranacea (Sladen, 1889) (new comb.)

Sladen 1889: ; A.M. Clark 1962: 68

Range: Antarctica between Marion Island and the Crozet Islands and west of the Crozet Islands, 2750-3200 m.

Fisher (1928) tenatively referred this species and *B. distincta* to *Brisingella* (now *Hymenodiscus*). Only detached arms of this species are known and an unambiguous placement seems unlikely pending collection and examination of less damaged material. It also seems unlikely that species based on such limited material will remain valid after more material is studied.

*monacantha H.L. Clark, 1920 (new comb.)

H.L. Clark 1920: 107

Range: off coast of Peru, W. pt. Aguja, 4064 m.

Only two detached arms (no disc) of this species are known. It seems unlikely that a species based on such limited material will remain valid after more material is studied.

ochotensis Djakonov, 1950 (new comb.)

Djakonov 1950a (1968): 103; 1950b: 29; Korovchinsky 1976b: 171;

Range: Sea of Okhotsk, 1,366 m.

pannychia Fisher, 1928 (new comb.)

Fisher 1928: 18; Hayashi 1943: 149 (as *Brisingella*)

Range: Alaska-Bowers Bank, Bering Sea, 1542 m.

pusilla Fisher, 1917 (new comb.)

Fisher 1917f: 427; 1928: 16 (as *Brisingella*)

Range: Off Southern California from San Diego to Santa Cruz Island, 602-2118 m.

submembranacea (Döderlein, 1927) (new comb.)

Döderlein 1927: 292 (as *Brisinga*)

A.M. Clark 1962: 68 (ref. prob. to *Brisingella*)

Range: off Princess Elizabeth and Wilhelm II islands, 2450-2926 m.

tenella (Ludwig, 1905) (new comb.)

Ludwig 1905a: 255; Macan 1938: 419 (as *Brisinga*)

Fisher 1917f: 524 (as *Brisingella*)

Range: Galapagos, 2418 m

verticellata (Sladen, 1889) (new comb.)

Sladen 1889: 604; Verrill 1894: 283; 1895: 211; H.L. Clark 1941: 64 (as *Brisinga*)

Fisher 1917f: 427; 1919: 524; Downey 1986: 15; Downey in Clark and Downey 1992: 469 (as *Brisingella*)

Range: Georges Bank to Cuba and Gulf of Mexico, 640-2570 m.

PARABRISINGA Hayashi, 1943

Hayashi 1943: 144 Spencer and Wright 1966: U78 ; Downey in Clark and Downey 1992: 462

Type Species: *Parabrisinga pellucida* Hayashi 1943

Parabrisinga is differentiated from *Hymenodiscus* solely on the basis of the gonad morphology which may have been mistakenly for the pyloric caeca and thus misinterpreted by Hayashi as serial rather than paired. That in addition to the small size of the material described and the similarity of other morphological characters leads me to believe that *P. pellucida* is probably a synonym of some other species of *Hymenodiscus*. A search with Japanese colleagues was unsuccessful in locating the holotype which is now presumed lost or destroyed.

*pellucida Hayashi, 1943

Hayashi 1943: 144

Range: "Probably" Misaki, Japan, 700 m.

ADDENDA

The three preceding parts of this Index went to press in the series Echinoderm Studies in the following years: Part 1. Paxillosida and Notomyotida in
1988; Part 2. Valvatida in 1992 and Part 3. Velatida and Spinulosida in 1995
Inevitably with a work of this kind there were some omissions and errors and
also there have been some subsequent publications which included relevant
taxonomic changes, additions and extensions of range. The following section should bring the whole index up to date as of 1999 to the best of my
knowledge [A.M.C.]

List of genus-group names (for the whole class, recent only).

Coronaster	Labidiasteridae
Ctenodiscus	Ctenodiscidae, not Goniopectinidae
Damnaster nov	Porcellanasteridae
Endogenasterias	Asteriidae
Enigmaster nov.	Goniasteridae
Eoleptasterias	Asteriidae
Gephyreaster	Goniasteridae, not Radiasteridae (Blake, 1987)
Glyphodiscus	Goniasteridae, not Goniopectinidae (Sastry, unpubl
Lysasterias	Asteriidae, not Porcellanasteridae
Nectria and Nectriaster	Oreasteridae, not Goniasteridae
Odontohenricia nov.	Echinasteridae
Paralophaster	Solasteridae
Plazaster	Labidiasteridae
Tegulaster	Asterinidae, not Asteriidae
Tomidaster	Goniasteridae
Tremaster	Asterinidae

ADDITIONS TO PART 1.

Family LUIDIIDAE

Luidia magellanica Leipoldt, 1895. Add '(with synonym *Asterias (Smilasterias) terweili* Goldschmidt, 1924)' according to Madsen (1956).
L. prionota Fisher, 1913. A synonym of *L. hardwicki* (Gray, 1840)
according to Liao & A.M. Clark (1995 [1996]).
L. prionota [as *prionata*]: Mah 1998b:66 (extension of range from Philippines and N Australia to Hawaii).

Family ASTROPECTINIDAE

ASTROPECTEN - from Downey *in* Clark & Downey (1992:25-44), unless otherwise stated.

armatus - Atlantic records only - non *A. armatus* Gray, 1840 = *A. brasiliensis* Müller & Troschel, 1842.

aurantiacus multispinosus da Costa, 1942:3. A synonym of *A. aranciacus* (Linnaeus, 1758).

carcharicus and hartmeyeri Döderlein, 1917. Synonyms of *A. vappa* Müller & Troschel, 1842 according to A.M.C. *in* Liao & Clark (1995[1996]).

caribemexicanensis Caso, 1990 **sp. nov.**
  Caso 1990: 108 (comparison with *A. nitidus* Verrill, 1915).
  Range: Caribbean side of Mexico.

granulatus: Jangoux, 1973, non *A. granulatus* Müller & Troschel, 1842, = *A. monacanthus* Sladen, 1883, according to Walenkamp (1990).

guineensis Koehler, 1911. A synonym of *A. gruveli* Koehler, 1911.

jarlii Madsen, 1950. A synonym of *A. cingulatus* Sladen, 1883.

longibrachius sp. nov. Jangoux & Aziz, 1988
  Jangoux & Aziz 1988:635.
  Range: Off Réunion Island, western Indian Ocean, 450-580 m.
  ? Astropecten. Very long-armed and with two rows of actinal plates. A.M.C. **(new obs.)**

michaelseni Koehler, 1914. A synonym of *A. irregularis pontoporeus* Sladen, 1883.

nitidus: Caso. 1990:109.

nitidus forcipatus Verrill, 1915. Not subspecifically distinct from *A. nitidus nitidus* Verrill.

nuttingi Verrill, 1915. A synonym of *A. alligator* Perrier, 1881.

riensis Döderlein, 1917. A synonym of *A. brasiliensis* Müller & Troschel, 1842.

sinicus Döderlein, 1917, see *Ctenopleura*

sphenoplax Bell, 1892, see *Persephonaster*

validispinosus sp. nov. Oguro, 1982
  Oguro 1982:72.
  Range: Palau and Yap, Caroline Is.

weberi Döderlein, 1917. A synonym of *A. irregularis pontoporeus* (above).

Bathybiaster herwigi Bernasconi, 1972. Referred to *Psilaster* by A.M.C. *in* Clark & Downey (1992:46,80).

loripes obesus Sladen, 1889. Raised to specific rank by A.M.C. *in* Clark & Downey (1992:46). *B. spinulatus* Koehler, 1917 (from Kerguelen) probably a synonym of *obesus* rather than *loripes* A.M.C. **(new obs.)**

Ctenophoraster donghaiensis sp. nov. Liao & Sun, 1989

Liao & Sun 1989:225.

Range: East China Sea (31°N, 128°E), 147 m.

Ctenopleura sinica **new comb.** (Döderlein, 1917) Liao *in* Liao & Clark 1995 [1996]:82.

Dipsacaster pretiosus (Döderlein, 1902) Liao *in* Liao & Clark 1995 [1996]:84 (extension of range from Honshu, Japan to East China Sea at Zhejiang and possibly southern China (from fish market).

Dytaster agassizi Perrier, 1894, biserialis Sladen, 1889 and rigidus Perrier, 1894. Synonyms of *D. grandis* (Verrill, 1884), also *nobilis* Sladen, 1889. A subspecies of *D. grandis* according to Downey *in* Clark & Downey (1992).

parvulus Koehler, 1909. A synonym of *D. semispinosus* (Perrier, 1894) according to Downey (1992).

cherbonnieri Sibuet, 1975
Stampanato & Jangoux 1997:28 (extension of range from Angola and Namibia to NW of Amsterdam Island, southern Indian Ocean).

Persephonaster exquisitus sp. nov. Jangoux & Aziz, 1988
Jangoux & Aziz 1988:638.
Range: Off R,union Island, western Indian Ocean (21°S, 55°E), 1970-2950 m.

? Persephonaster pulcher (Perrier, 1881): H.L. Clark, 1941. Probably conspecific with *Psilaster cassiope* Sladen, 1889 but proposed for suppression by A.M.C. *in* Clark & Downey (1992).

spinulosus H.L. Clark 1941. Confirmed as a synonym of *P. patagiatus* (Sladen, 1889) by A.M. Clark (1992).

Plutonaster agassizi agassizi (Verrill, 1880) Stampanato & Jangoux 1997:28 (extension of range from Atlantic, including S of Cape Town, to near St Paul Island, southern Indian Ocean).

Plutonaster abbreviatus Sladen, 1889 and P. granulosus Perrier, 1891. Synonyms of *P. agassizi notatus* Sladen, 1889, reduced to a subspecies, edwardsi (Perrier, 1882). Possibly conspecific with *P. agassizi notatus*, but proposed for suppression

intermedius (Perrier, 1881) and marginatus Sladen, 1889. Both synonyms of *P. agassizi agassizi* (Verrill, 1880)

patagiatus (Sladen, 1889). Referred back to *Persephonaster* from *Psilaster*

proteus H.L. Clark, 1923. A synonym of *P. bifrons* (Wyville Thomson, 1873), though possibly recognisable as a South African subspecies. All according to A.M. Clark *in* Clark & Downey (1992).

Persephonaster euryactis (Fisher, 1913) (included in part 1 as *Proserpinaster*) Imaoka et al. (ref. in part 3) 1991:47 (in japanese, apart from description; apparently treating Persephonaster misakiensis Goto, 1914 as a synonym since range given as extending from the Philippines to Tosa Bay, SE Japan.

Psilaster florae (Verrill, 1878). Reduced to a subspecies of *P. andromeda* Müller & Troschel, 1842)

herwigi (Bernasconi, 1972), new combination, referred from *Bathybiaster*

patagiatus Sladen, 1889, referred back to *Persephonaster*.
All according to A.M. Clark *in* Clark & Downey (1992).

Psilasteropsis humilis Koehler, 1907. A synonym of *Persephonaster sphenoplax* (Bell, 1892), referred from *Astropecten*, according to A.M. Clark *in* Clark & Downey (1992).

Tethyaster aulophorus (Fisher, 1911c) Liao & Clark 1995 [1996]:85 (extension of range from Philippines to Hainan Island, 128-174 m.)

T. magnificus (Bell, 1882). Reduced to a subspecies of *T. vestitus* (Say, 1825), according to A.M. Clark *in* Clark & Downey (1992).

Tethyaster tangaroae sp. nov. Rowe, 1989
Rowe 1989:268.
Range: off Norfolk Island, NE Tasman Sea, 390-423 m.

Family RADIASTERIDAE

GEPHYREASTER Fisher, 1910. To the Goniasteridae according to Blake (1987).

Radiaster notabilis (Fisher, 1913) Imaoka et al. 1991:50 (extension of range to S Japan).

Family PORCELLANASTERIDAE

ALBATROSSIA Ludwig, 1905 and ALBATROSSASTER Ludwig, 1907. Synonyms of

CAULASTER Perrier, 1882: Belyaev & Mironov 1996:892 (revived from the synonymy of *Porcellanaster* Wyville Thomson, 1877; with synonyms *Albatrossia* Ludwig, 1905 and *Albatrossaster* Ludwig, 1907).

eremicus (Sladen, 1889) **new comb.**: Belyaev & Mironov 1996:896 (revived from synonymy of *Porcellanaster ceruleus* Wyville Thomson, 1877, with synonym *Albatrossaster richardi* Koehler, 1909; range including Tasman Sea, SE Pacific and Antarctic).

pedunculatus Perrier, 1882: Belyaev & Mironov 1996:892 (size range described).

DAMNASTER gen. nov. H.E.S. Clark & McKnight, 1994
H.E.S. Clark & MacKnight 1994:1368.
Type species: *Damnaster tasmani* Clark & McKnight, 1994.

332

tasmani sp. nov. H.E.S. Clark & McKnight, 1994
Clark & McKnight 1994:1368.
Range: Tasman Sea, W of New Zealand (35-46°S, 156-167°E), 1647-4868 m.

HYPHALASTER Sladen, 1883: Belyaev & Mironov 1993:205.
australis sp. nov. Belyaev & Mironov, 1993
Belyaev & Mironov 1993:206.
Range: Southern Ocean S and SW from Heard Island, 1320-1820 (?2020) m.
inermis Sladen, 1883, sensu lato: Belyaev & Mironov 1993:210 (N and tropical Atlantic, Indian Ocean, mid-Pacific).
multispinus sp. nov. Belyaev & Mironov, 1993
Belyaev & Mironov 1993:214.
Range: NW Pacific SE from Kamchatka (48°N, 169°E), 2915-3015 m.

PORCELLANASTER Wyville Thomson, 1877: Belyaev & Mironov 1996: 887.
ceruleus Wyville Thomson, 1877: Belyaev & Mironov 1996:887 (synonymy restricted).
caulifer Sladen, 1883, *granulosus* Perrier, 1885 and *tuberosus* Sladen, 1883:
Belyaev & Mironov 1996:887 (revived from synonymy of *P. ceruleus*).
STYRACASTER Sladen, 1883
Belyaev & Moskalev 1986:1015 (key to the species).
armatus Sladen, 1883: Belyaev & Moskalev 1986:875 (extension of range to the Arabian Sea, and the central Pacific, S from Hawaii and depth down to 5420 m.).
chuni Ludwig, 1907: Belyaev & Moskalev 1986:876.
longispinus sp. nov. Belyaev & Moskalev, 1986
Belyaev & Moskalev 1986:772.
Range: SE from Kurile Is, NW Pacific (38°N, 166°E to 54°N, 160°E), 5040 (?4990) - 6300 (?6330) m.
paucispinus Ludwig, 1907: Belyaev & Moskalev 1986:873 (range extended to NW Pacific, near Kurile Is and depth down to 5200 m.).
robustus Koehler, 1908: Belyav & Moskalev 1986:873 (extension of range in Southern Ocean from Atlantic to Indian Ocean sector and depth down to 4540 m.).
simplipaxillatus sp. nov. Belyaev & Moskalev, 1986
Belyaev & Moskalev 1986:872.
Range: Western Indian Ocean, off the Seychelles, 3890 m.
transitivus sp. nov. Belyaev & Moskalev, 1986
Belyaev & Moskalev 1986:867.
Range: Mid-Pacific and off Peru, 4660-5840 m.

Family CTENODISCIDAE nov. Blake, 1987

[as Ctenodiscididae, misled by A.M.C.]
Ctenodiscus crispatus (Retzius, 1805). Additional synonym *Asterias arancia* Dewhurst, 1834.

Family GONIOPECTINIDAE

Goniopecten binghami Boone, 1928 and Pectinidiscus aliciae Carrera-Rodriguez & Tommasi, 1977. Synonyms of *Prionaster elegans* Verrill, 1899, according to A.M.C. *in* Clark & Downey (1992:111).

Family BENTHOPECTINIDAE

Benthopecten spinosus (Sladen) (tentatively) [see *B. spinosus* (Verrill) in Index pt. 1, p. 311, where inclusion of *Pararchaster occidentalis* Sladen, 1889 in synonymy should have led to extension of range to Japan] Imaoka et al. 1991:49.
Luidiaster oxyacanthus (Sladen) (in Index pt. 1, p. 315, as *Cheiraster (Christopheraster)*), Imaoka et al. 1991:48.

ADDITIONS TO PART 2

Family CHAETASTERIDAE

Chaetaster moorei Bell, 1894. A synonym of *Chaetaster vestitus* Koehler, 1910 according to Rowe *in* Rowe & Gates (1995:56).
Chaetaster vanzolinicus Tommasi, 1972. A synonym of *Chaetaster nodosus* Perrier, 1875, according to Downey *in* Clark & Downey (1992:146).

Family ODONTASTERIDAE

Acodontaster waitei (Koehler,1920). A synonym of *Acodontaster conspicuus* (Koehler, 1920), according to Stampanato & Jangoux 1993:175.
Asterodon granulosus Perrier, 1891. A synonym of *Asterodon singularis* (Müller & Troschel, 1843), according to Orovitz & Tablado 1990:31.
Odontaster setosus Verrill, 1899. A synonym of *Odontaster robustus* Verrill, 1899, though both of uncertain validity with respect to *O. hispidus* Verrill, 1880, according to A.M.Clark *in* Clark & Downey 1992:157.

Family GANERIIDAE

Ganeria hahni Perrier, 1891: Bernasconi, 1964. Maintained as a synonym of
Ganeria falklandica Gray, 1840, by A.M.Clark (1992:166).

Family ASTERINIDAE

Anseropoda insignis: Chave & Malahoff 1998:43, fig.
Asterina atyphoida H.L. Clark 1916: Marsh & Pawson 1993:281 (extension
of range from South Australia to Rottnest Island, Western Australia).
cepheus (Müller & Troschel, 1842) Liao & Clark 1995 [1996]:130 (restored
to specific rank distinct from *A. burtoni* Gray)).
orthodon Fisher, 1922: Liao *in* Liao & Clark 1995 [1996]:130 (specimens
from Hainan possibly distinct from those from Hong Kong).
lutea H.L. Clark, 1938, A. nuda H.L. Clark, 1921 and A. orthodon Fisher,
1922. Synonyms of *Asterina sarasini* (de Loriol, 1897) according to
Rowe *in* Rowe & Gates (1995:35).
perplexa H.L. Clark, 1938. A synonym of *Asterina inopinata* Livingstone,
1933 according to Rowe *in* Rowe & Gates (1995:35).
ASTERINIDES Verrill, 1913, with type species A. *folium* (Lütken, 1859)
Rowe *in* Rowe & Gates 1995:33; Campbell & Rowe 1997:130. Confirmed as generically distinct from *Asterina* Nardo, 1834.
Asterinopsis lymani (Perrier, 1881). A synonym of *Asterinopsis pilosa* (Perrier, 1881) according to A.M.C. *in* Clark & Downey (1992:190).
praetermissa Livingstone, 1933, see *Paranepanthia*
Nepanthia belcheri (Perrier, 1875) Liao *in* Liao & Clark 1995 [1996]:133
(extension of range to southernmost China).
Paranepanthia praetermissa (Livingstone, 1933). A synonym of *Paranepanthia rosea* H.L. Clark, 1938, according to Rowe *in* Rowe & Gates
(1995:39).
Patiriella exigua (Lamarck, 1816) Stampanato & Jangoux 1997:31.
gunni (Gray, 1840). Invalidated by *Asterias calcar* var. *hexagona* Lamarck,
1816, which needs rejection by the ICZN [**new obs.**]
nigra H.L. Clark, 1938. A synonym of *P. oliveri* (Benham, 1911) according
to Rowe *in* Rowe & Gates (1995:40).
obscura Dartnall, 1971. A synonym of *P. pseudoexigua* Dartnall, 1971
according to Rowe *in* Rowe & Gates (1995:41).
paradoxa sp. nov. Campbell & Rowe, 1997, Campbell & Rowe 1997:131.
Range: Dhofar, southern Oman.

Family PORANIIDAE

Poraniella Verrill, 1914, referred to the family Asteropseidae by A.M. Clark (1984:20 and *in* Clark & Downey (1992).

japonica Fisher, 1939. A synonym of *Poraniopsis inflatus* (Fisher, 1906) according to Oguro (1989).

inflata flexilis (Fisher, 1910) Ahearn 1995:29.

Spoladaster veneris (Perrier, 1879) Stampanato & Jangoux 1997:31 (extension of depth range to 1050 m.)

Family GONIASTERIDAE

Calliaster Gray, 1840, with synonym *Mabahissaster* Macan, 1938 according to Rowe *in* Rowe & Gates (1995:64).

childreni Gray, 1840: Liao & Clark 1995 [1996]:92 occurrence in S China doubtful).

regenerator Döderlein, 1922 and

spinosus H.L. Clark, 1916. Both referred to *Milteliphaster* by Rowe (in} Rowe & Gates (1995:66).

pedicellaris: Chave & Malahoff 1998: & fig.

zengi (Macan, 1938) **new comb.** Rowe & Gates 1995:64.

Calliderma spectabilis: Chave & Malahoff 1998:43 & fig.

Ceramaster patagonicus (Sladen, 1889), reduced to a subspecies of

grenadensis (Perrier, 1881) according to Downey *in* Clark & Downey (1992:237).

trispinosus H.L. Clark, 1923, reduced to a subspecies of

granularis (Retzius, 1783) according to Downey (1992:234).

ENIGMASTER McKnight & H.E.S. Clark, 1996

McKnight & Clark 1996:205.

Type species: *Enigmaster scalaris* McKnight & Clark, 1996

scalaris McKnight & H.E.S. Clark, 1996

McKnight & Clark 1996:205.

Range: Off Auckland Is, S of New Zealand, 520 m.

Hippasteria argentinensis Bernasconi, 1961 and *H. strongylactis* H.L. Clark, 1926, reduced to subspecies of *H. phrygiana* (Parelius, 1768) according to Downey (1992:248).

imperialis Goto, 1914: Mah 1998b:67 (extension of range from southern Japan to Hawaii).

Litonotaster rotundigramulum. A synonym of *Litonotaster intermedius* (Perrier, 1883) according to Downey (1992:251).

MABAHISSASTER Macan, 1938, with *M. zengi* Macan, 1938. A synonym of *Calliaster* Gray, 1840 according to Rowe *in* Rowe & Gates (1995:64)

Mediaster capensis H.L. Clark, 1923, reduced to a subspecies of *M. bairdi* Verrill, 1882) according to Downey (1992:252).

Milteliphaster regenerator (Döderlein, 1922) and *M. spinosus* (H.L. Clark, 1916)

new combs Rowe *in* Rowe & Gates (1995:66).

Nymphaster pentagonus H.L. Clark, 1916. A synonym of *Nymphaster moebii* (Studer, 1884) according to Rowe *in* Rowe & Gates (1995:67)

prehensilis (Perrier, 1885) (following Macan, 1938) and subspinosus (Perrier, 1881). Synonyms of *Nymphaster arenatus* (Perrier, 1881) according to Downey (1992:254).

Ogmaster capella (Müller & Troschel, 1842) Liao *in* Liao & Clark 1995 [1996]:94 (with possible synonym *Stellaster septemtrionalis* Oguro *in* Imaoka et al. 1991).

Peltaster nidarosiensis (Storm, 1881). A synonym of *Peltaster placenta* (Müller & Troschel, 1842) according to Downey (1992:258).

Pseudarchaster jordani and *myobrachius* both Fisher, 1906, probably conspecific according to Mah (1998b)

tessellatus Sladen, 1889, reduced to a subspecies of *gracilis* Sladen, 1889 by Downey (1992:263).

Sphaeriodiscus placenta: Fouda & Hellal 1987 (extension of range into northern Red Sea from Mediterranean).

Stellaster septemtrionalis sp. nov. Oguro *in* Imaoka et al., 1991

Imaoka et al. 1991:72; Liao *in* Liao & Clark 1995 [1996]:94 (possibly a synonym of *Ogmaster capella* (Müller & Troschel, 1842).

Range: East China Sea (27°N, 124°E), 105 m.

Tosia clugreta Walenkamp, 1976. A synonym of *Tosia parva* (Perrier, 1881) according to Downey (1992:268).

Family ASTERODISCIDIDAE

Asterodiscides: Imaoka et al. 1991:85 (tabular key to species).

elegans (Gray, 1847) Liao *in* Liao & Clark 1995 [1996]:112 (new record from Hainan Island confirming range to S China).

japonicus Imaoka et al., 1991

Imaoka et al. 1991:81.

Range: East China Sea (27°N, 124°E), 105 m.

tuberculosus: Chave & Malahoff 1998, fig. 105.

Family OREASTERIDAE

Anthenea chinensis Gray, 1840: Liao & Clark 1995 [1996]:99 (confirmed as a species distinct from *A. pentagonula* (Lamarck, 1816); potential neotype material described).

difficilis Liao *in* Liao & Clark, 1995 [1996](= *A. flavescens*: Chen, 1932 andA.M.C., 1982, non *flavescens* (Gray, 1840) and *A. pentagonula*: Chang & Liao, 1964, non *A. pentagonula*: Döderlein, 1915) Liao & Clark 1995 [1996]:99.

edmondi A.M. Clark, 1970. A synonym of *A. sidneyensis* Döderlein, 1915 (revived) according to Rowe *in* Rowe & Gates (1995:98).

viguieri Döderlein, 1915: Liao & Clark 1995 [1996]:102 (type locality Amboina (A.M.C.); confirmed from southern China but not north, also possibly conspecific with *chinensis* (Gray, 1840) of which *regalis* Koehler, 1910 may be another synonym (Liao)).

Goniodiscaster granuliferus (Gray, 1847) Liao *in* Liao & Clark, 1995 [1996]:106 (new record from Taiwan Strait confirming chinese type locality).

Monachaster umbonatus Macan, 1938. A synonym of *M. sanderi* (Meissner, 1892 according to Walenkamp (1990).

Nectria Gray, 1840: Zeidler, 1995:165, formal proposal of *Nectria ocellata* Perrier, 1875 as type species.

Pentaceraster cumingi: Chave & Malahoff 1998:44 & fig.

magnificus (Goto, 1914) A.M.C. *in* Liao & Clark 1995 [1996]:108 (includes *P. regulus*: A.M.C., 1982, non (Müller & Troschel, 1842)).

sibogae Döderlein, 1916: Liao *in* Liao & Clark 1995 [1996):109 (extension of range from Indonesia to Hainan Island, S China).

Family ASTEROPSEIDAE

Valvaster spinifera H.L. Clark, 1921. Confirmed as a synonym of *Valvaster-striatus* (Lamarck, 1816) Rowe *in* Rowe & Gates (1995:45).

Family MITHRODIIDAE

Mithrodia fisheri: Fouda & Hellal 1987 (extension of range to northern Red Sea).

Family OPHIDIASTERIDAE

BUNASTER Döderlein, 1895 reviewed, Marsh (1991).

DRACHMASTER Downey, 1970. A synonym of *Ophidiaster* L. Agassiz, 1836 according to Downey *in* Clark & Downey (1992:279).

FERDINA emended, Marsh & Price 1991:213.

sadhaensis nov. sp. Marsh & Campbell, 1991

Marsh & Campbell 1991:213.

Range: Southern Oman, 5-18 m.

Fromia nodosa: Fouda & Hellal 1987 (extension of range to northern Red Sea).

Gomophia gomophia (Perrier, 1875), *Gomophia mamillifera* (Livingstone, 1930) and *G. rosea* (H.L. Clark, 1921) **new combs** Rowe *in* Rowe & Gates (1995:83,84) (referred from *Nardoa*).

sphenisci (A.M. Clark, 1967), referred from *Nardoa*, Rowe & Gates 1995:84

Hacelia tuberculata nov. sp. Liao, 1985

Liao 1985:30.

Range: Diaoyudao, E China (24°45'N, 123°15'E), 100-200 m.

Leiaster speciosus von Martens, 1966: Liao *in* Liao & Clark 1995 [1996]:116 (extension of range to Xisha (Paracel) Is).

Linckia gracilis nov. sp. Liao, 1985

Liao 1985:32.

Range: Diaoyudao, E China, 100-200 m.

Nardoa lemonnieri Koehler, 1910. A synonym of *Nardoa galatheae* (Lütken, 1865) according to Rowe *in* Rowe & Gates (1995:88).

mamillifera Livingstone, 1930 and *N. rosa* H.L. Clark, 1921, see *Gomophia*

sphenisci, see *Gomophia*

tuberculata Gray, 1840. Additional synonyms *N. pauciforis* (von Martens, 1866) and *N. finschi* (de Loriol, 1891), confirmed, Rowe & Gates 1995:88.

Neoferdina japonica nov. sp. Oguro & Misaki, 1986

Oguro & Misaki 1986:60.

Range: Japan.

Ophidiaster bullisi (Downey, 1970) **new comb.** Downey (1992:279)

multispinus nov. sp. Liao & A.M. Clark, 1996 (nom. nov. for *O. armatus*: A.M. Clark, 1982b and Liao & Clark, 1995 [1996]:120, non *O. armatus* Koehler, 1910)

Liao & Clark 1996:37.

Range: Hainan Strait to Fujian Province, southern China, 5-55 m.

ophidianus: Fouda & Hellal 1987 (extension of range to northern Red Sea from Mediterranean).

pinguis H.L. Clark, 1941. A synonym of *Ophidiaster alexandri* Verrill, 1915 according to Downey (1992:280).

O. reyssi Sibuet, 1977: Downey 1992:282 (extension of range to the Mediterranean).

Paraferdina sohariae nov. sp. Marsh & Price, 1991
Marsh & Price 1991:66.
Range: Sri Lanka.
laccadivensis: Marsh & Price 1991 (extension of range to W Sumatra).

ADDITIONS TO PART 3

Family SOLASTERIDAE

CROSSASTER Müller & Troschel, 1840. Type species *Asterias papposa* Linnaeus, 1776). [Omitted in 1996]
multispinus H.L. Clark, 1916. A valid species distinct from *C. japonicus* Fisher, 1991 according to Rowe *in* Rowe & Gates (1995:113)

Family PTERASTERIDAE

Pteraster obesus myonotus Fisher, 1916: Mah 1998b:68 (extension of range from the Philippines and N Tasman Sea to Hawaii).

Family ECHINASTERIDAE

Echinaster (Othilia) echinophorus and *E.(O.) sentus*: Hendler et al., 1995:84-87 (with colour photographs of live specimens).
E. stereosomus Fisher, 1913: Liao *in* Liao & Clark 1995 [1996]:137 (extension of range from the Philippines and elsewhere to Hainan Island).
Metrodira subulata Gray, 1840 (retained in Metrodiridae) Liao & Clark 1995 [1996]:134 (extension of range from the Philippines and elsewhere to vicinity of Hainan Island and Taiwan Strait).

REFERENCES

Agassiz, A. 1866. On the embryology of *Asteracanthion berylinus* Ag. and a species allied to *A. rubens* M.T., *A. pallidus* Ag. *Proc. A. Acad. Arts Sci.* [1863] 6: 106-114.
Ahearn, C.G. 1995. Catalogue of the type specimens of seastars (Echinodermata: Asteroidea) in the National Museum of Natural History. *Smithson. Contr. Zool.* No. 572. 59 pp.
Aldrich, F.A. 1956. A comparative study of the identification characters of *Asterias forbesi* and *A. vulgaris* (Echinodermata: Asteroidea). *Notulae Philadelphia Acad. Nat. Hist.* 285: 1-3.

Alton, M. 1966. Bathymetric distribution of sea stars (Asteroidea) off the northern Oregon coast. *Jl Fish Res. Bd Canada* 23(11): 1673-1714.

Asbjornsen, P.C. 1856. Description d'un nouveau genre d'astéries. pp. 95-101; *Brisinga endecacnemos*. p. 101. In: M. Sars, Koren & Danielssen (eds), *Fauna Littoralis Norvegiae*. 2. Christiania.

Aziz, A. & Jangoux, M. 1985b. Four new species and one new subspecies of Asteroidea (Echinodermata) collected by the 'Siboga' Expedition in the Indo-Malayan region. *Bijdr. Dierk*. 55(2): 263-274.

Bell, F.J. 1881b. Contributions to the systematic arrangement of the Asteroidea. 1. The species of the genus *Asterias*. *Proc. zool. Soc. Lond*. 1881: 492-515.

Bell, F.J. 1891c. *Asterias rubens* and the British species allied thereto. *Ann. Mag. nat. Hist*. 7: 469-479.

Belyaev, G.M. 1982. The systematic position of the multirayed sea star from the sublittoral zone of Moneron Island. *Biol. Morya* 1982(6): 29-33.

Belyaev, G.M. & Mironov, A.N. 1993. Deep-sea starfishes of the genus *Hyphalaster*. *Trud. Inst. Okeanol. P.P Shirshov* 127: 215-217.

Belyaev, G.M. 1996. Starfishes *Porcellanaster* and *Caulaster* from the Atlantic and the Antarctic. *Zool. Zh*. 75: 886-899.

Belyaev, G.M. & Moskalev, L.I. 1986. Starfishes of the genus *Styracaster*. 1. *S. longispinus* sp. n. from the group of long-armed species. 2. New and rare short-armed species. 3. Composition, geographical and vertical distribution. *Zool. Zh*. 65(5,6): 771-780; 867-877; 1015-1023.

Bernasconi, I. 1956b. Algunos asteroideos de Antartida. *An. Soc. cient. Argent*. 161: 7-30. [Reprinted 1959 in *Contrçoes Inst. Antart. Argent*. No. 1.]

Bernasconi, I. 1961c. Nota sobre *Stichaster striatus* Müller & Troschel. *Physis, B. Aires* 21: 319-321.

Bernasconi, I. 1979. Asteriidae, Coscinasteriinae de la Argentino y Antartida. *Revta Mus. argent. Cienc. Nat. Bernardino Rivadavia Inst. Nac. Invest. Nienc. Nat*. (Hidrobiol.) 5(11): 241-246.

Blake, D.B. 1989. Asteroidea: Functional morphology, classification and phylogeny. *Echinoderm Studies* 3: 179-223.

Brandt, J.F. 1851. Bemerkungen über die Asteriden und Echiniden Okhotskischen, Kamtschatischen und Behringschen Meeres. pp. 27-33. In: Middendorff, A.T. von. *Reise in den Öussersten Norden und Osten Siberiens* 2(1). St Petersburg.

Brun, E. 1970. *Asterias hispida* Pennant, 1777, und *Uraster hispidus* (Pennant) Forbes, 1840 (Echinodermata, Asteridae): proposed suppression under the plenary powers in favour of *Leptasterias muelleri* (M. Sars, 1846). *Bull. Zool. Nom*. 26: 238-239.

Byrne, M., Morrice, M.G. & Wolf, B. 1997. Introduction of the northern Pacific asteroid *Asterias amurensis* to Tasmania: Reproduction and current distribution. *Mar. Biol*. 126: 673-685.

Campbell, A.C. & Rowe, F.W.E. 1997. A new species in the asterinid genus *Patiriella* (Echinodermata, Asteroidea) from Dhofar, southern Oman: A temperate taxon in a tropical locality. *Bull. Nat. Hist. Mus. Lond*. (Zool.) 62(2): 129-136.

Carus, J.V. 1885. *Prodromus Faunae Mediterraneae sive descriptio Animalium Maris Mediterranei Incolarum etc*. 524 pp. Stuttgart.

Caso, M.E. 1943. *Asteridos de Mexico*. Univ. Nation. Autonoma Mexico, Mexico, 136 pp.

Caso, M.E. 1961. Equinodermos de Mexico. Univ. Nation. Autonoma Mexico, Mexico, 388 pp.

Caso, M.E. 1990. Un nuevo Asteroidea del caribe mexicano *Astropecten caribemexi-canensis* sp. nov. y comparacion con la especie *Astropecten nitidus* Verrill. *An. Inst. Cienc. Mar Limnol. Univ. Aut. Mexico* 17(1): 107-130.

Chave, E.F. & Malahoff. A. 1998. *In deeper waters: Photographic studies of Hawaiian deep-sea habitats and life forms.* University of Hawai'i Press.

Cherbonnier, G. & Guille, A. 1967. Complément à la faune des échinodermes de la mer de Banyuls. *Vie Milieu* 18(2B): 317-330.

Chia, F.-S. 1966. Systematics of the six-rayed sea-star *Leptasterias* in the vicinity of San Juan Island, Washington. *Syst. Zool.* 15(4): 300-306.

Clark, A.H. 1916b. One new starfish and five new brittle stars from the Galapagos Islands. *Ann. Mag. nat. Hist.* 17: 115-122.

Clark, A.H. 1920. A new name for *Heliaster multiradiata* (Gray). *Proc. Biol. Soc. Wash.* 33: 183.

Clark, A.H. 1934. A new starfish from Puerto Rico. *Smithson. Misc. Collns* n.s. 91(14): 1-3.

Clark, A.H. 1946. Echinoderms from the Pearl Islands, Bay of Panama. *Smithson. Misc. Collns* 106(5): 1-11.

Clark, A.M. 1950. A new species of sea star from Norfolk Island. *Ann. Mag. Nat. Hist.* 3: 808-811.

Clark, H.E.S. 1962b. A new genus and species of asteroid from New Zealand. *Trans. R. Soc. N.Z.* (Zool.) 2: 45-46.

Clark, A.H. & McKnight, D.G. 1994. *Damnaster tasmani*, a new genus and species of Asteroidea (Echinodermata) from New Zealand. *Invert. Tax.* 8(6): 1367-1372.

Clark, H.L. 1907. The starfishes of the genus *Heliaster*. *Bull. Mus. comp. Zool. Harv.* 51(2): 23-76.

Costa, A.A. Elias da. 1941. *Un genre nouveau et une espèce nouvelle de la faune échinologique.* 11 pp., Porto.

Costa, A.A. Elias da. 1942. *Une sous-espèce nouvelle de la faune échinologique:* Astropecten auranticus multispinosus *sous-espèce nouvelle.* 8 pp., Guimaraes.

Costa, A.A. Elias da. 1952. *Nouvelle désignation pour le genre* Mortensenia *Elias da Costa, 1941, et pour l'espèce* Mortensenia lusitanica sp. n. 7 pp., Porto.

Dartnall, A.J., Pawson, D.L., Pope, E.C. & Smith, B.J. 1969. Replacement name for the preoccuppied genus name *Odinia* Perrier, 1885 (Echinodermata: Asteroidea). *Proc. Linn. Soc. N.S.W.* 93: 211.

Dewhurst, H.W. 1834. *The natural history of the order Cetacea and the oceanic inhabitants of the arctic regions.* xx+328 pp. London.

Djakonov, A.M. 1929a. Neue Seesterne aus dem Ochotskischen Meer. 1. *Leptasterias fisheri* sp. n. *C.r. Acad. Sci. Leningrad* 1929: 233-238.

Djakonov, A.M. 1929b. Neue Seesterne aus dem Ochotskischen Meer. 2. *Leptasterias orientalis* sp. n. *C. r. Acad. Sci. Leningrad* 1929: 277-281.

Djakonov, A.M. 1930b. Zur Frage der Artberechtigung der *mülleri-groenlandica*-Gruppe der Asteriden Gattung *Leptasterias* mir Beschreibung einer neuen Art aus dem Siberischen Eismeer. *Zool. Anz.* 91: 27-50.

Djakonov, A.M. 1938a. Monographische Bearbeitung der Asteriden des westlichen Nordpazifik. (Die Gattung *Leptasterias*). *Trud. zool. Inst. Akad. Nauk SSSR* 4(5): 749-914.

Downey, M.E. 1970a. Zorocallida, new order, and *Doraster constellatus*, new genus and species, with notes on the Zoroasteridae (Echinodermata Asteroidea). *Smithson. Contr. Zool.* No. 64. 18 pp.

Downey, M.E. 1971c. *Ampheraster alaminos*, a new species of the family Asteriidae (Echinodermata, Asteroidea) from the tropical western Atlantic. *Proc. Biol. Soc. Wash.* 84: 51-54.

Downey, M.E. 1972. *Midgardia xandaros*, new genus, new species. A large brisingid starfish from the Gulf of Mexico. *Proc. Biol. Soc. Wash.* 84(48): 421-426.

Downey, M.E. 1986. Revision of the Atlantic Brisingida (Echinodermata, Asteroidea) with description of a new genus and family. *Smithson. Contr. Zool.* 435: 57 pp.

Dungen et al. 1982. Catastrophic decline of a top carnivore in the Gulf of California rocky intertidal zone. *Science* 216: 989-991.

Fisher, W.K. 1917b. The asteroid genus *Coronaster*. *Proc. Biol. Soc. Wash.* 30: 23-26.

Fisher, W.K. 1917f. New genera and species of Brisingidae. *Ann. Mag. nat. Hist.* 20(8): 418-431.

Fisher, W.K. 1919b. North Pacific Zoroasteridae. *Ann. Mag. Nat. Hist.* 3: 387-393.

Fisher, W.K. 1922b. Notes on Asteroidea. 3. *Ann. Mag. Nat. Hist.* 10(9): 590-598.

Fisher, W.K. 1923. A preliminary synopsis of the Asteriidae, a family of sea stars. *Ann. Mag. Nat. Hist.* 12(9): 247-258; 595-607.

Fisher, W.K. 1924a. The genus *Sclerasterias* Perrier. *Bull. Inst. Oceanogr. Monaco* 444: 1-8.

Fisher, W.K. 1924b. A remarkable new sea star from Japan. *Proc. U.S. Natn. Mus.* 64: 1-6.

Fisher, W.K. 1926a. Notes on the Asteroidea. 4. *Ann. Mag. Nat. Hist.* 18:1 96-200.

Fisher, W.K. 1928a. Asteroidea of the North Pacific and adjacent waters. 2. Forcipulata (part). *Bull. U.S. Natn. Mus.* 76(2): 1-245.

Fisher, W.K. 1930. Asteroidea of the North Pacific and adjacent waters. 3. Forcipulata (concluded). *Bull. U.S. Natn. Mus.* 76(3): 1-356.

Fouda, M. & Hellal. 1987. *Fauna and flora of Egypt 2. The echinoderms of the North-western Red Sea. Asteroidea.* Al Azhar University, Department of Zoology. Egypt. iv+71 pp.

Galkin, S.V. & Korovchinsky, N.M. 1984. Vertical and geographical spread of the sea stars of the genus *Freyella* and notes about their ecology and origin. *Trud. Inst. Okeanol. Akad. Nauk SSSR* 119: 164-178. (In Russian)

Goldschmidt, 1924. On a new *Asterias* from Chile. *Ann. Mag. Nat. Hist.* (9)14: 499-502.

Grieg, J.A. 1927a. Evertebrater fra bankerne ved Spitzberge indsamlet av 'Tovik' og 'Armauer Hansen' somrene 1925 og 1926. *Bergens Mus. Aarb.* 1926(5): 1-28.

Grieg, J.A. 1927b. Echinoderms from the west coast of Norway. *Nyt. Mag. Naturvidensk.* 65:127-136.

Grieg, J.A. 1928. The Folden Fjord: Echinodermata. *Tromso Mus. Skr.* 1(7): 1-12.

Gruzov, E.N. 1964. A new deep water starfish *Astrocles djakonovi* sp. n. (Brisingidae) from the Okhotsk Sea. *Zool. Zh.* 43: 1394-1396 (In Russian).

Hayashi, R. 1943. Contributions to the classification of the sea-stars of Japan. 2. Forcipulata. *J. Fac. Sci. Hokkaido imp. Univ.* Zool. 9: 227-280.

Hayashi, R. 1975. A new sea-star of the Coscinasterinae from Toyama Bay. *Annot. Zool. Jap.* 48: 198-202.

Heddle, D. 1995. The descent of the Asteroidea and the reaffirmation of paxillosidan primitiveness, pp. 179-183. In: Emson, R. Smith, A. & Campbell, A. (eds), *Echinoderm Research 1995.* Rotterdam & Brookfield: Balkema.

Hendler, G., Miller, J.E., Pawson, D.L. & Kier, P.M. 1995. *Sea stars, sea urchins, and allies: Echinoderms of Florida and the Caribbean.* Smithsonian Institution

Press: Washington & London. 390 pp. [A comprehensive study of the species found down to 30 m., each described and illustrated, mostly in colour.]

Herring, P.J. 1974. New observations on the bioluminescence of Echinoderms. *J. Zool., Lond.* 172: 401-418.

Hutton, F.W. 1872b. *Catalogue of the Echinodermata of New Zealand, with diagnoses of the species*. Wellington. vi+17 pp.

James, D.B. 1996. Notes on the family Goniasteridae from the Indian Seas. *J. Mar. Biol. Ass. India* 38(1,2): 133-138. [Not seen]

Jangoux, M. & Massin, C. 1986. Catalogue commenté des types d'échinodermes actuels conservés dans les collections nationales Belges. *Bull. Inst. R. Sci. Nat. Belg.* 56: 83-97.

Johnston, G. 1836. Illustrations in British Zoology: *Asterias rubens* and *A. johnstoni* Gray; *Asterias aranciaca* and *A. endeca*; *Asterias papposa. Mag. Nat. Hist.* 9: 144-147; 298-300; 474-475.

Jullien, J. 1878. Description d'un nouveau genre de stelléride de la famille des astériadées. *Bull. Soc. Zool. Fr.* 3: 141-143.

Koehler, R. 1895c. Notes échinologiques. *Rev. Biol. N. Fr.* 7: 317-342.

Koehler, R. 1896b. Dragages profonds exécutés bord du 'Caudan' dans le Golfe de Gascogne (août -septembre, 1895): rapport préliminaire sur les échinodermes. *Rev. Biol. N. Fr.* 7: 439-496.

Koehler, R. 1909c. Échinodermes recueillis dans les mers arctiques par la mission arctique française, commandée par M. Bénard. *Bull. Mus. Hist. Nat. Paris* 1909: 121-123.

Koehler, R. Koehler, R. 1921b. Echinodermes (astéries, ophiures, échinides et crinoïdes) des dernières campagnes de la Princesse Alice et de l'Hirondelle. 2. *Bull. Inst. océanol. Monaco* No. 396. 8 pp.

Korovchinsky, N.M. 1976a. New species of abyssal starfishes of the genus *Freyella* (Brisingidae). *Zool. Zh.* 55(8): 1205-1215. (In Russian).

Korovchinsky, N.M. 1976b. New data on the deep-sea seastars (Asteroidea Brisingidae) from the North-Western Pacific. *Trud. Inst. Okeanol. Akad. Nauk SSSR* 99:165-177.

Korovchinsky, N.M. & Galkin, S.V. 1984. New data on the fauna of starfishes of the genus *Freyella* (Brisingidae). *Zool. Zh.* 63(8): 1187-1194 (In Russian).

Lambert, P. 1978a. British Columbia Marine Faunistic Survey Report: Asteroids from the Northeast Pacific. *Fish. Mar. Surv. Tech. Rep.* No 773. 23 pp.

Lambert, P. 1978b. New geographic and bathymetric records for some northeast Pacific Asteroids (Echinodermata: Asteroidea). *Syesis* 11: 61-64.

Liao, Y. 1985. Two new species of ophidiasterid sea-stars from the vicinity of Diaoyudao, East China Sea. *Chin. J. Oceanol. Limnol.* 3(1): 30-36.

Liao, Y. & Clark, A.M. 1995 [available 1996]. *The Echinoderms of southern China*. Science Press: Beijing, New York. 614 pp.

Liao, Y. 1996. On a new species of *Ophidiaster* (Echinodermata: Asteroidea) from southern China. *Bull. nat. Hist. Mus. Lond.* (Zool.) 62(1): 37-39.

Liao, Y. & Sun, S. 1989. A new species of astropectinid sea-stars from East China Sea. *Chin. J. Oceanol. Limnol.* 7(3): 225-232.

Lissner, A. & Hart, D. 1993. Chapter 6. Class Asteroidea. pp. 97-112. *In Taxonomic Atlas of the benthic Fauna of the Santa Maria Basin western Santa Barbara Channel.* 14. 306 pp.

Loriol, P. de. 1883. Catalogue raisonné des échinodermes recueillis par M. V. de Robillard à l'île Maurice. *Mem. Soc. Phys. Hist. nat. Genève.* 28(8): 1-64.

Loriol, P. de. 1894. Notes pour servir à l'étude des échinodermes. *Rev. suisse Zool.* 2: 467-497.

Ludwig, H. 1878. Zür Kentniss der Gattung *Brisinga*. *Zool. Anz.* 2: 21-234.

Ludwig, H. 1897c. Diagnosen der Seesterne des Mittelsmeeres. *Verh. Ver. Rheinland* 53: 281-309.

Ludwig, H. 1905b. Asterien und Ophiuren der schwedischen Expedition nach den Magalhaenslaendern 1895-1897. *Z. Wiss. Zool.* 82: 39-79.

Macan, T.T. 1938. Asteroidea. *Scient. Rep. John Murray Exped.* 4(9): 323-435. [Omitted from part 1]

McKnight, D.G. 1977b. A note on the order Zorocallida (Echinodermata: Asteroidea). *N.Z.O.I. Rec.* 3(18): 159-161.

McKnight, D.G. 1993. Records of echinoderms (excluding holothurians) from the Norfolk Ridge and Three Kings Rise north of New Zealand. Records of echinoderms from the Chatham Islands. *N.Z. J. Zool.* 20: 165-190; 191-200.

McKnight, D.G. & H.E.S. Clark. 1996. *Enigmaster scalaris*, n. gen., n. sp., a puzzling sea-star (Echinodermata, Asteroidea) from the Auckland Islands. *J. R. Soc. N.Z.* 26: 205-214.

Madsen, F.J. 1956b. Echinoidea, Asteroidea and Ophiuroidea from depths exceeding 6000 metres. *Galathea Rep.* 2:23-32.

Mah, C.L. 1998a. Preliminary phylogeny and taxonomic revision of the Brisingida. pp. 273-277. In: R. Mooi & M. Telford (eds) Proc. Ninth int. Echinoderm Conference. San Francisco 1996. Rotterdam/Brookfield: Balkema.

Mah, C.L. 1998b. New records, taxonomic notes and a checklist of Hawaiian starfish. *Occ. Pap. Bishop Museum* 55: 65-71. [Doubts cast on some old unconfirmed records not itemised here in Addenda.]

Mah, C.L. [in press] Redescription and taxonomic notes on the South Pacific brisingidan *Brisingaster robillardi* (Asteroidea), with new ontogenetic and phylogenetic information. *Zoosystema*.

Marenzeller, E. von. 1893c. Zoologische Ergebnisse der Tiefsee-Expeditionen im Östlichen Mittelmeere auf S.M. Schiff 'Pola'. 1. *Anz. Preuss. Akad. Wiss. Math.-naturwiss. Cl.* 8: 65-67.

Marenzeller, E. von. 1895b. Echinodermen gesammelt im stlichen Mittelmeere. 1893-1894. *Anz. Akad. Wiss. Math.-naturwiss. Cl.* 28:189-191.

Marion, A.F. 1883. Les progrès récents des sciences naturelles. *Rev. Scient.* 5: 129-136.

Marsh, L.M. 1991. A review of the echinoderm genus *Bunaster* (Asteroidea: Ophidiasteridae). *Rec. W. Aust. Mus.* 15(2): 419-433.

Marsh, L.M. & A.C. Campbell. 1991. A new species of *Ferdina* from Oman. *Bull. Br. Mus. nat. Hist.* (Zool.) 57(2): 213-219.

Marsh, L.M. & D.L. Pawson. 1993. Echinoderms of Rottnest Island. pp. 279-304. In: Wells, F.E. et al. *The Marine flora and fauna of Rottnest Island, Western Australia. Proc. Fifth Int. Mar. Biol. Workshop.* Western Australian Museum, Perth.

Marsh, L.M. & A.R.G. Price. 1991. Indian Ocean echinoderms collected during the Sindbad Voyage. *Bull. Br. Mus. Nat. Hist.* Zool. 57(1): 61-70.

Mein, B. 1992. Beitrag zür Kenntniss antarktischer Seestern (Asteroidea, Echinodermata). *Mitt. Hamb. Zool. Mus. Inst.* 89: 239-259.

Meyen, F.J.F. 1834. *Reise um die Erde* 1. 503 pp. [Not seen]

Millott, N. 1966. In: R.A. Boolootian. (Ed.) *Physiology of Echinodermata.* New York: Interscience Publishers.

Moyana & Larrain Prat, A.P. 1976. *Doraster qawoshqari* sp. nov., nuevo Asteroidea de Chile austral. (Echinodermata, Zorocallida, Zoroasteridae). *Bol. Soc. Biol. Concepcion* 50: 103-111.

Murdoch, J. 1885. Marine Invertebrates. In Report of the international Polar Expedition to Point Barrow, Alaska. Washington.

Norman, A.M. 1893. A month on the Trondhjem Fiord. 1. *Ann. Mag. Nat. Hist.* 6: 341-357.

Nutting, C.C. 1895. Narrative and preliminary report. *Bull. Labs Nat. Hist. St. Univ. Iowa* 3: 1-2.

Oguro, C. 1982. Notes on the sea-star fauna and a new species of *Astropecten* in Palau and Yap Islands. *Proc. Jap. Soc. Syst. Zool.* 23: 71-79.

Oguro, C. 1989. *Poraniopsis inflata* with new synonym (*P. japonica*) (Fisher). *Bull. biogeograph. Soc. Japan* 44: 49-50.

Oguro, C. & Misaki, H. 1986. A new sea-star, *Neoferdina japonica* from Japan. *Proc. Jap. Soc. Syst. Zool.* 34: 60-64.

O'Loughlin, P.M. & O'Hara, T.D. 1990. A review of the genus *Smilasterias* (Echinodermata: Asteroidea) with descriptions of two new species from south-eastern Australia, one a gastric brooder, and a new species from Macquarie Island. *Mem. Mus. Vict.* 50: 307-324.

Opinion 984. 1972. *Asterias hispida* Pennant, 1777, suppressed under the plenary Powers in favour of *Leptasterias muelleri* (M. Sars, 1846). *Bull. zool. Nom.* 29: 115-116.

Orovitz, B. & A. Tablado. 1990. *Asterodon singularis* (Müller & Troschel, 1843) unica especie del genero *Asterodon* Perrier, 1891 (Asteroidea:Odontasteridae). *Gayana Zool.* 54(1-2): 21-32.

Perrier, E. 1882b. Note sur les *Brisinga*. *CR hebd. Séanc. Acad. Sci. Paris* 95:61-63. (Trans. *Ann. Mag. nat. Hist.* 5(10): 261-263.)

Perrier, E. 1885a. Note préliminaire sur les échinodermes recueillis par le 'Travailleur' et le 'Talisman'. *Nouv. Archs Miss. Sci. Litt.* Rapp. Instr. (2)6: 1-154.

Perrier, E. 1885d. Sur les Brisingidae de la mission du `Talisman'. *CR hebd.* Séanc. *Acad. Sci. Paris* ‡101}: 441-444. [Translation in: *Ann. Mag. Nat. Hist.* (5)16: 312-314]

Perrier, E. 1891c. Sur les stellérides receuillis dans le Golfe de Gascogne, aux Açores et à Terre-Neuve pendant les campagnes scientifiques du yacht 'l'Hirondelle'. *CR hebd. Séanc. Acad. Sci. Paris* 112(21): 1225-1228.

Rathbun, R. 1887. Descriptions of the species of *Heliaster* (a genus of starfishes) represented in the U.S. National Museum. *Proc. U.S. Natn. Mus.* 10: 440-449.

Rowe, F.W.E. & J. Gates. 1995. Echinodermata. In A. Wells. (ed.), *Zoological Catalogue of Australia*. 33. 510 pp. Melbourne: CSIRO Australia. [Not seen].

Sars, G.O. 1875. *On some remarkable forms of animal life from the great deeps off the norwegian coast.* 2. Researches on the structure and affinity of the genus *Brisinga* based on the study of a new species, *Brisinga coronata*. 112 pp. Christiania: Christiania University.

Sars, M. 1846. Beobachtungen über die Entwickelung der Seesterne. pp. 47-62. *Fauna littoralis Norwegiae*. Christiania.

Shepherd, S.A. 1967b. A revision of the starfish genus *Uniophora* (Asteroidea: Asteriidae). *Trans. R. Soc. S. Aust.* 91: 3-14.

Shin, S. 1995a. Echinodermata from Chindo Island, Korea. *Kor. J. Syst. Zool.* 11(1): 115-124.

Shin, S. 1995b. A systematic study of the Asteroidea in the Eastern Sea, Korea. *Kor. J. Syst. Zool.* 11(2): 243-263.

Stampanato, S. & Jangoux, M. 1993. Les astérides (Echinodermata) de la Baie Breid (Côte de la Princese Ragnild, Quartier Enderby, Antarctique), avec la description d'une nouvelle espèce de *Solaster*. *Bull. Inst. R. Sci. Nat. Belg. Biol.* 63: 175-184.

Stampanato, S. & Jangoux, M. 1997. Les astérides (Echinodermata) récoltés autour des îles Saint-Paul et Amsterdam (Océan Indien Sud). *Zoosystema* 19(1): 27-33.

Stimpson, W. 1864. Synopsis of the marine Invertebrates collected by the late Arctic Expedition. *Proc. Acad. Nat. Sci., Philadelphia* 1863: 138-142.

Studer, T. 1885. Die Seesterne Südgeorgiens nach der Ausbeute der deutschen Polarstation, 1881 und 1883. *Jahrsber. Wiss. Anst.* Hamburg 2: 150.

Tablado, A. & Maytia, S. 1988. Presencia de *Perissasterias polyacantha* H.L. Clark, 1923 (Echinodermata, Asteroidea) em el Oceano Atlantico Sudoccidental. *Comun. Zool. Mus. Hist. Nat. Montevideo* 12(169): 1-11.

Thomson, C.W. 1873b. On the Echinoidea of the Porcupine deep-sea dredging expedition. *Phil. Trans. R. Soc.* 719: 59-71.

Tortonese, E. 1958. Euclasteroidea: nuevo ordine di Asteroidi (Echinodermi). *Doriana* 2(88): 1-3.

Tyler, P.A. & Zibrowius, H. 1992. Submersible observations of the invertebrate fauna on the continental slope southwest of Ireland(NE Atlantic Ocean). *Oceanol. Acta* 15(2): 211-226.

Verrill, A.E. 1871c. Marine fauna of Eastport, Maine. *Bull. Essex Inst.* 3: 2-6.

Verrill, A.E. 1873b. Report on the invertebrate animals of Vineyard Sound. *Rep. U.S. Fish Commn* 1871: 295-778.

Verrill, A.E. 1879a. *Preliminary checklist of the marine Invertebrata of the Atlantic coast from Cape Cod to the Gulf of St Lawrence*. 32 pp. New Haven: A.E. Verrill.

Verrill, A.E. 1880c. Notice of recent additions to the marine invertebrate fauna of the northeastern coast of America. *Amer. J. Sci.* 19: 137-140.

Verrill, A.E. 1884b. Notice of the remarkable marine fauna occupying the outer banks off the southern coast of New England, and of some additions to the fauna of Vineyard Sound. *Ann. Rep. U.S. Commn* 1882: 641-669 (1-29).

Viguier, J. 1878. Anatomie comparée du squelette des stellérides. *Archs Zool. exp. gén.* 7: 33-250.

Walenkamp, J. 1980. Systematics and zoogeography of Asteroidea (Echinodermata) from Inhaca Island, Mozambique. *Zool. Verh.* 70: 3-86.

Xantus, J. 1860. Descriptions of three new species of starfishes from Cape San Lucas. *Proc. Acad. nat. Sci. Philad.* 1860: 568.

Yamaoka, M. 1987. Fossil asteroids from the Miocene Morozaki group, Aichi Prefecture. central Japan. *Kaseki No Tomo* 135: 5-23.

Zeidler, W. 1995. Case 2951. *Nectria* Gray, 1840. (Echinodermata: Asteroidea): proposed designation of *Nectria ocellata* Perrier, 1875 as the type species. *Bull. Zool. Nom.* 52(2): 164-165.